Problems in Analysis

A SYMPOSIUM IN HONOR OF SALOMON BOCHNER

PRINCETON MATHEMATICAL SERIES

1. The Classical Groups, Their Invariants and Representation. By HERMANN WEYL.
2. Topological Groups. By L. PONTRJAGIN. Translated by EMMA LEHMER.
3. An Introduction to Differential Geometry with Use of the Tensor Calculus. By LUTHER PFAHLER EISENHART.
4. Dimension Theory. By WITOLD HUREWICZ and HENRY WALLMAN.
5. The Analytic Foundations of Celestial Mechanics. By AUREL WINTNER.
6. The Laplace Transform. By DAVID VERNON WIDDER.
7. Integration. By EDWARD JAMES MCSHANE.
8. Theory of Lie Groups: I. By CLAUDE CHEVALLEY.
9. Mathematical Methods of Statistics. By HARALD CRAMÉR.
10. Several Complex Variables. By SALOMON BOCHNER and WILLIAM TED MARTIN.
11. Introduction to Topology. By SOLOMON LEFSCHETZ.
12. Algebraic Geometry and Topology. Edited by R. H. FOX, D. C. SPENCER, and A. W. TUCKER.
14. The Topology of Fibre Bundles. By NORMAN STEENROD.
15. Foundations of Algebraic Topology. By SAMUEL EILENBERG and NORMAN STEENROD.
16. Functionals of Finite Riemann Surfaces. By MENAHEM SCHIFFER and DONALD C. SPENCER.
17. Introduction to Mathematical Logic, Vol. I. By ALONZO CHURCH.
18. Algebraic Geometry. By SOLOMON LEFSCHETZ.
19. Homological Algebra. By HENRI CARTAN and SAMUEL EILENBERG.
20. The Convolution Transform. By I. I. HIRSCHMAN and D. V. WIDDER.
21. Geometric Integration Theory. By HASSLER WHITNEY.
22. Qualitative Theory of Differential Equations. By V. V. NEMICKII and V. V. STEPANOV. Translated under the direction of SOLOMON LEFSCHETZ.
23. Topological Analysis, revised edition. By GORDON T. WHYBURN.
24. Analytic Functions. By NEVANLINNA, BEHNKE and GRAUERT, et al.
25. Continuous Geometry. By JOHN VON NEUMANN. Foreword by ISRAEL HALPERIN.
26. Riemann Surfaces. By L. AHLFORS and L. SARIO.
27. Differential and Combinatorial Topology. Edited by STEWART S. CAIRNS.
28. Convex Analysis. By R. T. ROCKAFELLAR.
29. Global Analysis. Edited by D. C. SPENCER and S. IYANAGA.
30. Singular Integrals and Differentiability Properties of Functions. By E. M. STEIN.
31. Problems in Analysis. Edited by R. C. GUNNING.

PROBLEMS IN ANALYSIS

A Symposium in Honor of Salomon Bochner

ROBERT C. GUNNING
GENERAL EDITOR

PRINCETON, NEW JERSEY
PRINCETON UNIVERSITY PRESS
1970

L. C. Card 76–106392

I. S. B. N. 0–691–08076–3

A. M. S. 1968: 0004

Printed in the United States of America
by Princeton University Press, Princeton, New Jersey

Foreword

A symposium on problems in analysis in honor of Salomon Bochner was held in Fine Hall, Princeton University, April 1–3, 1969, to celebrate his seventieth birthday, which took place on August 20, 1969. The symposium was sponsored by Princeton University and the United States Air Force Office of Scientific Research; the organizing committee consisted of W. Feller, R. C. Gunning, G. A. Hunt, D. Montgomery, R. G. Pohrer, and W. R. Trott.

This volume contains some of the papers delivered by the invited speakers at the symposium, together with a number of papers contributed by former students of Professor Bochner and dedicated to him on this occasion.

The papers were received by June 1, 1969.

Contents

Problems in Analysis

A SYMPOSIUM IN HONOR OF SALOMON BOCHNER

On the Group of Automorphisms of a Symplectic Manifold

EUGENIO CALABI[1]

1. Introduction

Let $\mathbf{X}$ be a connected, differential manifold of $2n$ dimensions. A *symplectic structure* on $\mathbf{X}$ is the geometrical structure induced by a differentiable exterior 2-form ω defined on $\mathbf{X}$, satisfying the following conditions:

(i) The form ω is closed: $d\omega = 0$;

(ii) It is everywhere of maximal rank; this means that the $2n$-form ω^n (nth exterior power of ω) is everywhere different from zero, or equivalently, the skew-symmetric $(2n) \times (2n)$ matrix of coefficients of ω, in terms of a basis for the cotangent space, is everywhere nonsingular.

A classical theorem, ordinarily attributed to Darboux, states that a $2n$-dimensional symplectic manifold (i.e., a manifold with a symplectic structure) can be covered by a local, differentiable coordinate system $\{U; (x)\}$ where $(x) = (x^1, \ldots, x^{2n})\colon U \to \mathbf{R}^{2n}$, in terms of which the local representation of the structural form ω becomes

$$(1.1)\qquad \begin{aligned} \omega|_U &= dx^1 \wedge dx^2 + dx^3 \wedge dx^4 + \cdots + dx^{2n-1} \wedge dx^{2n} \\ &= \sum_{j=1}^{n} dx^{2j-1} \wedge dx^{2j}; \end{aligned}$$

such a system of coordinates is called a *canonical* system.

The purpose of this study is to describe the group $\mathbf{G}$ of automorphisms of a symplectic manifold, i.e., the group of all differentiable automorphisms of $\mathbf{X}$ which leave the structural 2-form ω invariant, and the invariant subgroups of $\mathbf{G}$. The group $\mathbf{G}$ can also be characterized as mapping canonical coordinate systems into canonical systems.

[1] The research reported here was supported in part by the National Science Foundation.

Two normal subgroups of **G** are distinguished immediately as follows:

DEFINITION 1.1. Let $(\mathbf{X}, \omega)$ be a $2n$-dimensional symplectic manifold and let **G** be the group of all symplectic transformations of **X**. If **X** is not compact, we denote by $\mathbf{G}_0$ the subgroup of **G** consisting of all symplectic transformations of **X** that have compact support; that is to say, a symplectic transformation $\mathbf{g} \in \mathbf{G}$ belongs to $\mathbf{G}_0$ if and only if **g** equals the identity outside a compact region of **X**.

DEFINITION 1.2. Let $(\mathbf{X}, \omega)$ be a $2n$-dimensional symplectic manifold and let **G** be the group of all symplectic transformations of **X**. We denote by $\mathbf{G}_{0,0}$ the subgroup of **G** called the *minigroup* generated by the so-called locally supported transformations, defined as follows: a transformation $\mathbf{g} \in \mathbf{G}$ is called locally supported if there exists a canonical coordinate system $\{U; (x)\}$ defined in a contractible domain U with compact closure, such that the support of **g** lies in U.

The minigroup $\mathbf{G}_{0,0}$ and its corresponding Lie algebras are introduced here merely for expository convenience. In Section 3 it will be shown that the commutator subgroup of the arc-component of the identity in $\mathbf{G}_0$ coincides either with $\mathbf{G}_{0,0}$ or with a normal subgroup of codimension 1 in $\mathbf{G}_{0,0}$ (see Theorem 3.7, Section 3).

An elementary example of a locally supported transformation is the following: let $\{U; (x)\}$ be a canonical coordinate system in **X**; let its range $V = (x)(U) \subset \mathbf{R}^{2n}$ contain the ball $\left\{(t) \,\middle|\, \sum_{i=1}^{2n} (t^i)^2 < A\right\}$ for some $A > 0$; choose a real-valued differentiable function $\vartheta(r)$ of a real variable $r \geqq 0$ with support contained in a closed segment $[0, A']$ with $A' < A$. Then it is easily verifiable that the transformation **f** in $\mathbf{R}^{2n}$ (with the 2-form $\omega = \sum_{j=1}^{n} dt^{2j-1} \wedge dt^{2j}$),

$$(t) \to (t') = \mathbf{f}(t) \text{ with}$$
$$t'^{2j-1} = t^{2j-1} \cos \vartheta(r) - t^{2j} \sin \vartheta(r),$$
$$t'^{2j} = t^{2j-1} \sin \vartheta(r) + t^{2j} \cos \vartheta(r),$$
$$\left(r = \sum_{i=1}^{2n} (t^i)^2;\ 1 \leqq j \leqq n\right),$$

is a symplectic transformation which equals the identity for $r \geqq A'$. Therefore its restriction to V defines via the coordinate map (x) a symplectic transformation in U that can be trivially extended by the identity map in $\mathbf{X} - U$ to a locally supported symplectic transformation in **X**.

We shall state here the main results of this study in a preliminary form; more precise and stronger versions of these are repeated as theorems in the later sections.

STATEMENT 1. The groups $\mathbf{G}$, $\mathbf{G}_0$, and $\mathbf{G}_{0,0}$ are infinite dimensional Lie groups in terms of the Whitney C^∞ topology (in the case of $\mathbf{G}$) and the compactly supported C^∞ topology (in the case of $\mathbf{G}_0$ and $\mathbf{G}_{0,0}$).

STATEMENT 2. The minigroup $\mathbf{G}_{0,0}$ is a closed, normal subgroup of $\mathbf{G}_0$; the quotient group $\mathbf{G}_0/\mathbf{G}_{0,0}$ is locally isomorphic to the de Rham cohomology group $H_0^1(\mathbf{X}, \mathbf{R})$, the first cohomology group of $\mathbf{X}$ with real coefficients and compact support.

STATEMENT 3. If $\mathbf{X}$ is not compact, the completion $\mathbf{G}_{1,0}$ of $\mathbf{G}_{0,0}$ in the compact-open topology of presheaves (i.e., the group obtained by adjoining to $\mathbf{G}_{0,0}$ the infinite products of sequences of $\mathbf{g}_\nu \in \mathbf{G}_{0,0}$, where for each compact $K \subset \mathbf{X}$ only finitely many of the $\mathbf{g}_\nu$ differ from the identity in K) is a closed, normal subgroup of $\mathbf{G}$; the quotient group $\mathbf{G}/\mathbf{G}_{1,0}$ is locally isomorphic to the de Rham group $H^1(\mathbf{X}, \mathbf{R})$, i.e., to the first cohomology group with closed support.

STATEMENT 4. The group $\mathbf{G}_{1,0}$ has no connected, closed, normal subgroups other than the identity; in particular its commutator subgroup is an open subgroup. The same is true of $\mathbf{G}_{0,0}$, of course, if $\mathbf{X}$ is compact (in which case $\mathbf{G}_{0,0} = \mathbf{G}_{1,0}$). On the other hand, if $\mathbf{X}$ is not compact, the commutator subgroup $\mathbf{G}'_{0,0}$ of $\mathbf{G}_{0,0}$ is normal in $\mathbf{G}_0$, has codimension equal to 1, relative to $\mathbf{G}_{0,0}$, and has no connected, closed, normal subgroups other than the identity.

The next two sections deal with the Lie group structure of the groups $\mathbf{G}$ and $\mathbf{G}_0$, emphasizing the relationship with the corresponding Lie algebras; the main tools used here are due to J. Moser [1]. In Section 3 we prove the four main statements just given at the Lie algebra level, and in Section 4 we expand the results at the group level, trying as far as we have succeeded to obtain results on the global structure of these groups and their closed, normal subgroups.

2. Infinite dimensional Lie groups of differentiable transformations

We shall summarize in this section some of the known facts about infinite dimensional Lie groups or pseudogroups of differentiable transformations, especially with regard to their relationships with the corresponding Lie algebras of tangent vector fields.

We denote by $\mathbf{G}$, $\mathbf{H}$, and so forth, groups of differentiable transformations of a manifold; the corresponding pseudogroups of local transformations are denoted by G, H, and so forth; the associated Lie algebras of globally defined vector fields will be denoted by German capitals $\mathfrak{G}$, $\mathfrak{H}$, and so forth; the corresponding presheaves of local vector fields will be denoted by small German letters, such as $\mathfrak{g}$, $\mathfrak{h}$, etc.

An infinite dimensional Lie group $\mathbf{G}$ of global, differentiable transformations in a manifold $\mathbf{X}$ (or alternately a pseudogroup G of local transformations) is an infinite dimensional, differential manifold (naturally we do not exclude from this notion finite dimensional manifolds), in the following sense: for any finite dimensional, differential manifold $\mathbf{M}$ a map $\varphi: \mathbf{M} \to \mathbf{G}$ is defined to be differentiable, if and only if the corresponding evaluation map $\tilde{\varphi}: \mathbf{M} \times \mathbf{X} \to \mathbf{X}$, where $\tilde{\varphi}(t, \mathbf{x}) = (\varphi(t))(\mathbf{x})$, is differentiable (in the case of a pseudogroup, one requires also that the domain of definition of $\tilde{\varphi}$ be open in $\mathbf{M} \times \mathbf{X}$).

For our present purposes, it seems irrelevant to fix any topology on $\mathbf{G}$; we regard it rather as a *compactoid*, that is to say, we allow ourselves to consider the equivalence class of topologies that are compatible with the category of differentiable maps $\varphi: \mathbf{M} \to \mathbf{G}$ just defined. The Lie algebra $\mathfrak{G}$ associated to $\mathbf{G}$ is then the set of tangent vector fields (respectively presheaves of local vector fields) each defined as the equivalence class of differentiable paths in $\mathbf{G}$ originating at the identity with the obvious equivalence relation. The question that ordinarily arises here is how to recapture the arc component of the identity in $\mathbf{G}$ from the sheaf of germs $\mathfrak{g}$ of Lie algebras determined by $\mathfrak{G}$.

Any local cross section ϑ in $\mathfrak{g}$ (that is to say, each local vector field belonging to $\mathfrak{g}$) defines a one-parameter subpseudogroup of differentiable transformations by integrating the vector field to the corresponding autonomous flow. The composition of such one-parameter local flows defines a pseudogroup $\Gamma_{\mathbf{G}}$ in the pseudogroup G associated to $\mathbf{G}$. If $\mathbf{G}$ is finite dimensional, it is known classically that $\Gamma_{\mathbf{G}}$ defines an open subpseudogroup of G. In the infinite dimensional case the same conclusion holds, if $\mathbf{G}$ acts real analytically, at least by considering first local cross sections of $\mathfrak{g}$ and collating the resulting local transformations. In the C^{∞} case for infinite dimensional Lie groups, several authors have shown examples to the effect that one-parameter subpseudogroups are not dense in a neighborhood of the identity, and the details of some of these examples give a strong indication that even the composition of the elements of such one-parameter systems may not fill out any neighborhood of the identity in $\mathbf{G}$. Some authors have suggested a method based on affine connections; this method permits one to attach to each local vector field in $\mathfrak{G}$ a differentiable *path* in $\mathbf{G}$ originating at the identity but not constituting, in general, a one-parameter pseudogroup, so that the union of such paths fill out a neighborhood of the identity in $\mathbf{G}$. This method is satisfactory in the case of differentiably acting pseudogroups definable in terms of a first-order, integrable differential system in the coordinate transformations but would require higher order connections for pseudogroups of a more complicated nature.

The correspondence between presheaves of Lie algebras of local vector fields and pseudogroups of local differentiable maps can be established in a natural way only as a local one-to-one correspondence between differentiable paths. Thus the fundamental theorem on existence, uniqueness, and continuity with respect to initial data for ordinary differential equations establishes a one-to-one, locally biregular correspondence between differentiable, one-parameter families of local vector fields in $\mathfrak{g}$ (i.e., paths in $\mathfrak{g}$) and local flows in $\mathbf{X}$ belonging to G (i.e., paths in G): any topological structure in the stalks of the sheaf of germs of one-parameter families of vector fields in $\mathbf{X}$ belonging to $\mathfrak{g}$ (provided that its definition includes minimal regularity conditions) yields a well defined topology on the stalks of the corresponding sheaf of germs of G-flows. The corresponding topology on the sets of germs of elements of G is then obtained by passage to the quotient, assigning to each path in G (originating at the germ of the identity) the germ of the terminal element $\in G$ of the path. Thus one obtains from the sheaf of germs of Lie algebras $\mathfrak{g}$ a sheaf G of germs of diffeomorphisms. The group $\mathbf{G}$ of global cross sections in G can be obtained without any difficulty if the manifold $\mathbf{X}$ is compact.

In the case where $\mathbf{X}$ is not compact, the corresponding sheaves $\mathfrak{g}$ of germs of Lie algebras of vector fields and G of germs of pseudogroups of transformations lead to the corresponding global Lie algebras and groups in many ways, of which two are the most important: the ones with unrestricted support and the ones with compact support. They are obtained from the topologies of the corresponding stalks by "globalizing" them, in the former case either by the compact-open extension of a uniform topology on the stalks (compact-open topology) or by a Whitney topology and in the latter case by a uniform topology over uniformly compactly supported cross sections. It is worthwhile noting that the global groups obtained from the pathwise-connected sheaf of groups is not necessarily connected, as we know well in the case of the celebrated group of all differentiable, orientation-preserving automorphisms of spheres.

3. The Lie algebra of symplectic vector fields

We apply the concepts reviewed in the previous section to the case of the Lie algebra associated with the group $\mathbf{G}$ or $\mathbf{G}_0$ of automorphisms of a symplectic manifold $(\mathbf{X}, \omega)$.

If φ is any exterior p-form and ξ is a vector field in a manifold, we denote by $\xi \llcorner \varphi$ the interior product of ξ with φ; this is the $(p-1)$-form (identically zero if $p = 0$) whose value at a $(p-1)$-vector $\eta_1 \wedge \cdots \wedge \eta_{p-1}$ is given by

$$(3.1) \qquad (\xi \llcorner \varphi)(\eta_1 \wedge \cdots \wedge \eta_{p-1}) = \varphi(\xi \wedge \eta_1 \wedge \cdots \wedge \eta_{p-1}).$$

The Lie derivative of φ with respect to ξ is then obtained from the formula

$$[\xi, \varphi] = d(\xi \llcorner \varphi) + \xi \llcorner d\varphi.$$

It is well known that any differentiable path in the sheaf of germs of diffeomorphisms, originating at the germ of the identity at any $\mathbf{x} \in \mathbf{X}$, leaves the form φ invariant along the orbit of $\mathbf{x}$, if and only if the path in the sheaf of germs of vector fields given by the differential of the given path yields a one-parameter family of germs of vector fields ξ satisfying $[\xi, \varphi] = 0$. Thus the Lie algebra of the symplectic group $\mathbf{G}$ or $\mathbf{G}_0$ is described locally by the vector fields ξ satisfying, since ω is closed,

$$d(\xi \llcorner \omega) = 0. \tag{3.2}$$

We now introduce the Lie algebras corresponding to the groups $\mathbf{G}$, $\mathbf{G}_0$, and $\mathbf{G}_{0,0}$ previously given in Definitions 1.1 and 1.2.

PROPOSITION 3.1. *The Lie algebras* $\mathfrak{G}$, $\mathfrak{G}_0$, *and* $\mathfrak{G}_{0,0}$ *corresponding, respectively, to the groups* $\mathbf{G}$, $\mathbf{G}_0$, *and* $\mathbf{G}_{0,0}$ *are given by the global vector fields* ξ *on* $\mathbf{X}$ *satisfying* (3.2) *and, in addition,*

(i) *satisfying no further conditions in the case of* $\mathfrak{G}$;
(ii) *having compact support in the case of* $\mathfrak{G}_0$;
(iii) *generated, in the case of* $\mathfrak{G}_{0,0}$, *by vector fields* ξ_ν, *where each* ξ_ν *has compact support contained in a contractible domain* U_ν *admitting a canonical coordinate system* $(x^1, \ldots, x^{2n})$, *that is to say, such that there exists a function* u_ν *with compact support in* U_ν *satisfying*

$$\xi_\nu \llcorner \omega = du_\nu,$$

or equivalently

$$\xi_\nu = \sum_{j=1}^{n} \left(\frac{\partial u_\nu}{\partial x^{2j}} \frac{\partial}{\partial x^{2j-1}} - \frac{\partial u_\nu}{\partial x^{2j-1}} \frac{\partial}{\partial x^{2j}} \right). \tag{3.3}$$

The proof of this proposition is clear.

Since the form ω is everywhere of maximal rank, the bundle map from the tangent to the cotangent vector bundle defined by assigning to ξ the 1-form $\xi \llcorner \omega$ is bijective. Thus, for instance, the Lie algebra $\mathfrak{G}$ is isomorphic, as a vector space over the real numbers, to the set of all closed Pfaffian forms on $\mathbf{X}$. This isomorphism induces a Lie algebra structure on the set of all Pfaffian forms called the "Poisson bracket"; more precisely, for any 1-form α we denote by $\alpha^\#$ the uniquely defined tangent vector ξ such that $\xi \llcorner \omega = \alpha$ (at the bundle, sheaf, local, or global level); then the Poisson bracket of two Pfaffian forms α and β is defined to be

$$\{\alpha, \beta\} = [\alpha^\#, \beta^\#] \llcorner \omega.$$

The Lie algebra of all C^∞ 1-forms (globally defined on **X**) with the Poisson bracket is isomorphic under the map $\{\alpha \to \alpha^\#\}$ to the Lie algebra of all tangent vector fields; the vector subspace consisting of all closed 1-forms is a Lie subalgebra. It follows from (3.2) that this subalgebra is isomorphic to the algebra $\mathfrak{G}$ of all vector fields ξ such that $[\xi, \omega] = 0$. The subalgebras $\mathfrak{G}_0$ and $\mathfrak{G}_{0,0}$ of $\mathfrak{G}$ are similarly characterized. We shall now define some vector subspaces of these Lie algebras that will, in fact, turn out to be ideals.

Definition 3.2. We denote by $\mathfrak{G}'$ the vector subspace of $\mathfrak{G}$ consisting of all germs of vector fields $\xi \in \mathfrak{G}$ such that $\xi \mathbin{\llcorner} \omega$ is exact; that is to say, $\mathfrak{G}'$ consists of all the vector fields $(du)^\#$ where u is an arbitrary differentiable function on **X**. Similarly we denote by $\mathfrak{G}'_0$ the vector subspace of $\mathfrak{G}_0$ consisting of all the vector fields $(du)^\#$ where u is an arbitrary function on **X** with compact support. Finally, we denote by $\mathfrak{G}''_0$ the vector subspace of $\mathfrak{G}'_0$ consisting of the vector fields $\xi = (du)^\#$, where u is a function with compact support on **X** satisfying, in addition,

$$\int_{\mathbf{X}} u\omega^n = 0. \tag{3.4}$$

Remarks. In the case of the unrestricted Lie algebra $\mathfrak{G}$, one cannot define the algebra that corresponds to $\mathfrak{G}''_0$ in the case of $\mathfrak{G}_0$; clearly, if **X** is a compact manifold, then $\mathfrak{G}_0 = \mathfrak{G}$ and, in defining $\mathfrak{G}'_0 = \mathfrak{G}'$ the function u is determined by the vector field $\xi = (du)^\#$ only up to an additive constant; therefore one can always choose u so as to satisfy (3.4); thus we have

$$\mathfrak{G}'_0 = \mathfrak{G}''_0 = \mathfrak{G}'.$$

On the other hand, if **X** is not compact, the function u generating an element $\xi \in \mathfrak{G}'_0$ is uniquely determined by ξ, since it is required by definition to have compact support; therefore the condition (3.4) becomes meaningfully restrictive, and indeed

$$\dim_{\mathbf{R}} (\mathfrak{G}'_0/\mathfrak{G}''_0) = 1.$$

Similarly, using the theorems of de Rham, for all symplectic manifolds

$$\dim_{\mathbf{R}} (\mathfrak{G}/\mathfrak{G}') = \dim_{\mathbf{R}} H^1(X, \mathbf{R}) = b_1(\mathbf{X}),$$

where $b_1(\mathbf{X})$ denotes the first Betti number of **X** with respect to real coefficients (and homology with compact support) while in the case of noncompact **X**

$$\dim_{\mathbf{R}} (\mathfrak{G}_0/\mathfrak{G}'_0) = b_1^0(\mathbf{X}),$$

where b_1^0 denotes the first Betti number for homology with closed support.

We now show that the vector subspaces $\mathfrak{G}'$, $\mathfrak{G}'_0$, and $\mathfrak{G}''_0$ are indeed Lie algebra ideals of $\mathfrak{G}$ and $\mathfrak{G}_0$, respectively.

PROPOSITION 3.3 (R. S. Palais). *The commutator ideal of* $[\mathfrak{G}, \mathfrak{G}] = \mathfrak{G}^2$ *is contained in* $\mathfrak{G}'$, *while the commutators* $\mathfrak{G}_0^2 = [\mathfrak{G}_0, \mathfrak{G}_0]$ *and* $\mathfrak{G}_0^3 = [\mathfrak{G}_0, \mathfrak{G}_0^2]$ *are contained, respectively, in* $\mathfrak{G}'_0$ *and* $\mathfrak{G}''_0$.

PROOF. Let $\xi = \alpha^\#$, $\eta = \beta^\#$ be arbitrary elements of $\mathfrak{G}$, that is to say, let α and β be closed 1-forms. Then $\{\alpha, \beta\} = [\xi, \eta] \llcorner \omega$ is not only closed but indeed exact, for, with no assumption on ξ, η, and ω, we have the following identities:

$$\begin{aligned}[\xi, \eta] \llcorner \omega &= [\xi, \eta \llcorner \omega] - \eta \llcorner [\xi, \omega] \\ &= d(\xi \llcorner (\eta \llcorner \omega)) + \xi \llcorner d(\eta \llcorner \omega) - \eta \llcorner [\xi, \omega] \\ &= d(\xi \llcorner (\eta \llcorner \omega)) + \xi \llcorner [\eta, \omega] - \eta \llcorner [\xi, \omega] - \xi \llcorner (\eta \llcorner d\omega) \\ &= dw + \xi \llcorner [\eta, \omega] - \eta \llcorner [\xi, \omega] - \xi \llcorner (\eta \llcorner d\omega),\end{aligned} \tag{3.5}$$

where

$$w = \xi \llcorner (\eta \llcorner \omega);$$

thus, under the assumptions $d\omega = 0$, $[\xi, \omega] = [\eta, \omega] = 0$, we have

$$[\xi, \eta] \llcorner \omega = dw.$$

This shows that $[\mathfrak{G}, \mathfrak{G}] \subset \mathfrak{G}'$ and, obviously, also $[\mathfrak{G}_0, \mathfrak{G}_0] \subset \mathfrak{G}'_0$, if ξ and η (and hence w) have compact support.

In order to show that $\mathfrak{G}^3 \subset \mathfrak{G}''_0$, we assume that $\xi = \alpha^\#$, $\eta = \beta^\#$ belong to $\mathfrak{G}_0$; if w is defined by (3.6), we have an elementary identity.

$$\begin{aligned} w\omega^n = (\xi \llcorner (\eta \llcorner \omega))\omega^n &= -n(\xi \llcorner \omega) \wedge (\eta \llcorner \omega) \wedge \omega^{n-1} \\ &= -n\alpha \wedge \beta \wedge \omega^{n-1}.\end{aligned} \tag{3.6}$$

It follows that, if either of the two closed forms with compact support α or β is exact (say $\alpha = du$, where u has compact support),

$$\int_{\mathbf{X}} w\omega^n = -n \int_{\mathbf{X}} d(u\beta \wedge \omega^{n-1}) = 0.$$

This completes the proof of this proposition.

The next results show that the continuation of the commutator sequences yields no new ideals. For this purpose we recall the definition of the Lie algebra $\mathfrak{G}_{0,0}$ of the minigroup (Definition 1.2 and Proposition 3.1). Clearly, since $\mathfrak{G}_{0,0} \subset \mathfrak{G}'_0$, we have $[\mathfrak{G}_{0,0}, \mathfrak{G}_{0,0}] \subset \mathfrak{G}'_{0,0} \subset \mathfrak{G}''_0$, where $\mathfrak{G}'_{0,0}$ is defined to be $\mathfrak{G}_{0,0} \cap \mathfrak{G}'_0$; in other words $\mathfrak{G}'_{0,0}$ is generated by the vector fields $(du)^\#$, where u is a function which has compact support in a contractible domain U admitting a canonical coordinate system and satisfies (3.4).

LEMMA 3.4. *The Lie algebras* $\mathfrak{G}_{0,0}$ *and* $\mathfrak{G}'_{0,0}$ *coincide, respectively, with* $\mathfrak{G}'_0$ *and* $\mathfrak{G}''_0$. *The algebra* $\mathfrak{G}'$ *in the case of a noncompact manifold is generated by infinite sums with locally finite supports of elements of* $\mathfrak{G}''_0 = \mathfrak{G}'_{0,0}$.

PROOF. Let $\xi \in \mathfrak{G}'_0$. This means that $\xi = (du)^{\#}$, where u is a differentiable function on $\mathbf{X}$ with compact support. Let $\{U_\nu\}_{\nu \in J}$ be a locally finite, open cover of $\mathbf{X}$ by contractible open sets U_ν, each of which admits a canonical coordinate system $(U_\nu, {}_\nu(x))$ and let (φ_ν) be a differentiable partition of unity, where each φ_ν has compact support contained in U_ν; set $u_\nu = \varphi_\nu u$; then $(du_\nu)^{\#} \in \mathfrak{G}_{0,0}$ and all but a finite number of them vanish, since u has compact support. This shows that $(du)^{\#} \in \mathfrak{G}'_0$ can be expressed as a sum of elements in $\mathfrak{G}_{0,0}$. Now suppose that $\xi = (du)^{\#} \in \mathfrak{G}''_0$: since in this case $\mathbf{X}$ is assumed not to be compact and is connected, condition (3.4) is equivalent to the existence of a $(2n-1)$-form ψ on $\mathbf{X}$ with compact support, such that $d\psi = u\omega^n$. Let $\psi_\nu = \varphi_\nu \psi$ be a decomposition of ψ by the partition of unity (φ_ν), and define u_ν by the condition

$$u_\nu \omega^n = d\psi_\nu = d(\varphi_\nu \psi).$$

Then clearly $(du_\nu)^{\#} \in \mathfrak{G}'_{0,0}$, showing that $\mathfrak{G}''$ is generated additively by $\mathfrak{G}'_{0,0}$.

The final assertion about $\mathfrak{G}'$ now has to be proved only in the case where $\mathbf{X}$ is not compact. In this case an element $\xi = (du)^{\#}$ is defined by a function u on $\mathbf{X}$ whose support is unrestricted. Since $\mathbf{X}$ is connected and not compact, the $2n$-dimensional real cohomology group of $\mathbf{X}$ with unrestricted support is trivial; therefore the $2n$-form $u\omega^n$ can be represented by $d\psi$, where ψ is a $(2n-1)$-form. Decomposing ψ by means of the partition of unity φ_ν as in the case of $\mathfrak{G}''_0$, we show that ξ can be expressed as an infinite sum of elements $\xi_\nu \in \mathfrak{G}'_{0,0}$ with compact, locally finite supports; this completes the proof of the assertion.

LEMMA 3.5. *The Lie algebra* $\mathfrak{G}'_{0,0}$ *coincides with its own commutator.*

PROOF. Let $\xi = (du)^{\#}$ be a vector field in $\mathfrak{G}_{0,0}$ defined from a function u with compact support $K \subset U$ where U is a contractible domain in which canonical coordinates $(x) = (x^1, \ldots, x^{2n})$ can be defined. Expressing u in terms of these coordinates, the vector field ξ can be described by (3.3). In terms of these same coordinates, we have

$$\omega^n = n!\, dx^1 \wedge \cdots \wedge dx^{2n}.$$

Since $\xi \in \mathfrak{G}_{0,0}$, the function u satisfies (3.4); in other words we have

$$\int_{(x)(U)} u(x)\, dx' \cdots dx^{2n} = 0. \tag{3.7}$$

We shall show that there exists a set of $4k$ functions (v_i, w^i) $(i = 1, 2, \ldots, 2n)$ each with compact support contained in U and satisfying, like the function u, equation (3.7) and satisfying

$$du = \sum_{i=1}^{2n} \{dv_i, dw^i\}$$

or equivalently

$$\xi = (du)^{\#} = \sum_{i=1}^{2n} [(dv_i)^{\#}, (dw^i)^{\#}].$$

This means that u is determined from the other functions (using 3.3 and 3.5) by the formula

$$u(x) = \sum_{i=1}^{2n} \sum_{j=1}^{n} \left(\frac{\partial v_i}{\partial x^{2j}} \frac{\partial w^i}{\partial x^{2j-1}} - \frac{\partial v_i}{\partial x^{2j-1}} \frac{\partial w^i}{\partial x^{2j}} \right). \tag{3.8}$$

Since the function u satisfies (3.4), there are, by virtue of Gauss' theorem, $2n$ functions $z^1, \ldots, z^{2n}$ in $\mathbf{X}$ each with compact support in U such that

$$u(x) = \sum_{i=1}^{2n} \frac{\partial z^i(x)}{\partial x^i}. \tag{3.9}$$

As a matter of fact, the functions z^i can be chosen so that their support is contained in an arbitrarily small, connected, open neighborhood of the compact support K of the function u; in particular we let each of the functions z^i have support contained in a compact K_1 with $K \subset K_1 \subset U$.

Now consider the functions $\bar{v}_i = -(-1)^i z^{(i-(-1)^i)}$ and $\bar{w}^i = x^i$; disregarding the bars, they satisfy (3.8). In fact we see that, for each i,

$$\sum_{j=1}^{n} \left(\frac{\partial \bar{v}_i}{\partial x^{2j}} \frac{\partial \bar{w}^i}{\partial x^{2j-1}} - \frac{\partial \bar{v}_i}{\partial x^{2j-1}} \frac{\partial \bar{w}^i}{\partial x^{2j}} \right) = \frac{\partial z^{(i-(-1)^i)}}{\partial x^{(i-(-1)^i)}}.$$

Consequently, summing both members of this equation, we have a solution of (3.8); however, the functions $\bar{v}_i$ do not, in general, satisfy (3.4), nor do $\bar{w}_i = x^i$ even have compact support. This can be remedied in a second stage, as we shall now proceed to do.

Choose a non-negative valued, differentiable function $h(x)$ which is identically equal to 1 in K_1 and whose support is a compact domain $K_2 \subset U$. We can then replace the functions $\bar{w}^i = x^i$ by $h(x) \cdot x^i$; the new choice of $\bar{w}^i$ will not affect equation (3.8) and the resulting functions $h \cdot x^i$ have compact support, so that $(d(h \cdot x^i))^{\#} \in \mathfrak{G}_{0,0}$. Next, pick two more non-negative, nonzero, differentiable functions h' and h'' with compact supports K_3 and K_4 mutually disjunct and contained in $U - K_2$. Since

$\int_U h'\omega^n$ and $\int_U h''\omega^n$ are both positive, there exist real constants c_i' and c_i'' $(i = 1, 2, \ldots, 2n)$ such that

$$\int_U (\bar{v}_i - c_i'h')\omega^n = \int_U (h \cdot \bar{w}^i - c_i''h'')\omega^n = 0 \qquad (1 \leqq i \leqq 2n).$$

Thus we can set

$$v_i = \bar{v}_i - c_i h'; \quad w^i = h(x)x^i - c_i''h'' \qquad (1 \leqq i \leqq 2n).$$

The replacement of the $4n$ functions $\bar{v}_i$ and $\bar{w}^i$ by v^i and w^i, respectively, does not affect the computation of the Poisson brackets $\{dv_i, dw^i\}$, so that (3.8) is still satisfied, while each of the functions v_i and w^i satisfies, as does the function u, equation (3.4). Hence

$$(dv_i)^{\#} \text{ and } (dw_0^i)^{\#} \subset \mathfrak{G}_{0,0}'$$

and

$$(du)^{\#} = \sum_{i=1}^{2n} [(dv_i)^{\#}, (dw^i)^{\#}].$$

This completes the proof of Lemma 3.5.

Before stating the first main result on the Lie algebras $\mathfrak{G}$ and $\mathfrak{G}_0$, we need one more notion, that of the *symplecting pairing*.

DEFINITION 3.6. In a symplectic manifold $(\mathbf{X}, \omega)$ we define the *symplectic pairing* to be the alternating, bilinear map of $H_0^1(\mathbf{X}, \mathbf{R})$ (first cohomology group with compact support and real coefficients) into the real numbers, denoted by $(\mathbf{u}, \mathbf{v})$ as follows: let α, β be closed 1-forms representing by means of de Rham's theorem the cohomology classes $\mathbf{u}$, $\mathbf{v}$, respectively. Then the symplectic pairing is given by

$$(3.10) \qquad (\mathbf{u}, \mathbf{v}) = \int_{\mathbf{X}} \alpha \wedge \beta \wedge \omega^{n-1}.$$

It is well known that the symplectic pairing is nonsingular in the case of the symplectic structure subordinate to a compact Kähler manifold, while it is identically zero in the case of the natural symplectic structure of the cotangent bundle of any differential manifold or, more generally, in any (necessarily noncompact) symplectic manifold with $n \geqq 2$, where ω^{n-1} is cohomologous to zero.

THEOREM 3.7. *Let* $(\mathbf{X}, \omega)$ *be a symplectic manifold, let* $\mathfrak{G}$ *be the Lie algebra of all differentiable, symplectic, global vector fields, and let* $\mathfrak{G}_0$ *be the Lie subalgebra of* $\mathfrak{G}$ *consisting of vector fields with compact support. Then the commutator algebra* $\mathfrak{G}^2 = [\mathfrak{G}, \mathfrak{G}]$ *coincides with* $\mathfrak{G}'$ *and* $\mathfrak{G}/\mathfrak{G}^2$ *is isomorphic to* $H^1(\mathbf{X}, \mathbf{R})$*; furthermore, if* $\mathbf{X}$ *is not compact, the commutator* *algebra* $\mathfrak{G}_0^2 = [\mathfrak{G}_0, \mathfrak{G}_0]$ *coincides with either* $\mathfrak{G}_0'$ *or* $\mathfrak{G}_0''$ *depending on whether the symplectic pairing of* $H_0^1(\mathbf{X}, \mathbf{R})$ *is, respectively, nontrivial or identically zero; the third derived algebra* $\mathfrak{G}_0^3 = [\mathfrak{G}_0, \mathfrak{G}^2]$ *coincides in all cases with* $\mathfrak{G}_0''$. *The Lie algebras* $\mathfrak{G}'$ *and* $\mathfrak{G}_0''$ *are each equal to their own commutator algebras.*

PROOF. From Proposition 3.3 we have the inclusion relations

$$[\mathfrak{G}, \mathfrak{G}] \subset \mathfrak{G}', \qquad [\mathfrak{G}_0, \mathfrak{G}_0] \subset \mathfrak{G}_0', \qquad [\mathfrak{G}_0, \mathfrak{G}_0'] \subset \mathfrak{G}_0''.$$

In Lemma 3.5 it was established that $[\mathfrak{G}_0'', \mathfrak{G}_0''] \supset \mathfrak{G}_0''$. Since $\dim_{\mathbf{R}} (\mathfrak{G}_0'/\mathfrak{G}_0'') = 1$, this means that $\mathfrak{G}_0^2$ coincides with either $\mathfrak{G}_0'$ or $\mathfrak{G}_0''$. It follows from equation (3.6) in the proof of Proposition 3.3 that $\mathfrak{G}_0^2 = \mathfrak{G}_0'$ if and only if the symplectic pairing is nontrivial, and otherwise $\mathfrak{G}_0^2 = \mathfrak{G}_0''$ (this will be the case, in particular, if $H_0^1(\mathbf{X}, \mathbf{R}) = 0$ or if ω^{n-1} is cohomologous to zero). The corresponding conclusions in the case of $\mathfrak{G}$ can be obtained by showing that $\mathfrak{G}'$ is equal to its own commutator subalgebra. This can be done by the following argument. There exists an open cover $\{U_\nu\}_{\nu \in J}$ of $\mathbf{X}$ by contractible canonical coordinate domains with the following additional property: there exists a partition of the indexing system J into a finite collection of subsets, $J = \bigcup_{\mu=1}^{N} J_\mu$ such that, for each μ, the open sets $(U_\nu)_{\nu \in J_\mu}$ are pairwise disjunct. Let (φ_ν) be a partition of unity subordinate to $\{U_\nu\}_{\nu \in J}$ and let $e_\mu = \sum_{\nu \in J_\mu} \varphi_\nu$; then $(e_\mu)_{\mu=1}^N$ is a finite partition of unity; any vector field $\xi \in \mathfrak{G}'$ can be expressed as $(du)^\#$ for some function u such that (after adding a suitable constant to u, if $\mathbf{X}$ is compact) $u\omega^n = d\psi$ for some $(2n - 1)$-form ψ, and let $\psi_\mu = e_\mu \cdot \psi$. Let u_μ be defined by the equation $u_\mu \omega^n = d\psi_\mu$ and consider the vector fields $\xi_\mu = (du_\mu)^\#$. By using the arguments of Lemma 3.5 in each U_ν for $\nu \in J_\mu$ we see that, since the domains $\{U_\nu\}_{\nu \in J_\mu}$ are pairwise disjunct, ξ_μ ($\mu = 1, 2, \ldots, N$) can be expressed as a sum of commutators of $2n$-pairs of vector fields with support in $\bigcup_{\nu \in J_\mu} U_\nu$; the sum of these N sums of commutators reproduces the given $\xi \in \mathfrak{G}'$ represented as a finite sum of commutators of vector fields belonging to $\mathfrak{G}'$. This completes the proof of the asserted commutator relations. This concludes the proof of the theorem.

An immediate application of Theorem 3.7, to be used in the next section, is that we can obtain, using de Rham's theorem, a complete set of Abelian representations of the Lie algebras $\mathfrak{G}$ and $\mathfrak{G}_0$ as follows. For any differentiable, compact, integral 1-cycle γ and any $\xi \in \mathfrak{G}$, we define

$$J(\xi)(\gamma) = \int_\gamma \xi \,\llcorner\, \omega. \tag{3.11}$$

Similarly, if $\xi \in \mathfrak{G}_0$ then, for any integral, locally finite, differentiable, integral 1-cycle γ we define $J(\gamma)(\xi)$ by the same formula. It is clear that the value of the integral depends only on the homology class of γ: it establishes the homomorphism of $\mathfrak{G}$ (respectively $\mathfrak{G}_0$) into hom $({}_0H_1(X, \mathbf{R}), \mathbf{R}) = H^1(X, \mathbf{R})$ [respectively into hom $(H_1(X, \mathbf{R}), \mathbf{R}) = H_0^1(X, \mathbf{R})$].

We have seen that these two homomorphisms are surjective and are Lie algebra homomorphisms, if we regard the cohomology groups as Abelian Lie algebras. We can also construct geometrically a Lie algebra representation J' of $\mathfrak{G}_0'$ with kernel $\mathfrak{G}_0''$ onto the additive group $\mathbf{R}$ as follows. If $\mathbf{X}$ is not compact, for each point $\mathbf{x} \in \mathbf{X}$ let $\gamma_{\mathbf{x}}$ be an infinite, locally finite, differentiable 1-chain whose boundary $\partial\gamma_{\mathbf{x}}$ equals $\mathbf{x}$. If $\xi \in \mathfrak{G}_0'$, define the function $\psi(\xi): \mathbf{X} \to \mathbf{R}$ by the equation

$$\psi(\xi)(\mathbf{x}) = \int_{\gamma_{\mathbf{x}}} \xi \llcorner \omega. \tag{3.12}$$

Since $\gamma_{\mathbf{x}}$ is defined modulo the group of 1-cycles and $\xi \in \mathfrak{G}_0'$, the function $\psi(\xi)$ is independent of the choice of the chain; in fact it coincides with the function u with compact support such that $(du)^{\#} = \xi$. The representation of $\mathfrak{G}_0'$ into $\mathbf{R}$ with kernel $\mathfrak{G}_0''$ then is obviously given by the integral

$$J'(\xi) = \int_{\mathbf{X}} u\omega^n = \int_{\mathbf{X}} \psi(\xi)\omega^n.$$

We have at the moment no satisfactory, faithful geometrical representation of the non-Abelian nilpotent Lie algebra $\mathfrak{G}_0/\mathfrak{G}_0''$ in the cases where $[\mathfrak{G}_0, \mathfrak{G}_0] = \mathfrak{G}_0'$, i.e., where the symplectic pairing is not identically zero. The next result concerns the structure of Lie algebra ideals of $\mathfrak{G}$ and $\mathfrak{G}_0$ that give rise to normal subgroups. It is clear that any normal subgroup $\mathbf{H}$ of any (finite or infinite dimensional) Lie group $\mathbf{G}$ of differentiable transformations has as its Lie algebra an ideal in the Lie algebra of $\mathbf{G}$. The converse, however, is false, unless the germs of the actions of $\mathbf{G}$ at each point are uniquely determined by their local power series expansion; this would happen, for instance, when $\mathbf{G}$ is finite dimensional or if its action is real analytic. In the case at hand, however, when $\mathbf{G}$ is the group of all automorphisms of a symplectic manifold $(\mathbf{X}, \omega)$, we know that there are nontrivial elements of $\mathbf{G}$ that act trivially in a nonempty open set, and nontrivially in another. In particular let U be an open subdomain of $\mathbf{X}$ with $U \neq \varnothing$, $\overline{U} \neq \mathbf{X}$; then the Lie subalgebra $\mathfrak{G}_U$ of $\mathfrak{G}$ consisting of all vector fields $\xi \in \mathfrak{G}$ that vanish identically in U is locally transitive in $X - \overline{U}$; clearly it is an ideal in $\mathfrak{G}$; however, the group $\mathbf{G}_U$ of global transformations of $\mathbf{X}$ defined by $\mathfrak{G}_U$ is not a normal subgroup of $\mathbf{X}$, since U is not invariant under $\mathbf{G}$. This fact justifies the restrictive condition imposed below.

Definition 3.8. Let $\mathbf{H}$ be a Lie group of differentiable transformations of a manifold $\mathbf{X}$ and $\mathfrak{H}$ its Lie algebra of vector fields. A Lie algebra ideal

$\mathfrak{H}' \subset \mathfrak{H}$ is called stable, if it is invariant under the adjoint representation of $\mathbf{H}$ in $\mathfrak{H}$.

It is easy to verify that the local isomorphism classes of normal Lie subgroups of $\mathbf{H}$ are in one-to-one correspondence with the stable ideals of the Lie algebra of $\mathbf{H}$.

THEOREM 3.9. *Let* $(\mathbf{X}, \omega)$ *be a* $2n$*-dimensional symplectic manifold, and let* $\mathbf{G}$ *be the group of all global, symplectic automorphisms of* $(\mathbf{X}, \omega)$*; let* $\mathbf{G}_0$ *be the subgroup of* $\mathbf{G}$ *consisting of the transformations with compact support and* $\mathbf{G}_0''$ *its third derived group. Then every nontrivial ideal of the Lie algebra* $\mathfrak{G}$ *of* $\mathbf{G}$ *(or* $\mathfrak{G}_0$ *of* $\mathbf{G}_0$*) that is stable under the adjoint action of* $\mathbf{G}_0$ *(indeed even of* $\mathbf{G}_0''$ *alone) contains the Lie algebra* $\mathfrak{G}_0''$ *of* $\mathbf{G}_0''$.

The proof of this theorem requires a lemma which is the analogue of the lemma of Palais–Cerf in the case of symplectic transformations. It is formulated here in a form somewhat stronger than strictly necessary because it is of independent interest, and the proof given here is independent of that of the Palais–Cerf Lemma for the sake of completeness.

LEMMA 3.10. *The identity component* $\mathbf{G}_0''$ *of the third derived group of* $\mathbf{G}_0$ *is strongly transitive on* $\mathbf{X}$ *in the following sense. Let* $\bar{\mathbf{g}}$ *be a germ of a symplectic transformation at a point* $\mathbf{x}_0 \in \mathbf{X}$ *with target* $\bar{\mathbf{g}}(\mathbf{x}_0) = \mathbf{x}_1$*, let* γ *be a simple, differentiable path from* $\mathbf{x}_0$ *to* $\mathbf{x}_1$ *parametrized by* t $(0 \leqq t \leqq 1$, $\gamma(0) = \mathbf{x}_0$, $\gamma(1) = \mathbf{x}_1)$ *and let* $|\gamma|$ *be the corresponding point set. Then for any neighborhood* U *of* $|\gamma|$ *there exists a global transformation* $\mathbf{g} \in \mathbf{G}_0''$ *that represents the given germ* $\bar{\mathbf{g}}$ *at* $\mathbf{x}_0$ *and equals the identity outside of* U*. Indeed* $\mathbf{g}$ *can be achieved by a* $\mathbf{G}_0''$*-flow that is trivial outside* U.

PROOF. The proof consists of constructing a sequence of three successive differentiable flows in $\mathbf{X}$ (trivial outside of the given neighborhood U of $|\gamma|$); the composition of successive differentiable flows into a differentiable flow is elementary. The first stage will bring the point $\mathbf{x}_0$ to $\mathbf{x}_1$ along the path γ; the second one, leaving $\mathbf{x}_1$ fixed, will rotate the tangent space $\mathbf{X}_{\mathbf{x}_1}$ of $\mathbf{X}$ at $\mathbf{x}_1$ so that the composition of this flow with the first stage will match the first order jet of the given germ $\bar{\mathbf{g}}$ of a symplectic map; the third stage will be a symplectic flow, trivial as the previous ones outside U which leaves $\mathbf{x}_1$ and each tangent vector at $\mathbf{x}_1$ fixed and, after composing it with the two previous stages, matches the germ $\bar{\mathbf{g}}$ itself.

Stage 1. The flow carrying $\mathbf{x}_0$ to $\mathbf{x}_1$ can be actually imbedded in a one-parameter group, by constructing a symplectic vector field $\xi \in \mathfrak{G}_0''$ on $\mathbf{X}$ with compact support in U whose restriction to $|\gamma|$ coincides with the tangent vectors $d\gamma(t)$ along the path. Since ξ has compact support, it generates a one-parameter subgroup of $\mathbf{G}_0''$ acting globally on $\mathbf{X}$ and trivially outside U, whose action on $\mathbf{x}_0$ is an extension to all of $\mathbf{R}$ of the parameter domain of the path γ. The tangent vector $d\gamma(t)$ at each point $\gamma(t)$ of $|\gamma|$ yields, by contraction with ω, a 1-form $d\gamma(t) \llcorner \omega = \alpha(t)$ which is orthogonal to $d\gamma(t)$. Let u be a function on $\mathbf{X}$ with the following properties:

(i) At each point $\gamma(t)$ of $|\gamma|$, $du(\gamma(t)) = \alpha(t)$.
(ii) The function u has compact support contained in U.
(iii) The integral $\int_{\mathbf{X}} u\omega^n$ vanishes.

[Note that property (i) implies that u is constant along $|\gamma|$.] It is clear that such a function u exists and that the vector field $\xi = (du)^{\#}$ has all the required properties to achieve the flow carrying $\mathbf{x}_0$ to $\mathbf{x}_1$. This shows that $\mathbf{G}_0''$ is pointwise transitive on $\mathbf{X}$.

Stage 2. We show now that $\mathbf{G}_0''$ is transitive on the bundle of symplectic tangential frames. We are given a germ $\bar{\mathbf{g}}_1$ of a symplectic transformation at a point $\mathbf{x}_1$ with $\bar{\mathbf{g}}_1(\mathbf{x}_1) = \mathbf{x}_1$ and we want to exhibit a $\mathbf{G}_0''$-flow $(\mathbf{g}(t))_{0 \leqq t \leqq 1}$ in $\mathbf{X}$ leaving $\mathbf{x}_1$ and everything outside a neighborhood of U of $\mathbf{x}_1$ pointwise fixed, such that $\mathbf{g}(0)$ is the identity and $d\mathbf{g}(1)(\mathbf{x}_1) = d\bar{\mathbf{g}}_1(\mathbf{x}_1)$. The fibre map $d\bar{\mathbf{g}}(\mathbf{x}_1)$ of the tangent space $\mathbf{X}_{\mathbf{x}_1}$ at $\mathbf{x}_1$ belongs to the $2n$-dimensional, real symplectic group $Sp(\mathbf{X}_{\mathbf{x}_1}, \omega(\mathbf{x}_1))$ which is known to be arcwise connected. Let $S(t)$ $(0 \leqq t \leqq 1)$ be a differentiable path in this group from the identity to $d\bar{\mathbf{g}}_1(\mathbf{x}_1)$ and let $s'(t) = \dfrac{dS(t)}{dt} \circ S^{-1}(t)$ its logarithmic derivative, so that $s'(t)$ belongs to the Lie algebra $\mathfrak{sp}\ (\mathbf{X}_{\mathbf{x}_1}, \omega(\mathbf{x}_1))$. This means that the bilinear form $\omega(\mathbf{x}_1) \circ s'(t)$ on $\mathbf{X}_{\mathbf{x}_1}$ defined by $(\omega(\mathbf{x}_1) \circ s'(t))(v, v') = \omega(\mathbf{x}_1)(s'(t)(v), v')$ is symmetric for each t $(0 \leqq t \leqq 1)$. We construct now a differentiable function $u\colon [0, 1] \times \mathbf{X} \to \mathbf{R}$, i.e., a differentiable one-parameter of functions $(u_t)_{0 \leqq t \leqq 1}$ on $\mathbf{X}$ $(u_t(x) = u(t, x))$ with the following properties for each value of the parameter t:

(i) The function u_t has compact support, contained in U.
(ii) The differential du_t of u_t vanishes at $\mathbf{x}_1$.
(iii) The Hessian $\mathbf{H}_{\mathbf{x}_1}[u_t]$ of u_t at $\mathbf{x}_1$ equals the symmetric bilinear form $\omega(\mathbf{x}_1) \circ s'(t)$.
(iv) The integral $\int_{\mathbf{X}} u_t\omega^n$ vanishes.

The construction of the family of functions (u_t) with the four properties just described is a simple matter. We now consider the one-parameter family of vector fields $\xi_t = (du_t)^{\#}$ $(0 \leqq t \leqq 1)$ and look at the differential equation.

$$\text{(3.13)} \qquad \frac{\partial \mathbf{g}(t, \mathbf{x})}{\partial t} = \xi(t)(\mathbf{g}(t, \mathbf{x})); \qquad \mathbf{g}(0, \mathbf{x}) = \mathbf{x}.$$

Property (i) of u_t and the definition of $\xi(t)$ imply that $(\mathbf{g}_t)_{0 \leqq t \leqq 1}$ ($\mathbf{g}_t(\mathbf{x}) = \mathbf{g}(t, \mathbf{x})$) has a global solution on $[0, 1] \times \mathbf{X}$, describing a $\mathbf{G}'_0$-flow in $\mathbf{X}$ that is trivial outside U; property (ii) implies that $\mathbf{x}_1$ is fixed throughout the flow. Property (iii) means that, at the fixed point $\mathbf{x}_1$ of this flow, the one-parameter family $(d\mathbf{g}_t(\mathbf{x}_1))$ of actions of $(\mathbf{g}_t)$ on the tangent space $\mathbf{X}_{\mathbf{x}_1}$ at $\mathbf{x}_1$ satisfies the same differential equation with the same initial condition as the path $S(t)$ in the linear symplectic group $Sp(\mathbf{X}_{\mathbf{x}_1}, \omega(\mathbf{x}_1))$; therefore $d\mathbf{g}_1(\mathbf{x}_1) = S(1) = d\bar{\mathbf{g}}(\mathbf{x}_1)$; the second stage of the process is then completed merely by remarking that property (iv) of u_t implies that the flow is actually in the commutator subgroup $\mathbf{G}''_0$ of $\mathbf{G}'_0$.

Stage 3. In order to complete the proof of the lemma, we now suppose that the given germ $\bar{\mathbf{g}}$ of a symplectic transformation leaves $\mathbf{x}_1$ fixed and acts as the identity on the tangent bundle $\mathbf{X}_{\mathbf{x}_1}$ of $\mathbf{x}_1$ and try to extend it to a global, symplectic $\mathbf{G}''_0$-flow on $\mathbf{X}$, trivial outside a given neighborhood U of $\mathbf{x}_1$.

Let $\{U', (x)\}$ be a local, canonical coordinate system in a neighborhood U' of $\mathbf{x}_1$, $U' \subset U$, with $x^i(\mathbf{x}_1) = 0$ $(1 \leqq i \leqq 2n)$ and including in its range the closed ball $\sum_{i=1}^{2n} (x^i)^2 \leqq a_0^2$ for some $a_0 > 0$; let U_r, for any positive $r \leqq a_0$, denote the subdomain of U' defined by $\sum_{i=0}^{2n} (x^i)^2 < r^2$ and let the germ $\bar{\mathbf{g}}$ be represented by a local symplectic map $x^i \to y^i = \bar{g}^i(x^1, \ldots, x^{2n})$ $(1 \leqq i \leqq 2n)$ in a neighborhood U_{a_1} of $\mathbf{x}_1$ $(0 < a_1 \leqq a_0)$. We have, thus, by assumption:

$$\sum_{j=1}^{n} dy^{2j-1} \wedge dy^{2j} = \sum_{j=1}^{n} dx^{2j-1} \wedge dx^{2j} \tag{3.14}$$

and

$$\bar{g}^i(x) = x^i + O(|x|^2), \quad \text{where } |x|^2 = \sum_{i=1}^{2n} (x^i)^2. \tag{3.15}$$

Let $h(s)$ be a monotone decreasing C^∞ function of $s \geqq 0$ which equals identically 1 for $0 \leqq s \leqq b$, is strictly positive for $b < s < 2b$, and vanishes for all $s \geqq 2b$, for some positive $b < \frac{1}{2}a_1$, sufficiently small for later purposes. Consider the map of $\mathbf{f}$: $[0, 1] \times \mathbf{X} \to \mathbf{X}$ defined as follows:

$$\mathbf{f}(t, \mathbf{x}) = \mathbf{x} \quad \text{if either } t = 0 \text{ or } \mathbf{x} \notin U_{2b}; \tag{3.16}$$

otherwise, in terms of the local coordinates $(x^1, \ldots, x^{2n})$, i.e., setting $f_t^i(x^1(\mathbf{x}), \ldots, x^{2n}(\mathbf{x})) = x^i(\mathbf{f}(t, \mathbf{x}))$,

(3.17)

$$f_t^i(x) = (t \cdot h(|x|))^{-1} \bar{g}^i(t \cdot h(|x|)x^1, \ldots, t \cdot h(|x|)x^{2n}), \qquad (i = 1, \ldots, 2n).$$

One verifies first that, because of (3.15), equations (3.16) and (3.17) match each other in a C^∞ way; furthermore, if the constant b is sufficiently small, the map $\mathbf{f}_t: \mathbf{X} \to \mathbf{X}$ ($\mathbf{f}_t(\mathbf{x}) = \mathbf{f}(t, \mathbf{x})$) is a diffeomorphism for each t (with $\mathbf{f}_0$ equal to the identity), while f_1 coincides with $\bar{g}$ in U_b and each $\mathbf{f}_t$ is symplectic in U_b. Thus the diffeotopy $(\mathbf{f}_t)_{0 \leqq t \leqq 1}$ is symplectic for each t except in the interior of $U_{2b} - U_b$ for $t > 0$. Furthermore, one verifies directly that in U_b we have $\dfrac{\partial f^i}{\partial t} = O(|x|^2)$ uniformly in t, so that the flow acts as the identity on the tangent space at $\mathbf{x}_1$. In order to replace $(\mathbf{f}_t)$ by a diffeotopy that is symplectic everywhere in $\mathbf{X}$ and for all t, we shall construct another diffeotopy $(\mathbf{g}'_t)_{0 \leqq t \leqq 1}$ acting as the identity for $t = 0$, and, for $t > 0$, outside $\bar{U}_{b'} - \bar{U}_{b/2}$ for some b' $(2b < b' < a_1)$ so that the composite diffeotopy $(g_t)_{0 \leqq t \leqq 1}$ with $g_t = f_t \circ (g'_t)^{-1}$ satisfies $\omega \circ \mathbf{g}_t = \omega$ for each t everywhere in $\mathbf{X}$. Let $\omega_t = \omega \circ \mathbf{f}_t$; then $\omega_t = \omega_0 = \omega$ except in the spherical shell $U_{2b} - \bar{U}_b$. Let b_1 be arbitrary constant with $2b < b_1 < a_1$; then $\omega_t - \omega$ is closed and cohomologous to 0 with compact support in the open submanifold $U_{b_1} - \bar{U}_{b/2}$. One can therefore find a differentiable, one-parameter family of 1-forms α_t with compact support in $U_{b_1} - \bar{U}_{b/2}$ satisfying for each t $(0 \leqq t \leqq 1)$

$$d\alpha_t = \frac{\partial \omega_t}{\partial t} \quad \text{and} \quad \alpha_0 = 0.$$

Consider the corresponding one-parameter family of vector fields $(\eta_t)_{0 \leqq t \leqq 1}$ uniquely determined by the equation $\eta_t \llcorner \omega_t = \alpha_t$ and set up the differential equation $\dfrac{\partial \mathbf{g}'_t}{\partial t} = \eta_t \circ \mathbf{g}'_t$ with $\mathbf{g}'_0 =$ identity.

An easy calculation shows that, if we set $\omega'_t = \omega \circ \mathbf{g}'_t$, we have

$$\frac{\partial \omega'_t}{\partial t} = [\eta_t, \omega'_t] = \frac{\partial \omega_t}{\partial t}, \qquad \omega'_0 = \omega_0 = \omega.$$

Thus we have defined another flow $(\mathbf{g}'_t)_{0 \leqq t \leqq 1}$ which has the same effect on ω as $(\mathbf{f}_t)$ in $U_{b_1} - U_{b/2}$, i.e., $\omega_t = \omega \circ \mathbf{f}_t = \omega \circ \mathbf{g}'_t$ and leaves the complement pointwise fixed. Consequently, if we set

$$\mathbf{g}_t = \mathbf{f}_t \circ (\mathbf{g}'_t)^{-1} \qquad (0 \leqq t \leqq 1),$$

we obtain a flow that coincides with $\mathbf{f}_t$ in $U_{b/2} \cup (\mathbf{X} - U_{b_1})$ and leaves ω invariant. Since this flow belongs clearly to the minigroup $\mathbf{G}_{0,0}$ which, by Lemma 1 of Theorem 3.4 coincides with $\mathbf{G}'_0$, there exists a differentiable one-parameter family of real valued, C^∞ functions $(v_t)_{0 \leqq t \leqq 1}$ with support in U_{b_1} such that $(\mathbf{g}_t)$ is the solution of the differential equation

$$\frac{\partial \mathbf{g}_t}{\partial t} = (dv_t)^{\#} \circ \mathbf{g}_t.$$

In order to replace this flow by a $\mathbf{G}_0''$-flow, we need only to replace (v_t) by a modified family $(v_t + w_t)$, where w_t is a family of differentiable functions with support in $U_a - \overline{U}_{b_1}$ such that

$$\int (v_t + w_t)\omega^n = 0 \qquad (0 \leqq t \leqq 1).$$

This concludes the proof of the lemma.

PROOF OF THEOREM 3.9. Suppose that $\mathfrak{J}$ is a nontrivial stable Lie algebra ideal of $\mathfrak{G}_0''$ other than 0 and let ξ_0'' be a nonzero vector field in $\mathfrak{J}$. Let U_0 be a neighborhood of a point $\mathbf{x}_0 \in \mathbf{X}$, in which ξ is everywhere $\neq 0$. Then there exists, in a smaller neighborhood U_1 of $\mathbf{x}_0$, a canonical coordinate system $(x^1, \ldots, x^{2n})$ such that $x^i(x_0) = 0$ and $\xi|_{U_1} = \dfrac{\partial}{\partial x^1}\cdot$ The existence of such a coordinate system can be verified readily from any of the known proofs of Darboux's theorem on the existence of canonical coordinates. In other words $\xi|_{U_1} = (dx^2)^{\#}$. Now suppose that η is any given vector field in $\mathfrak{G}_0''$. The theorem is proved if we can show that η is an element of the $\mathbf{G}_0''$-stable ideal of vector fields generated by ξ.

The condition $\eta \in \mathfrak{G}_0''$ means that the uniquely determined function v with compact support such that $(dv)^{\#} = \eta$ satisfies $\int v\omega^n = 0$; therefore there exists a $(2n - 1)$-form ψ with compact support K in X such that $d\psi = v\omega^n$. We choose, using Lemma 3.10, a finite, open cover $\{V_\nu, {}_\nu(x)\}_{\nu\in J}$ of a connected, open neighborhood of K by canonical coordinate domains $({}_\nu(x) = ({}_\nu x^1, \ldots, {}_\nu x^{2n}))$, each with the same range $\sum_{i=1}^{2n} ({}_\nu x^i)^2 < a^2$, with a sufficiently small, so as to satisfy the following additional property: for each $\nu \in J$ and each index i $(1 \leqq i \leqq 2n)$ the local symplectic map $\mathbf{f}_{\nu,i}\colon V_\nu \to U_1$ defined by

$$x^1 \circ \mathbf{f}_{\nu,i} = {}_\nu x^i$$
$$x^2 \circ \mathbf{f}_{\nu,i} = -(-1)^i{}_\nu x^{(i-(-1)^i)}$$

and

$$x^{2j+1} \circ \mathbf{f}_{\nu,i} = {}_\nu x^{(i+2j)}$$
$$x^{2j+2} \circ \mathbf{f}_{\nu,i} = -(-1)^i{}_\nu x^{(i+2j-(-1)^i)}, \qquad (2 \leqq j \leqq n-1),$$

(where the indices $(i + 2j)$, $(i + 2j - (-1)^i)$ are to be interpreted as integers modulo $(2n)$) can be extended to a global $\mathbf{G}_0''$-automorphism of $\mathbf{X}$, which we denote again by $\mathbf{f}_{\nu,i}$. Let $(\varphi_\nu)_{\nu\in J}$ be a partition of unity in K subordinate to the covering (V_ν) and set up the corresponding partitions $\psi = \sum_\nu \psi_\nu$, $v = \sum_\nu v_\nu$, $\eta = \sum_\nu \eta_\nu$ defined by setting, respectively, $\psi_\nu = \varphi_\nu\psi$, $v_\nu\omega^n = d\psi_\nu$, $\eta_\nu = (dv_\nu)^{\#}$. Then each $\eta_\nu \in \mathfrak{G}_0''$. We can prove the theorem by showing that each $\eta_\nu \in \mathfrak{J}$.

We first look at the corresponding $(2n-1)$-form ψ_ν. In terms of the local coordinates ${}_\nu(x)$ we have

$$\psi_\nu = \sum_{i=1}^{2n} (-1)^{i+1}\psi_\nu^i({}_\nu(x))\, d({}_\nu x^1) \wedge \cdots \wedge d({}_\nu x^{i-1}) \wedge d({}_\nu x^{i+1}) \wedge \cdots \wedge d({}_\nu x^{2n}).$$

Each of the $2n$ components $\psi_\nu^i({}_\nu(x))$ of ψ_ν (extended by the constant zero outside of V_ν) gives rise to a vector field $\zeta_\nu^i = (d\psi_\nu^i({}_\nu(x)))^\# \in \mathfrak{G}_0'$; for each $\nu \in J$ the relationship between the $2n$ vector fields ζ_ν^i $(1 \leqq i \leqq 2n)$ and η_ν can be described by the equation

$$\eta_\nu = \sum_{i=1}^{2n} \left[\frac{\partial}{\partial({}_\nu x^i)}, \zeta_\nu^i\right].$$

We now apply to each term the mapping $\mathbf{f}_{\nu,i}$, which sends $\dfrac{\partial}{\partial({}_\nu x^i)}$ into the restriction of local vector field $\dfrac{\partial}{\partial x^i}$ to some subdomain of U_1, i.e., into a restriction of the given vector field ξ, and ζ_ν^i into a vector field $d\mathbf{f}_{\nu,i}(\zeta_{\nu,i}) = \zeta_{\nu,i}^i$ with compact support in $\mathbf{f}_{\nu,i}(V_\nu) \subset U_1$. In other words we have the representation of

$$\eta = \sum_{\nu\in J} \sum_{i=1}^{2n} [d(f_{\nu,i}^{-1})(\xi), \zeta_\nu^i],$$

where each $\zeta_\nu^i \in \mathfrak{G}_0'$. We can modify each ζ_ν^i by adding to it a suitably chosen vector field $\eta_{\nu,i}' \in \mathfrak{G}_0'$ with support disjunct[2] from the compact support of $d(f_{\nu,i}^{-1}) \circ \xi$, so that the sum $\zeta_\nu^i + \eta_{\nu,i}'$ belongs to $\mathfrak{G}_0''$; this modification of ζ_ν^i does not affect the Lie product. This shows that the given vector field $\eta \in \mathfrak{G}_0''$ belongs to the $\mathbf{G}_0''$-stable ideal generated by ξ, thereby completing the proof of the theorem.

An analogous result on the commutator algebra $\mathfrak{G}'$ without restriction at infinity would be desirable. This can be achieved in an elementary way by making suitable assumptions of completeness in an appropriate topology, using the results just established, as follows.

COROLLARY. *Let* $(\mathbf{X}, \omega)$ *be a noncompact symplectic manifold,* $\mathfrak{G}$ *the Lie algebra of all differentiable automorphisms of* $(\mathbf{X}, \omega)$ *and* $\mathfrak{G}'$ *its commutator algebra. Then any nonzero ideal in* $\mathfrak{G}'$ *that is both: (i) invariant under the adjoint action of* $\mathbf{G}'$, *(ii) either closed under infinite sums with locally finite support or complete under a compact-open extension of a function-space topology on the stalks of the sheaf of germs, must be all of* $\mathfrak{G}'$.

[2] The case where $\mathbf{X}$ is compact and the support of ξ is all of $\mathbf{X}$ does not present any essential difficulty, since ξ could be replaced without loss of generality by its Lie product (assumed $\neq 0$) with a vector field in $\mathfrak{G}_0''$ with smaller support.

The proof of this corollary can be omitted.

Summarizing the conclusions of this section, we see that the finite-dimensional representations of **G** and $\mathbf{G}_0$ are very limited. They were described completely at the end of the proof of Theorem 3.7.

4. Representation of the automorphism groups

In this section we shall show how and to what extent the finite-dimensional representations of the Lie algebras $\mathfrak{G}$, $\mathfrak{G}_0$, and $\mathfrak{G}_0'$ can be "lifted" to representations of the corresponding Lie groups **G**, $\mathbf{G}_0$, and $\mathbf{G}_0'$, respectively. Since these representations are either Abelian or (in the case of $\mathfrak{G}_0/\mathfrak{G}_0''$ when **X** is not compact and the symplectic pairing in $H_0^1(\mathbf{X}, \mathbf{R})$ is nontrivial) nilpotent, it is clear that they give rise naturally to representations of the respective universal covering groups (analytic groups, in the terminology of Chevalley) $\tilde{\mathbf{G}}$, $\tilde{\mathbf{G}}_0$, and $\tilde{\mathbf{G}}_0'$ into quotient groups of the corresponding Lie groups $H^1(\mathbf{X}, \mathbf{R})$, $H_0^1(\mathbf{X}, \mathbf{R}) + \mathbf{R}$ (this is the semidirect product defined by means of the symplectic pairing) and **R**. However, since we have at present little or no knowledge about either the group of components (zero-dimensional homotopy group) of **G**, $\mathbf{G}_0$, or $\mathbf{G}_0'$, such information about their universal covering is not especially interesting except as an intermediate step in attacking the representation problem of the groups themselves.

In the remarks after the proof of Theorem 3.7 we exhibited the geometric realization of some Abelian representations of the Lie algebras $\mathfrak{G}$, $\mathfrak{G}_0$, and $\mathfrak{G}_0'$, using de Rham's theorem. We shall apply these geometric realizations to obtain the corresponding Abelian representations at the first of the analytic groups.

DEFINITION 4.1. Given a Lie group **L** (finite- or infinite-dimensional) we denote by $\Omega(\mathbf{L})$ the right path-semigroup of **L**; this is the semigroup whose elements are equivalence classes under right translations of piecewise differentiable, oriented paths in **L**, composed by composition of paths, after a right translation to match points of transition from each path to the next. We denote by $\Omega_0(\mathbf{L})$ the normal sub-semigroup of **L** represented by closed paths that are homotopically trivial.

It is clear that the quotient semigroup $\Omega(\mathbf{L})/\Omega_0(\mathbf{L})$ is naturally isomorphic to the universal covering group $\tilde{\mathbf{L}}$ of the identity component of **L**.

LEMMA 4.2. *Given a (finite- or infinite-dimensional) Lie group* **L** *of differentiable transformations of a manifold* **X** *and a differentiable representation* Φ *of* $\Omega(\mathbf{L})$ *into a Lie group* **H**, *this representation is trivial on* $\Omega_0(\mathbf{L})$ *(and thereby induces a representation of* $\tilde{\mathbf{L}}$*), if and only if the associated* *linear map $d\Phi$ of the Lie algebra $\mathfrak{L}$ of* **L** *into that of* **H** *is a Lie algebra homomorphism.*

PROOF. It is understood that a representation $\Phi: \Omega(\mathbf{L}) \to \mathbf{H}$ is differentiable, if and only if any (piecewise) differentiable, finite-dimensional parametric family of paths in **L** has as its image in **H** under Φ a corresponding (piecewise) differentiable family of elements in **H**. Given a differentiable representation Φ, then to any (piecewise) differentiable path σ: $[0, a] \to \mathbf{L}$ in **L** we associate the one-parameter family $\sigma_{|[0,t]}$: $[0, t] \to \mathbf{L}$ $(0 \leqq t \leqq a)$ of paths. The image of this family is a (piecewise) differentiable path in **H**, originating at the identity, and whose initial tangent

$$\dot{\Phi}_0 = \left.\frac{d\Phi(\sigma_{|[0,t]})}{dt}\right|_{t=0}$$

depends only on the initial tangent

$$\dot{\sigma}_0 = \left.\frac{d\sigma_{|[0,t]}}{dt}\right|_{t=0}$$

of the given path σ. The map of Lie algebras thus obtained is clearly linear. The conclusion of the proof is now reduced to a repetition of the standard proof of the first fundamental theorem of Lie groups, i.e., to the existence and uniqueness theorems for initial value problems in ordinary differential equations.

We shall now construct some specific representations, when **L** is either the group $\mathbf{G}$, $\mathbf{G}_0$, or $\mathbf{G}_0'$ of automorphisms of a symplectic manifold $(\mathbf{X}, \omega)$, of $\Omega(\mathbf{L})$, respectively, into the following Abelian Lie groups, respectively, $H^1(\mathbf{X}, \mathbf{R})$, $H_0^1(\mathbf{X} \cdot \mathbf{R})$, and **R** (written additively), satisfying the conditions of the lemma.

We consider, in their respective order, the groups: (i) **G**, (ii) $\mathbf{G}_0$, and (iii) $\mathbf{G}_0'$ (in case **X** is not compact) of automorphisms of a symplectic manifold $(\mathbf{X}, \omega)$ and let $\sigma = \{\sigma(t)\}$ describe a differentiable path in each of these groups, parametrized by t $(0 \leqq t \leqq 1)$ with $\sigma(0) =$ identity. In each case we consider also, respectively: (i) any finite singular (piecewise differentiable) 1-cycle γ with integer coefficients in **X**; (ii) any locally finite, integral, singular 1-cycle γ in **X**; (iii) for any given point **x** any choice of a singular, locally finite, infinite 1-chain $\gamma_{\mathbf{x}}$ in **X** whose boundary $\partial\gamma_{\mathbf{x}}$ is the 0-cycle **x**. In each case we introduce the finite 2-chains $\sigma(\gamma)$ [respectively $\sigma(\gamma_{\mathbf{x}})$ in case (iii)] obtained by the homotopy operator on **X** represented by σ and calculate the functionals $\overline{\Phi}(\sigma, \gamma)$ in cases (i) and (ii) and $\overline{\Phi}(\sigma, \gamma_{\mathbf{x}})$ in case (iii), defined respectively by

$$(4.1) \qquad \overline{\Phi}(\sigma, \gamma) = \int_{\sigma(\gamma)} \omega, \qquad \overline{\Phi}(\sigma, \gamma_{\mathbf{x}}) = \int_{\sigma(\gamma_{\mathbf{x}})} \omega.$$

PROPOSITION 4.3. *Let* $(\mathbf{X}, \omega)$ *be a symplectic manifold and consider, respectively:* (*i*) *the automorphism group* $\mathbf{G}$ *of* $(\mathbf{X}, \omega)$, (*ii*) (*if* $\mathbf{X}$ *is not compact*) *the compactly supported automorphism group* $\mathbf{G}_0$, *and* (*iii*) (*if* $\mathbf{X}$ *is not compact*) *the subgroup* $\mathbf{G}_0'$ *of* $\mathbf{G}_0$. *Let* $\overline{\Phi}(\sigma, \gamma)$ *and* $\overline{\Phi}(\sigma, \gamma_{\mathbf{x}})$ *be defined, in their respective cases, by* (4.1). *Then the following statements are valid. In cases* (*i*) *and* (*ii*) *the functional* $\overline{\Phi}(\sigma, \gamma)$ *depends only on the homotopy class of the path* σ (*relative to its end points*) *and on the real homology class of* γ [*with compact support in case* (*i*), *locally compact support in case* (*ii*)]. *In case* (*iii*) *the functional* $\overline{\Phi}(\sigma, \gamma_{\mathbf{x}})$ *depends only on* $\mathbf{x}$, *independent of the choice of* $\gamma_{\mathbf{x}}$; *the resulting function* $\Phi(\sigma, \mathbf{x})\colon \Omega(\mathbf{G}_0') \times \mathbf{X} \to \mathbf{R}$ *has, for each fixed* $\sigma \in \Omega(\mathbf{G}_0')$, *uniformly compact support in* $\mathbf{X}$ *and its integral*

$$\overline{\Psi}(\sigma) = \int_{\mathbf{X}} \Phi(\sigma, \mathbf{x})\omega^n \tag{4.2}$$

depends only on the homotopy class of the path $\sigma \in \Omega(\mathbf{G}_0')$ *relative to its end points; in other words* $\overline{\Phi}$ *and* $\overline{\Psi}$ *define, in case* (*i*), *a homomorphism* χ *of the universal covering group* $\tilde{\mathbf{G}}$ *of* $\mathbf{G}$ *onto* $H^1(\mathbf{X}, \mathbf{R})$; *in case* (*ii*), *an epimorphism* $\chi\colon \tilde{\mathbf{G}}_0 \to H_0^1(\mathbf{X}, \mathbf{R})$; *in case* (*iii*), *a homomorphism* $\chi\colon \tilde{\mathbf{G}}_0' \to \mathbf{R}$ *that is an epimorphism if and only if* $\mathbf{X}$ *is not compact.*

PROOF. It is clear that in case (i) the 2-chain $\sigma(\gamma)$ is finite; in case (ii) the 1-cycle γ may be, and in case (iii) the 1-chain $\gamma_{\mathbf{x}}$ has to be infinite, but since the chain homotopy operator represented by σ has compact support, the resulting 2-chains $\sigma(\gamma)$ and $\sigma(\gamma_{\mathbf{x}})$ are null at infinity, so that the functionals $\overline{\Phi}(\sigma, \gamma)$ and $\overline{\Phi}(\sigma, \gamma_{\mathbf{x}})$ are well defined.

We shall consider temporarily only case (ii). The integral $\int_{\sigma(\gamma)} \omega$ can be evaluated by reducing it to two successive, 1-dimensional integrals as follows. Denote by $\sigma_{|[0,s]}$ the path in $\mathbf{G}_0$ restricted to the interval $[0, s]$ of the parameter t of σ, and by σ^t the element $\sigma(t) \in G_0$ at the value t; let $\dot{\sigma}^t$ be the vector field $\dfrac{d(\sigma^t)}{dt}$ in $\mathbf{X}$ and by $\xi(t)$ the vector field $\dot{\sigma}^t \circ (\sigma^t)^{-1}$ (i.e., such that the path σ is the solution of the ordinary differential equation $\dfrac{d\sigma^t}{dt} = \xi(t) \circ \sigma^t$ with initial value $\sigma^0 =$ identity). Then both $\dot{\sigma}^t$ and $\xi(t)$ belong to the Lie algebra $\mathfrak{G}_0$. We calculate the derivative of $\overline{\Phi}(\sigma_{|[0,t]}, \gamma)$ with respect to t as follows:

$$\frac{d}{dt}\overline{\Phi}(\sigma_{|[0,t]}, \gamma) = \int_{\sigma^t\gamma} \dot{\sigma}^t \llcorner \omega = \int_\gamma \xi(t) \llcorner (\omega \circ (\sigma^t)^{-1}) = \int_\gamma \xi(t) \llcorner \omega, \tag{4.3}$$

where the last equality stems from the invariance of ω under the automorphism $(\sigma^t)^{-1}$. The integrand has compact support, so that the integral is well defined. $\overline{\Phi}(\sigma, \gamma)$ can thus be calculated by successive integrations as follows:

$$\overline{\Phi}(\sigma, \gamma) = \int_0^1 \left(\int_{\sigma^t\gamma} \dot{\sigma}^t \llcorner \omega \right) dt = \int_0^1 \left(\int_{\gamma} \xi(t) \llcorner \omega \right) dt. \tag{4.4}$$

The analogues of this formula in cases (i) and (iii) are of the same type and require the same proof. Since, in the three respective cases and for each t, the vector field $\xi(t)$ belongs to the Lie algebra $\mathfrak{G}$, $\mathfrak{G}_0$, and $\mathfrak{G}_0'$, respectively, the 1-form $\xi(t) \llcorner \omega$ is, respectively, closed, closed with compact support, and the differential of a function $u(t, x)$ with compact support in $[0, 1] \times \mathbf{X}$. Therefore the integral $\int_\gamma \xi(t) \llcorner \omega$ [resp. $\int_{\gamma_\mathbf{x}} \xi(t) \llcorner \omega$ in case (iii)] depends only on σ and, in case (i), on the compact homology class of the 1-cycle γ; in case (ii), on the (unbounded) homology class of γ; and, in case (iii), on the boundary $\mathbf{x}$ of $\gamma_\mathbf{x}$; in this case it merely reproduces the function $\{u(t, x)\}$ with compact supports whose differential with respect to x is $\xi(t) \llcorner \omega$. We denote by $\Phi'(\sigma, [\gamma])$ the resulting function resulting from Φ by passage to the quotient, where $[\gamma]$ denotes the homology class of γ [compact in case (i), noncompact in case (ii)], and by u_t the corresponding function of σ and x in case (iii), where $u_t(x) = u(t, x)$.

The maps $\Phi\colon \Omega(\mathbf{G}) \to H^1(\mathbf{X}, \mathbf{R})$ in case (i) and $\Phi\colon \Omega(\mathbf{G}_0) \to H_0^1(\mathbf{X}, \mathbf{R})$ in case (ii) defined by $\Phi(\sigma)([\gamma]) = \Phi'(\sigma, [\gamma])$ [where γ is the corresponding dual homology class] are clearly homomorphisms of the semigroup $\Omega(\mathbf{G})$ [resp. $\Omega(\mathbf{G}_0)$] into the corresponding cohomology groups. They are differentiable since they can be computed by integration over finite parametric families of paths in $\Omega(\mathbf{G})$ [or $\Omega(\mathbf{G}_0)$]. The differential of Φ at the identity reduces simply to the de Rham representation of the symplectic vector field $\xi(0) \in \mathfrak{G}$ (resp. $\mathfrak{G}_0$) by an element in $H^1(\mathbf{X}, \mathbf{R})$ [resp. $H_0^1(\mathbf{X}, \mathbf{R})$] as described in equation (3.7), after the proof of Theorem 3.4. This representation is a Lie algebra homomorphism, as a consequence of Proposition 3.3. Hence by Lemma 4.2 it induces a representation of the analytic (simply connected) group $\tilde{\mathbf{G}}$ into $H^1(\mathbf{X}, \mathbf{R})$ [resp. $\tilde{\mathbf{G}}_0$ into $H_0^1(\mathbf{X}, \mathbf{R})$], which proves the assertion in cases (i) and (ii). We remark that in these two cases the proof of the invariance of the map Φ on $\Omega(\mathbf{G})$ [or $\Omega(\mathbf{G}_0)$] under homotopy of paths could be obtained directly, without using Lemma 4.2.

In case (iii) we have a map $\bar{\chi}\colon \Omega(\mathbf{G}_0') \to \overline{\mathbf{R}}$ defined by

$$\bar{\chi}(\sigma) = \int_\mathbf{X} \left(\int_0^1 u_t(\mathbf{x})\, dt \right) \omega^n = \int_\mathbf{X} \left\{ \int_0^1 \left[\int_{\gamma_\mathbf{x}} \xi(t) \llcorner \omega \right] dt \right\} \omega^n.$$

Again one verifies in a straightforward way that this map is additive on $\Omega(\mathbf{G}_0')$ and differentiable, and, since the associated map of $\mathfrak{G}_0'$ (the differential of $\bar{\chi}$ with respect to t at $t = 0$) is a Lie algebra homomorphism, the map $\bar{\chi}$ is invariant under homotopy of paths in $\Omega(\mathbf{G}'_0)$, and therefore defines by passage to quotient a map

$$\chi\colon \tilde{\mathbf{G}}'_0 \to \mathbf{R},$$

where $\tilde{\mathbf{G}}'_0 = \Omega(\mathbf{G}'_0)/\Omega_0(\mathbf{G}'_0)$ is the universal covering group $\tilde{\mathbf{G}}'_0$ of the identity component in $\mathbf{G}'_0$. It follows trivially from deRham's theorem that the representations of $\tilde{\mathbf{G}}$, $\tilde{\mathbf{G}}_0$, and $\tilde{\mathbf{G}}'_0$ into $H^1(\mathbf{X}, \mathbf{R})$, $H^1_0(\mathbf{X}, \mathbf{R})$, and $\mathbf{R}$, respectively, are surjective (except the third case, if applied to a compact manifold). This completes the proof of Proposition 4.3.

It would be interesting to investigate how the representation $\chi\colon \tilde{\mathbf{G}}'_0 \to \mathbf{R}$ must be modified to obtain a corresponding one of $\mathbf{G}'_0$, i.e., what open subgroups of $\mathbf{G}'_0$ have representations χ_1 into quotient groups of $\mathbf{R}$ that are locally equivalent to χ. This, however, has not been possible except in the case $n = 1$ of open surfaces $\mathbf{X}$ with an oriented area element ω; in this case one knows from the theory of Teichmüller–Ahlfors that the group $\mathbf{G}_0$ is homotopically equivalent to the group of automorphisms of the fundamental group of $\mathbf{X}$. In particular, the identity component of $\mathbf{G}_0$ is simply connected, so that the representation χ_1 of this group onto $\mathbf{R}$ exists and coincides with χ. It has been pointed out that, in the special case of the Cartesian plane $\mathbf{R}^2$ with $\omega = dx \wedge dy$, the representation χ of the group $\mathbf{G}_0 = \mathbf{G}'_0$ of measure-preserving diffeomorphisms with compact support onto $\mathbf{R}$ had been found by Levi Civita. (I have been unable, however, to trace the reference.) This representation, intuitively, measures for each such diffeomorphism $\mathbf{g}$, how much $\mathbf{g}$ "twists" the plane in the mean, in the clockwise sense.

We shall now give the partial modification of Proposition 4.3 that defines Abelian representations of the groups $\mathbf{G}$ and $\mathbf{G}_0$.

DEFINITION 4.4. Let $(\mathbf{X}, \omega)$ be a symplectic manifold. We denote by $\Lambda = \Lambda_\omega$ the additive subgroup of $\mathbf{R}$ made up of the numbers $\omega[w]$ where w runs over all *integral*, compact, 2-dimensional homology classes, $w \in {}_0H_2(\mathbf{X}, \mathbf{Z})$. If Λ is discrete, we say that $(\mathbf{X}, \omega)$ is a *Hodgean* manifold.

THEOREM 4.5. *Let* $(\mathbf{X}, \omega)$ *be a Hodgean manifold; let* ${}_0\mathbf{G}$ *denote the open subgroup of the automorphism group* $\mathbf{G}$ *of* $(\mathbf{X}, \omega)$ *whose elements act as the identity on* ${}_0H_1(\mathbf{X}, \mathbf{Z})$; *if* $\mathbf{X}$ *is not compact, let* ${}_0\mathbf{G}_0$ *denote the open subgroup of* G_0 *whose elements act as the identity on* $H_1(\mathbf{X}, \mathbf{Z})$ (*noncompact homology*). *Then the representation* $\chi\colon \tilde{\mathbf{G}} \to H^1(\mathbf{X}, \mathbf{R})$ [*resp.* $\tilde{\mathbf{G}}_0 \to H^1_0(\mathbf{X}, \mathbf{R})$] *of Proposition* 4.3 *is locally equivalent to a representation* ${}_0\chi\colon \mathbf{G} \to$ hom $({}_0H_1(\mathbf{X}, \mathbf{Z}), \mathbf{R}/\Lambda)$ [*resp.* ${}_0\chi\colon \mathbf{G}_0 \to$ hom $(H_1(\mathbf{X}, \mathbf{Z}), \mathbf{R}/\Lambda)$], *where* Λ *is the additive subgroup of* $\mathbf{R}$ *consisting of the integral periods of* ω.

PROOF. The two parallel statements (which are distinct if $\mathbf{X}$ is not compact) concerning $\mathbf{G}$ and $\mathbf{G}_0$ are entirely analogous in content and proof, so that we shall confine our statements mainly to the first case.

Let $\mathbf{g} \in {}_0\mathbf{G}$ and let γ be any compact, 1-dimensional, singular differentiable cycle in $\mathbf{X}$ with integer coefficients, its homology class being denoted by $[\gamma] \in {}_0H_1(\mathbf{X}, \mathbf{Z})$. Then $\mathbf{g}(\gamma) - \gamma$ is a cycle which, by the definition of ${}_0\mathbf{G}$, bounds an integral 2-chain τ, which is compact (since either γ, in the first case, or $\mathbf{g}$, in the second case, has compact support). τ is unique modulo the group ${}_0Z_2(\mathbf{X}, \mathbf{Z})$ of compact, integral 2-cycles, so that the integral $\int_\tau \omega$ is uniquely determined by $\mathbf{g}$, ω and γ, modulo the discrete group $\Lambda \subset \mathbf{R}$. In addition, if γ is homologous to zero, i.e., $\gamma = \partial\tau_0$, then we can take $\tau = \mathbf{g}(\tau_0) - \tau_0$ and, since ω is invariant under g, $\int_\tau \omega = \int_{(g(\tau_0) - \tau_0)} \omega = 0$, so that $\int_\tau \omega$ depends only on the homology class of γ. We thus obtain a map ${}_0\chi$ by passage to quotients, ${}_0\chi\colon {}_0\mathbf{G} \to \operatorname{hom}({}_0H_1(\mathbf{X}, \mathbf{Z}), \mathbf{R}/\Lambda)$, defined by

$$(4.5) \qquad {}_0\chi(\mathbf{g})[\gamma] = \int_\tau \omega; \qquad \partial\tau = \mathbf{g}(\gamma) - \gamma.$$

Now suppose that $\mathbf{g}$ is (differentiably) *homotopic* to the identity under a homotopy $h\colon I \times \mathbf{X} \to \mathbf{X}$. This is a considerably weaker assumption than saying that $\mathbf{g}$ lies in the component of the identity in $\mathbf{G}$ (in the case of $\mathbf{G}_0$ we assume the homotopy to be with compact support). Let $h_t\colon \mathbf{X} \to \mathbf{X}$ be the differentiable map defined by $h_t(x) = h(t, x)$, $h_0 =$ identity, $h_1 = \mathbf{g}$; then $h_t(\gamma) - \gamma$ is homotopic to the identity and the resulting chain homotopy h_* defines a 2-chain $\tau = h_*\gamma$ for which we can define

$${}_1\chi(\mathbf{g})([\gamma]) = \int_{h_*\gamma} \omega \in \mathbf{R}.$$

This map ${}_1\chi$ has values in the additive group $\mathbf{R}$ and is a lift of the homomorphism χ restricted to the open subgroup ${}_1\mathbf{G} \subset \mathbf{G}$ of automorphisms that are homotopic to the identity ((identity component of $\mathbf{G}$) $\subset {}_1\mathbf{G} \subset {}_0\mathbf{G} \subset \mathbf{G}$). In this case it is easy to see that by restricting χ to the identity component of $\mathbf{G}$ we obtain the representation characterized in Proposition 4.3.

This completes the proof of the theorem.

An analogous result in the case of open subgroups of $\mathbf{G}_0'$ has not been found, since the argument used in order to obtain the real-valued representation of its covering group $\mathbf{G}_0'$ requires in an essential way Lemma 4.2, i.e., that the homotopy h from the identity to $\mathbf{g}$ be within the group $\mathbf{G}_0'$ itself.

5. Conclusions and problems

The most interesting conclusions of this paper in terms of further developments are probably related to Lemma 3.10, the analogue of the Palais–Cerf Lemma. It would be interesting to describe a set of sufficient conditions for a pseudogroup Γ of differentiable transformations to have the Palais–Cerf property, i.e., that each germ of an element of Γ can be extended to a global transformation of $\mathbf{X}$ by an element of the group $\mathbf{G}$ of global cross-sections of Γ over $\mathbf{X}$ (or, better, an element in the identity component of $\mathbf{G}$). The corresponding question of the extension of a germ to global transformations belonging to a derived group $\mathbf{G}^{(k)}$ of $\mathbf{G}$ appears to be much more complicated (see, for example, the case where Γ is the pseudogroup preserving a positive volume element ω_n in an n-manifold). It is conjectured that the existence of a syzygean resolution of the sheaf Θ_Γ of germs of Γ-vector fields,

$$0 \leftarrow \Theta_\Gamma \leftarrow \Phi^{-1} \leftarrow \Phi^{-2} \leftarrow \cdots,$$

with each Φ^{-j} a fine sheaf, may be of significance here. In particular the length of such a resolution may describe the dimension of the primary obstruction to the extension of a cross-section in Θ_Γ over a compact $K \subset \mathbf{X}$ to all of $\mathbf{X}$. In the case of the symplectic group, Θ_Γ is isomorphic by Proposition 3.1 to the sheaf of germs of closed 1-forms, so that the syzygean resolution can be described by

$$0 \leftarrow \Theta_\Gamma \leftarrow \mathscr{A}^0 \leftarrow (\mathbf{R}) \leftarrow 0$$

($\mathscr{A}^p$ = sheaf of germs of p-forms), which has length 1. In the case where Γ consists of the germs of diffeomorphisms preserving a nowhere-vanishing n-form ω_n, the corresponding resolution is

$$0 \leftarrow \Theta_\Gamma \leftarrow \mathscr{A}^{n-1} \overset{d}{\leftarrow} \mathscr{A}^{n-2} \leftarrow \cdots \leftarrow \mathscr{A}^0 \leftarrow \mathbf{R} \leftarrow 0,$$

which has length $n - 1$. The primary obstructions to the extension of cross-sections in Θ_Γ from K to $\mathbf{X}$ are represented in these two special cases by the relative cohomology groups $H^2(\mathbf{X}, K; \mathbf{R})$ and $H^n(\mathbf{X}, K; \mathbf{R})$ ($n = \dim \mathbf{X}$), respectively.

UNIVERSITY OF PENNSYLVANIA
AND
INSTITUT DES HAUTES ETUDES SCIENTIFIQUES,
BURES-SUR-YVETTE

REFERENCE

1. MOSER, JÜRGEN, "On the volume elements on a manifold," *Trans. Amer. Math. Soc.*, *120*, No. 2 (1965), pp. 286–294.

On the Minimal Immersions of the Two-sphere in a Space of Constant Curvature

SHIING-SHEN CHERN[1]

1. Introduction

This study arose from reading Calabi's interesting paper [2] on the minimal immersions of the two-sphere S^2 into an N-dimensional sphere S^N. Using an idea of H. Hopf, Calabi observed the strong implications of the fact that the submanifold is a two-sphere. His treatment made essential use of the global coordinates on S^N. We shall give a different approach by supposing only that the ambient space is a Riemannian manifold of constant sectional curvature. Some general remarks on the higher osculating spaces are intended to make the development more natural and to prepare the way for future applications to the case of submanifolds of higher dimensions.

2. Higher osculating spaces of a submanifold and fundamental forms

Let X be a Riemannian manifold of dimension N and

$$x: M \to X \tag{1}$$

be an immersion of a differentiable manifold M of dimension n into X. In this section we will agree on the following ranges of indices:

$$\begin{aligned} 1 &\leqq A, B, C, D \leqq N, \\ 1 &\leqq i, j \leqq n, \\ n + 1 &\leqq \alpha, \beta \leqq N. \end{aligned} \tag{2}$$

[1] Work done under partial support of NSF grant GP-8623.

Let e_A be a local orthonormal frame field. The Levi–Civita connection defines the covariant differentials

$$De_A = \sum_B \omega_{AB} e_B, \tag{3}$$

where

$$\omega_{AB} + \omega_{BA} = 0. \tag{4}$$

(See [3] for details.) If ω_B is a coframe field dual to e_A, so that we have

$$ds^2 = \sum_A \omega_A^2, \tag{5}$$

the structure equations of the space are

$$\begin{aligned} d\omega_A &= \sum_B \omega_B \wedge \omega_{BA}, \\ d\omega_{AB} &= \sum_C \omega_{AC} \wedge \omega_{CB} + \Omega_{AB}, \end{aligned} \tag{6}$$

where

$$\Omega_{AB} = -\tfrac{1}{2} \sum_{C,D} R_{ABCD} \omega_C \wedge \omega_D. \tag{7}$$

The space X is said to be of constant curvature c, if and only if

$$\Omega_{AB} = -c\omega_A \wedge \omega_B. \tag{8}$$

Throughout this paper we will suppose that X is of constant curvature.

To study the geometry of the immersed submanifold M we restrict ourselves (locally) to orthonormal frame fields over M such that e_i are tangent vectors at $x \in M$, so that e_α are normal vectors to M. Then we have

$$\omega_\alpha = 0. \tag{9}$$

By the first structure equation in (6) and Cartan's lemma, we can put

$$\omega_{i\alpha} = \sum_j h_{\alpha ij} \omega_j, \qquad h_{\alpha ij} = h_{\alpha ji}. \tag{10}$$

A curve C through x is a smooth function $x(s)$, $|s| < L$, with $x(0) = x$. We can suppose the parameter s to be its length. By covariant differentiations along C we get the vector fields

$$\frac{Dx}{ds}, \frac{D^2x}{ds^2}, \frac{D^3x}{ds^3}, \ldots \tag{11}$$

At $s = 0$ the vectors

$$\frac{Dx}{ds}, \frac{D^2x}{ds^2}, \ldots, \frac{D^mx}{ds^m}$$

are said to span the *osculating space of order m*, the osculating space of order one being the tangent line. The *osculating space $T_x^{(m)}$ of order m of M at $x \in M$* is defined to be the space spanned by all the osculating spaces of order m of the curves through x and lying on M. We then have

$$T_x^{(1)} \subset T_x^{(2)} \subset \cdots \subset T_x^{(m)} \subset \cdots, \tag{12}$$

where $T_x^{(1)}$ is the tangent space to M at x and we shall write $T_x^{(1)} = T_x$. Let

$$\dim T_x^{(m)} = n + p_1(x) + \cdots + p_{m-1}(x), \qquad m = 1, 2, \ldots. \tag{13}$$

A point $x \in M$ is called a *regular point of order m*, if

$$p_a(x) = \text{const}, \qquad a = 1, \ldots, m-1,$$

in a neighborhood of x.

We can write

$$Dx = \sum_i \omega_i e_i, \tag{14}$$

so that, by (3) and (10),

$$\begin{aligned} D^2x &= \sum_{i,\alpha} \omega_i \omega_{i\alpha} e_\alpha + \text{terms in } e_i \\ &= \sum_{i,j,\alpha} h_{\alpha ij} \omega_i \omega_j e_\alpha + \text{terms in } e_i. \end{aligned} \tag{15}$$

The quadratic differential forms

$$(e_\alpha, D^2x) = \sum_{i,j} h_{\alpha ij} \omega_i \omega_j \tag{16}$$

are called the *second fundamental forms* of M. The integer $p_1(x)$ defined previously is equal to the number of linearly independent vectors among $\sum_\alpha h_{\alpha ij} e_\alpha$, $i \leqq j$, and is also the number of linearly independent second fundamental forms. In the generic case we have $p_1(x) = n(n+1)/2$.

Suppose now that x is a regular point of order $m - 1 \geqq 2$, so that $p_1(y), \ldots, p_{m-2}(y)$ are constants when $y \in M$ is in a neighborhood of x. We shall use the following ranges of indices:

$$\begin{gathered} 1 \leqq \lambda_0 = i \leqq n, \\ n + 1 \leqq \lambda_1 \leqq n + p_1, \\ n + p_1 + 1 \leqq \lambda_2 \leqq n + p_1 + p_2, \\ \cdots \\ n + p_1 + \cdots + p_{m-2} + 1 \leqq \lambda_{m-1} \leqq N. \end{gathered} \tag{17}$$

Let e_A be a local orthonormal frame field, such that $e_{\lambda_0}, e_{\lambda_1}, \ldots, e_{\lambda_b}$ span $T_x^{(b+1)}$, $b = 0, 1, \ldots, m-2$. We then have

$$\omega_{\lambda_{b-1}\lambda_{b+1}} = 0, \qquad b = 1, \ldots, m-2. \tag{18}$$

By exterior differentiation of (18) and making use of the second equation in (6) and the equation (8), we obtain

$$\sum_{\lambda_b} \omega_{\lambda_{b-1}\lambda_b} \wedge \omega_{\lambda_b\lambda_{b+1}} = 0, \qquad b = 1, \ldots, m-2. \tag{19}$$

This allows us to introduce by recurrence the quantities $h_{\lambda_b i_1 \cdots i_{b+1}}$ defined by the equations

$$\sum_{\lambda_b} h_{\lambda_b i_1 \cdots i_{b+1}} \omega_{\lambda_b \lambda_{b+1}} = \sum_{i_{b+1}} h_{\lambda_{b+1} i_1 \cdots i_{b+2}} \omega_{i_{b+2}}, \tag{20}$$

and the $h_{\lambda_b i_1 \cdots i_{b+1}}$ are symmetric in the Latin indices $i_1, \ldots, i_{b+1}$. We find

$$\begin{aligned} (e_{\lambda_{m-1}}, D^m x) &= \sum_{i, \lambda_1, \ldots, \lambda_{m-2}} \omega_i \omega_{i\lambda_1} \omega_{\lambda_1\lambda_2} \cdots \omega_{\lambda_{m-2}\lambda_{m-1}} \\ &= \sum_{i_1, \ldots, i_m} h_{\lambda_{m-1} i_1 \cdots i_m} \omega_{i_1} \cdots \omega_{i_m}, \end{aligned} \tag{21}$$

which are differential forms of degree m and are to be called the mth *fundamental forms* of M in X. By (20) it follows that, if the mth fundamental forms vanish at a point, those of higher order vanish at the same point.

For later applications it will be convenient to write the structure equations (6) in relation with the equations of parallel displacement (3). It suffices to take the second covariant differential in the sense of the exterior calculus. We then have

$$\hat{D}^2 e_A = \sum_B \Omega_{AB} e_B = -c\omega_A \wedge \sum_B \omega_B e_B, \tag{22}$$

where the notation $\hat{D}^2$ is used to distinguish it from the D^2 used previously. It follows that for our frame field attached to a submanifold M we have

$$\begin{aligned} \hat{D}^2 e_i &= -c\omega_i \wedge \sum_j \omega_j e_j, \\ \hat{D}^2 e_\alpha &= 0. \end{aligned} \tag{23}$$

3. Minimal surfaces

We consider the special case that M is a two-dimensional manifold immersed as a minimal surface in X (so that $n = 2$ in the formulas of the last section). The latter condition is expressed analytically by

$$\sum_i h_{\alpha ii} = 0. \tag{24}$$

From (20) it follows that

$$\sum_j h_{\lambda_b j j i_3 \cdots i_{b+1}} = 0, \qquad b = 1, 2, \ldots. \tag{25}$$

Because of the symmetry of $h_{\lambda_b i_1 \cdots i_{b+1}}$ in its Latin indices, the same relation holds when contracted with respect to any two Latin indices. At a generic point we have, therefore,

$$p_1(x) = p_2(x) = \cdots = 2. \tag{26}$$

It will be convenient in this case to make use of the complex notation and the isothermal coordinates on M. We put

$$\varphi = \omega_1 + i\omega_2, \tag{27}$$

so that the metric on M is

$$ds^2 = \varphi\bar{\varphi}. \tag{28}$$

From the first equation of (6) we find

$$d\varphi = -i\omega_{12} \wedge \varphi. \tag{29}$$

The Gaussian curvature K of M is defined by the equation

$$d\omega_{12} = -\frac{i}{2} K\varphi \wedge \bar{\varphi}. \tag{30}$$

The existence of local isothermal coordinates means that we can write, locally,

$$\varphi = \lambda dz. \tag{31}$$

Substituting into (29), we obtain

$$(d\lambda + i\omega_{12}\lambda) \wedge dz = 0, \tag{32}$$

so that $d\lambda + i\omega_{12}\lambda$ is a multiple of dz. This remark will be useful later on.

From (21) and (25) we see that the mth fundamental forms at a generic point of a minimal surface can be written

$$(e_\lambda, D^m x) = \sum_{i_1, \ldots, i_m} h_{\lambda i_1 \cdots i_m} \omega_{i_1} \cdots \omega_{i_m} = \mathrm{Re}\,(\bar{H}_\lambda^{(m)} \varphi^m), \qquad 2m - 1 \leqq \lambda \leqq N, \tag{33}$$

where

$$H_\lambda^{(m)} = h_{\lambda \underbrace{1 \cdots 1}_{m}} + i h_{\lambda \underbrace{1 \cdots 1}_{m-1} 2} \tag{34}$$

4. A theorem on an elliptic differential system

We wish to prove the following theorem:

THEOREM. *Let* $w_\alpha(z)$ *be complex-valued functions which satisfy the differential system*

$$\frac{\partial w_\alpha}{\partial \bar z} = \sum_\beta a_{\alpha\beta} w_\beta, \qquad 1 \leqq \alpha, \beta \leqq p, \tag{35}$$

in a neighborhood of $z = 0$, *where* $a_{\alpha\beta}$ *are complex-valued* C^1-*functions. Suppose the* w_α *do not all vanish identically in a neighborhood of* $z = 0$. *Then:*

(1) *the common zeroes of* w_α *are isolated;*
(2) *at a common zero of* w_a *the ratios* $w_1 : \cdots : w_p$ *tend to a limit.*

LEMMA 1. *Under the hypotheses of the theorem let* $w_\alpha = o(|z|^{r-1})$ *at* $z = 0$, $r \geqq 1$. *Then* $\lim_{z \to 0} w_\alpha(z) z^{-r}$ *exists.*

PROOF. Let D be a disk of radius R about $z = 0$, and let $\zeta \in D$, $\zeta \neq 0$. Write

$$z = x + iy, \qquad \zeta = \xi + i\eta.$$

Let $\rho(z) = \max_\alpha |w_\alpha|$. From (35) we obtain

$$d\left\{\frac{w_\alpha\, dz}{z^r(z - \zeta)}\right\} = \frac{1}{z^r(z - \zeta)} \sum_\beta a_{\alpha\beta} w_\beta\, d\bar z \wedge dz.$$

By Stokes' Theorem it follows that

$$-2\pi i w_\alpha(\zeta)\zeta^{-r} + \int_{\partial D} \frac{w_\alpha\, dz}{z^r(z - \zeta)} = \iint_D \frac{\sum_\beta a_{\alpha\beta} w_\beta}{z^r(z - \zeta)}\, d\bar z \wedge dz. \tag{36}$$

Taking absolute value, we obtain

$$2\pi\rho(\zeta)|\zeta^{-r}| \leqq \int_{\partial D} \frac{\rho(z)|dz|}{|z^r(z - \zeta)|} + A \iint_D \frac{\rho(z)\, dx\, dy}{|z^r(z - \zeta)|}, \tag{37}$$

where $A > 0$ is a constant depending only on the system (35). With $z_0 \in D$, $z_0 \neq 0$, we integrate this inequality with respect to $d\xi\, d\eta / |\zeta - z_0|$ over D. Remarking that

$$\frac{1}{|(z - \zeta)(z_0 - \zeta)|} = \frac{1}{|z - z_0|}\left|\frac{1}{z - \zeta} + \frac{1}{\zeta - z_0}\right|,$$

$$\iint_D \frac{dx\, dy}{|z - \zeta|} < 2R, \tag{38}$$

the integration gives

$$2\pi \iint_D \frac{\rho(\zeta)\, d\xi\, d\eta}{|\zeta^r(\zeta - z_0)|} \leqq 4R \int_{\partial D} \frac{\rho(z)|dz|}{|z^r(z - z_0)|} + 4AR \iint_D \frac{\rho(z)\, dx\, dy}{|z^r(z - z_0)|}$$

or

$$(2\pi - 4AR) \iint_D \frac{\rho(z)\, dx\, dy}{|z^r(z - z_0)|} \leqq 4R \int_{\partial D} \frac{\rho(z)|dz|}{|z^r(z - z_0)|}.$$

We choose R so small that $2\pi - 4AR > 0$. Then the integral at the left-hand side of the preceding equation is bounded as $z_0 \to 0$. It follows from (37) that $\rho(\zeta)|\zeta^{-r}|$, and hence $|w_\alpha(\zeta)\zeta^{-r}|$, is bounded as $\zeta \to 0$. By (36) we see that $\lim_{\zeta \to 0} w_\alpha(\zeta)\zeta^{-r}$ exists.

LEMMA 2. *Under the hypotheses of the theorem suppose* $w_\alpha = o(|z|^{r-1})$, *all* r. *Then* $w_\alpha \equiv 0$ *in a neighborhood of* $z = 0$.

PROOF. Suppose, to the contrary, that $\rho(z_0) \neq 0$, $|z_0| < R$. We multiply (37) by $d\xi\, d\eta$ and integrate over D. This gives

$$2\pi \iint_D \rho(\zeta)|\zeta^{-r}|\, d\xi\, d\eta \leqq 2R \int_{\partial D} \rho(z)|z^{-r}|\, |dz| + 2AR \iint_D \rho(z)|z^{-r}|\, dx\, dy,$$

or

$$(2\pi - 2AR) \iint_D \rho(z)|z^{-r}|\, dx\, dy \leqq 2R \int_{\partial D} \rho(z)|z^{-r}|\, |dz|.$$

There exist two positive constants a and b independent of r such that the left-hand side of the above inequality is $\geqq a|z_0|^{-r}$ and the right-hand side is $\leqq bR^{-r}$. Combining them, we obtain

$$\frac{a}{b} \leqq \left|\frac{z_0}{R}\right|^r.$$

But this leads to a contradiction as $r \to \infty$.

From the two lemmas the theorem follows immediately. In fact, suppose $w_\alpha(0) = 0$ and suppose $w_\alpha(z)$ do not all vanish in a neighborhood of $z = 0$. By Lemma 2 there is a largest integer $r \geqq 1$ such that $w_\alpha = o(|z|^{r-1})$ at $z = 0$. By Lemma 1, $\lim_{z \to 0} w_\alpha(z)z^{-r}$ exists. These limits cannot be all zero, as otherwise we would have $w_\alpha = o(|z|^r)$. This implies the two statements in the theorem.

5. Minimal immersions of the two-sphere

Consider now the minimal immersion

$$S^2 \to X, \tag{39}$$

where, as before, X is a Riemannian manifold of dimension N and of constant curvature c. It will be convenient to make use of complex vectors, and we put

$$E_1 = e_1 + ie_2. \tag{40}$$

By (23) we have

$$\hat{D}^2 E_1 = -\frac{c}{2}\varphi \wedge \bar{\varphi} E_1. \tag{41}$$

On the other hand, using (10) and the conditions (24), we can write

$$DE_1 = -i\omega_{12}E_1 + \bar{\varphi}\sum_\alpha H_\alpha^{(2)} e_\alpha, \tag{42}$$

where

$$H_\alpha^{(2)} = h_{\alpha 11} + ih_{\alpha 12}. \tag{43}$$

The surface S^2 being supposed to be oriented, the frame e_1, e_2 in the tangent plane is defined up to a rotation, and E_1 up to the change

$$E_1 \to E_1^* = e^{i\tau}E_1, \tag{44}$$

where τ is real. Under the change (44), φ and $H_\alpha^{(2)}$ are changed according to

$$\begin{aligned} \varphi \to \varphi^* &= e^{i\tau}\varphi, \\ H_\alpha^{(2)} \to H_\alpha^{(2)*} &= e^{2i\tau}H_\alpha^{(2)}, \end{aligned} \tag{45}$$

from which it follows that

$$\bar{H}_\alpha^{(2)}\varphi^2 = \bar{H}_\alpha^{(2)*}\varphi^{*2}.$$

Thus the form

$$\sum_\alpha (\bar{H}_\alpha^{(2)})^2\varphi^4 \tag{46}$$

is globally defined on the surface S^2, being independent of the choice of the frame field.

Taking the exterior derivative of (42) and collecting the terms in e_α, we obtain, in view of (41),

$$\left(dH_\alpha^{(2)} + \sum_\beta H_\beta^{(2)}\omega_{\beta\alpha} + 2i\omega_{12}H_\alpha^{(2)}\right) \wedge \bar{\varphi} = 0,$$

or

$$d\bar{H}_\alpha^{(2)} + \sum_\beta \bar{H}_\beta^{(2)}\omega_{\beta\alpha} - 2i\omega_{12}\bar{H}_\alpha^{(2)} \equiv 0, \quad \text{mod } \varphi. \tag{47}$$

From (47) we derive

$$d\sum_\alpha (\bar{H}_\alpha^{(2)})^2 - 4i\omega_{12}\left(\sum_\alpha (\bar{H}_\alpha^{(2)})^2\right) \equiv 0, \quad \text{mod } \varphi. \tag{48}$$

We use the local isothermal coordinate z introduced in (31) and write the form (46) as $f(z)\,dz^4$. It follows from (32) and (48) that $f(z)$ is holomorphic, so that $f(z)\,dz^4$ is an Abelian form of degree 4 on S^2. This is possible only when $f(z) = 0$ or

$$\sum_\alpha (H_\alpha^{(2)})^2 = 0, \tag{49}$$

or, by (43),

$$\sum_\alpha h_{\alpha 11}^2 = \sum_\alpha h_{\alpha 12}^2, \qquad \sum_\alpha h_{\alpha 11} h_{\alpha 12} = 0. \tag{50}$$

Geometrically this means that the vectors $\sum_\alpha h_{\alpha 11} e_\alpha$ and $\sum_\alpha h_{\alpha 12} e_\alpha$ in the osculating space of the second order are perpendicular to each other and are of the same length

$$k_1 = \left(\sum_\alpha h_{\alpha 11}^2\right)^{1/2} \geqq 0.$$

The point is singular if and only if $H_\alpha^{(2)} = 0$ or $k_1 = 0$. From (47) and the theorem in §4 we see that, if $H_\alpha^{(2)} \not\equiv 0$, the singular points are isolated and that at each such singular point the "osculating space of the second order" is well defined and varies continuously with the point.

Suppose $k_1 \not\equiv 0$. We can choose the vectors e_3, e_4 of the frame field so that, if

$$E_2 = e_3 + i e_4, \tag{51}$$

equation (42) can be written

$$DE_1 = -i\omega_{12} E_1 + k_1 \bar{\varphi} E_2. \tag{52}$$

We extend the scalar product over complex vectors, so that it will be complex bilinear in both arguments. Then

$$\begin{aligned} (E_1, \bar{E}_1) &= (E_2, \bar{E}_2) = 2, \\ (E_1, E_1) &= (E_1, E_2) = (E_2, E_2) = 0. \end{aligned} \tag{53}$$

Since D preserves the scalar product, we obtain from (3), (52), and (53),

$$DE_2 = -k_1 \varphi E_1 - i\omega_{34} E_2 + \Phi$$

where Φ is a linear combination of the vectors e_μ, $5 \leqq \mu \leqq N$. Suppose the point under consideration to be a regular point, i.e., $k_1 > 0$. Consideration of the terms in e_μ in $\hat{D}^2 E_1$ gives

$$\bar{\varphi} \wedge \Phi = 0,$$

so that Φ is of the form

$$\Phi = \frac{1}{k_1}\bar{\varphi}\left(\sum_\mu H_\mu^{(3)} e_\mu\right).$$

It should be remarked that with this notation, $\mathrm{Re}\,(\bar{H}_\mu^{(3)}\varphi^3)$ are the third fundamental forms of the surface.

The form of degree 6:

$$\sum_\mu (\bar{H}_\mu^{(3)})^2 \varphi^6 \tag{54}$$

is independent of the choice of the frame field and, defined to be zero at a singular point, is well defined over the whole surface S^2. With the local isothermal coordinate z it can be written as $g(z)\,dz.^6$ We shall show that $g(z)$ is a holomorphic function. This follows immediately from the structure equations. In fact, consideration of the term in E_2 in $\hat{D}^2 E_1$ gives

$$\{dk_1 + ik_1(2\omega_{12} - \omega_{34})\} \wedge \bar{\varphi} = 0.$$

We can therefore put

$$\omega_{34} = 2\omega_{12} + \theta_1 \tag{55}$$

where θ_1 is a real one-form. Substituting back, we obtain

$$(dk_1 - ik_1\theta_1) \wedge \bar{\varphi} = 0$$

or

$$\theta_1 \equiv -id \log k_1, \quad \mod \bar{\varphi}. \tag{56}$$

Since $D^2 E_2 = 0$, the terms involving e_μ give, on using (55) and (56),

$$d\bar{H}_\mu^{(3)} + \sum_\nu \bar{H}_\nu^{(3)}\omega_{\nu\mu} - 3i\omega_{12}\bar{H}_\mu^{(3)} \equiv 0, \quad \mod \varphi, \qquad 5 \leqq \nu \leqq N, \tag{57}$$

from which it follows that

$$d\sum_\mu (\bar{H}_\mu^{(3)})^2 - 6i\omega_{12}\left(\sum_\mu (\bar{H}_\mu^{(3)})^2\right) \equiv 0, \quad \mod \varphi. \tag{58}$$

Using (32), we derive, as before, that $g(z)$ is holomorphic. Thus the form (54) is an Abelian form of degree 6 and it must be zero. We can therefore define

$$E_3 = e_5 + ie_6 \tag{59}$$

such that

$$DE_2 = -k_1\varphi E_1 - i\omega_{34}E_2 + k_2\bar{\varphi}E_3.$$

Continuing in this way, we can define a local frame field e_A at a regular point of S^2, such that, if

(60) $$E_s = e_{2s-1} + ie_{2s}, \qquad 1 \leqq s \leqq m,$$

we have

(61) $$DE_s = -k_{s-1}\varphi E_{s-1} - i\omega_{2s-1,2s}E_s + k_s\bar{\varphi}E_{s+1}, \qquad 1 \leqq s \leqq m,\ k_0 = k_m = 0.$$

The integer m is the smallest integer such that DE_m is a linear combination of E_{m-1} and E_m. Equations (61) will be called the *Frenet–Boruvka Formulas* for the minimal immersion of a two-sphere in X. Under the change (44), E_s will be changed according to

(62) $$E_s \to E_s^* = e^{si\tau}E_s.$$

Generalizing (55) we set

(63) $$\omega_{2s-1,2s} = s\omega_{12} + \theta_{s-1}, \qquad 2 \leqq s \leqq m.$$

Then the one-forms θ_{s-1}, $2 \leqq s \leqq m$, remain invariant under a change of the frame field in the tangent plane.

Computing $\hat{D}^2E_s$ and using (41) and

(64) $$\hat{D}^2E_s = 0, \qquad s > 1,$$

we obtain

(65) $$k_1^2 = \tfrac{1}{2}(c - K) \geqq 0$$

and

(66) $$\begin{aligned} &\{dk_s + ik_s(\theta_s - \theta_{s-1})\} \wedge \varphi = 0, && \theta_0 = 0, \\ d\theta_s + i\{k_s^2 - k_{s+1}^2 &- \tfrac{1}{2}(s+1)K\}\varphi \wedge \bar{\varphi} = 0, && 1 \leqq s \leqq m-1. \end{aligned}$$

These are the integrability conditions of (61), and they can be simplified. In fact, relative to the complex structure on S^2 we use the operators ∂, $\bar{\partial}$, and

(67) $$d^c = i(\bar{\partial} - \partial).$$

From the fact that $d \log k_s$ and $\theta_s - \theta_{s-1}$ are real one-forms, we obtain from the first equation of (66),

$$\theta_s - \theta_{s-1} = d^c \log k_s, \qquad 1 \leqq s \leqq m-1,$$

which gives

(68) $$\theta_s = d^c \log (k_1 \cdots k_s), \qquad 1 \leqq s \leqq m-1.$$

To a real-valued smooth function u let its Laplacian Δu be defined by

$$dd^c u = \frac{i}{2}\Delta u \varphi \wedge \bar{\varphi}. \tag{69}$$

Then, in view of (68), the second equation of (66) gives

$$\tfrac{1}{2}\Delta \log (k_1 \cdots k_s) + k_s^2 - k_{s+1}^2 - \tfrac{1}{2}(s+1)K = 0 \tag{70}$$

or

$$\tfrac{1}{2}\Delta \log k_s - k_{s-1}^2 + 2k_s^2 - k_{s+1}^2 - \tfrac{1}{2}K = 0. \tag{71}$$

We summarize our results in the following theorem:

THEOREM. *Let $S^2 \to X$ be a minimal immersion of the two-sphere into a Riemannian manifold of constant curvature. There is an integer m such that the osculating spaces of order m (and dimension $2m$) are parallel along the surface. The singular points are isolated. At a regular point a complete system of local invariants is given by the quantities $k_s > 0$, $1 \leqq s \leqq m-1$, which satisfy the conditions* (71).

The simplest case is when the Gaussian curvature K is constant. If $K = c$, the tangent plane is parallel along the surface and the surface is totally geodesic. If $K < c$, it follows from (65) and then recursively from (71) that k_s are all constants. The same relations give

$$K = \frac{2}{m(m+1)}c, \tag{72}$$

and all the k_s can be expressed in terms of c and m. The surface is thus determined up to an isometry in the space.

6. The case when the ambient space is the N-sphere

Consider the special case, studied by Calabi, of the minimal immersion

$$S^2 \to S^N, \tag{73}$$

where S^N is the sphere of radius 1 in R^{N+1} (so that $c = 1$). The preceding theorem has the consequence that the surface must lie on an even-dimensional great sphere S^{2m}. Without loss of generality we suppose $N = 2m$. In this case the tangent vectors of S^{2m} can be realized in R^{2m+1}, and the vectors E_s by vectors in the complex number space C^{2m+1}. If x is a point on the surface, we can write

$$dx = \tfrac{1}{2}\bar{\varphi}E_1 + \tfrac{1}{2}\varphi\bar{E}_1. \tag{74}$$

Moreover, the Levi–Civita connection is defined by orthogonal projection into the tangent spaces of S^{2m}, and we have

$$\begin{aligned} DE_1 &= -\varphi x + dE_1, \\ DE_s &= dE_s, \qquad 1 < s. \end{aligned} \tag{75}$$

We interpret E_s as homogeneous coordinate vectors in the complex projective space $P_{2m}(C)$. Equation (61) with $s = m$ shows that E_m is an algebraic curve. It will be called the *generating curve* of the minimal surface. From the other equations of (61) we see that $E_m E_{m-1}$ is the line tangent to the generating curve at E_m, etc., and $E_m E_{m-1} \cdots E_1$ is its osculating space of order $m - 1$. Since

$$(E_r, E_s) = 0, \qquad 1 \leqq r, s \leqq m,$$

the generating curve has the property that its osculating spaces of order $m - 1$ belong completely to the nonsingular hyperquadric Q_{2m-1} in $P_{2m}(C)$, the equation of Q_{2m-1} being

$$z_0^2 + z_1^2 + \cdots + z_m^2 = 0, \tag{76}$$

where $z_0, z_1, \ldots, z_m$ are the homogeneous coordinates in $P_{2m}(C)$. It is a classical result (see [4], p. 235) that the linear subspaces of dimension $m - 1$ on Q_{2m-1} form an irreducible algebraic variety of (complex) dimension $m(m + 1)/2$.

To construct the minimal surface from the generating curve, we take its conjugate complex curve $\bar{E}_m$. The corresponding osculating spaces of order m are conjugate complex and therefore meet in a real point x. The minimal surface is the locus of such points of intersection. Boruvka showed that if the generating curve is a normal curve, the resulting minimal surface has constant Gaussian curvature and is given by the spherical harmonics (see [1]). In general, the surface described by x may have isolated points without tangent plane, so that we obtain only a generalized surface.

By (65) we can relate the area of a minimal S^2 with a projective invariant of the generating curve, thus giving a geometrical interpretation of Calabi's result that this area is an integral multiple of 2π.

University of California
Berkeley

REFERENCES

1. Boruvka, O., "Sur les surfaces représentées par les fonctions sphèriques de première espèce," *J. de Math. Pures et Appl.*, 1933, pp. 337–383.

2. Calabi, E., "Minimal immersions of surfaces in Euclidean spheres," *J. Diff. Geom.*, *1* (1967), pp. 111–125.
3. Chern, S., *Minimal Submanifolds in a Riemannian Manifold.* Mimeographed lecture notes, University of Kansas, 1968.
4. Hodge, W. V. D., and D. Pedoe, *Methods of Algebraic Geometry*, Vol. II. New York: Cambridge University Press, 1952.

Intersections of Cantor Sets and Transversality of Semigroups

HARRY FURSTENBERG

1. Introduction

Let X be a compact metric space endowed with a notion of dimension for subsets of X (analogous to Hausdorff dimension for subsets of a manifold). Two closed subsets $A, B \subset X$ will be called *transverse* if

$$(1) \qquad \dim A \cap B \leqq \begin{cases} \dim A + \dim B - \dim X & \text{when this is } \geqq 0 \\ 0 & \text{otherwise.} \end{cases}$$

Let S_1 and S_2 be two semigroups (or groups) acting on X (i.e., each S_i is a semigroup of continuous transformations of X into itself). S_1 and S_2 will be called *transverse* if whenever A is a closed S_1-invariant set and B a closed S_2-invariant set, A and B are transverse. Finally, two transformations T_1 and T_2 of $X \to X$ are *transverse* if the semigroups they generate are transverse.

The investigations to be described arose in an attempt to verify that certain pairs of endomorphisms of a compact Abelian group are transverse. Before discussing these we shall say something of the background of the problem. If S is a semigroup acting on a space X one can frequently verify that for almost all points $x \in X$ (in an appropriate sense) the orbit Sx of x is dense in X. When the exceptional set is not empty, one would like to have conditions to ensure that a given point is not exceptional. The following is a hypothetical situation where this is realized. We assume that

A. (i) S' is another semigroup acting on X, and for every $x \in X$

$$(2) \qquad \dim \overline{Sx} + \dim \overline{S'x} \geqq \dim X.$$

(ii) If A is a closed S-invariant set with $\dim A = \dim X$, then $A = X$. Under **A**(i), if the semigroup S' attaches a "small" orbit to a point x, then the semigroup S will have a "large" orbit at x. In particular, if $\dim \overline{S'x} = 0$, then necessarily $\dim \overline{Sx} = \dim X$ and, by (ii), Sx is dense in X. We have then a sufficient condition which ensures that a point has a dense orbit under S.

Now the condition **A**(ii) is actually met in many situations. Condition **A**(i) is, however, a very special one, and the notion of transversality was introduced as a less delicate notion which might provide evidence for the validity of (2). The relationship between the two is this. Suppose A is a closed S-invariant set, B a closed S'-invariant set with $\dim A + \dim B < \dim X$. According to **A**(i), $A \cap B$ must be empty. If, on the other hand, the semigroups are transverse, what can be inferred is that $\dim A \cap B = 0$. In many cases transversality of S and S' may be seen to imply that (2) is valid for all x outside a set of dimension 0. Actually we have demanded too much in **A**(i), and the best one can hope for is that (2) be valid for all x outside of a small "constructible" set. For example, there are usually points x for which both Sx and $S'x$ are finite. These periodic points can be explicitly determined.

Let us consider still another hypothesis regarding transformation semigroups S_1 and S_2:

B. If A is a proper closed subset of X invariant under both S_1 and S_2, then $\dim A = 0$.

Note that if S_1 and S_2 are transverse and, in addition, **A**(ii) is valid, then B will be verified. For

$$\dim A = \dim A \cap A \leqq \begin{cases} 2 \dim A - \dim X, & \text{if } 2 \dim A \geqq \dim X \\ 0, & \text{if } 2 \dim A < \dim X. \end{cases}$$

The first alternative leads to $\dim A = \dim X$ whence $A = X$ and the second alternative leads to $\dim A = 0$.

2. Examples

For a first example we take $X =$ the circle group $K = \mathbf{R}/\mathbf{Z}$. The integers act as endomorphisms on K, and if S is any multiplicative semigroup of integers, it may be thought of as acting on K. In [1] we obtained the following theorems:

THEOREM 1. *Let p and q be two integers > 1, with both not powers of the same integer, and denote by S_p and S_q the semigroups generated respectively by p and q. If A is a proper closed subset of K invariant under both S_p and S_q then A is finite.*

Theorem 1 may be regarded as a theorem in diophantine approximation. For if S is a semigroup containing both S_p and S_q and α is an irrational, then by Theorem 1, $\overline{S\alpha} = K$. In particular we have

THEOREM 2. *If S is a multiplicative semigroup of integers* 0 *not consisting only of powers of a single integer and α is irrational, then for every $\epsilon > 0$ there exist $s \in S$ and $r \in \mathbf{Z}$ with* $\left|\alpha - \frac{r}{s}\right| < \frac{\epsilon}{s}$.

The condition that the semigroup not consist only of powers of a single integer is necessary for the conclusion. For if $S \subset \{a^n\}$, $a > 0$, and we take $\alpha = \sum_1^\infty a^{-2n} + \sum_1^\infty a^{-n^2}$, then α is irrational and cannot be "well approximated" by fractions with denominator in S.

Theorem 1 is a sharpening of **B**. However it is also a consequence of **B**. For it can be shown without much difficulty (see [1]) that if A has the property in question, then the set of differences $A - A$ is either all of K or it has 0 as an isolated point. The first alternative is impossible if $\dim A = 0$, and the second implies that A is finite. Thus Theorem 1 would be a consequence of the following conjecture:

Conjecture 1. The semigroups S_p and S_q of Theorem 1 are transverse.

Related to this is the following:

Conjecture 2. For every irrational $x \in K$,

$$\dim \overline{S_p x} + \dim \overline{S_q x} \geqq 1.$$

Of course when x is rational then $S_p x$ and $S_q x$ are finite.

If we associate with a real number its expansion to the base p, then multiplying by p corresponds to shifting the expansion. A subset of K invariant under S_p corresponds to a family of one-sided infinite sequences with entries $0, \ldots, p - 1$ and shift invariant. In particular, if we consider the set of all sequences whose entries belong to a proper subset of $\{0, \ldots, p - 1\}$, these will determine a closed S_p-invariant set. We will call such a set a *Cantor p-set*. Conjecture 1 implies that if p and q are not powers of the same integer, then each Cantor p-set is transverse to each Cantor q-set.

We note that the Cantor p-set corresponding to the subset $\{a_1, \ldots, a_r\} \subset \{0, \ldots, p - 1\}$ has Hausdorff dimension $\log r/\log p$ (see [1], p. 35). Moreover if A is any proper closed S_p-invariant set, then A is contained in some proper Cantor p^m-set for some m. It follows that $\dim A < \dim K$ unless $A = K$. This verifies condition **A**(ii).

An interesting consequence of Conjecture 2 is the following:

Conjecture 2′. Suppose that p and q are not powers of the same integer. Then in the expansion of p^n to the base pq every digit and every combination of digits occurs as soon as n is sufficiently large.

For concreteness let us take $p = 2$ and $q = 5$ so that $pq = 10$. Suppose that some combination of digits $b_1, \ldots, b_l$ was missing from the expansions of infinitely many 2^n. We do not lose generality by supposing that $b_1 \neq 0$, $b_l \neq 0$. Choose a subsequence of these n which increases very rapidly, say $\{n_i\}$. Form the number $\xi = \sum_1^\infty 5^{-n_i}$. If $n_{i+1} - n_i \to \infty$ then $\overline{S_5\xi}$ is countable and by Conjecture 2, $\dim \overline{S_{10}\xi} = 1$. Hence $S_{10}\xi$ is dense in K. We consider the decimal expansion of ξ. First, to obtain the decimal expansion of 5^{-n_i} we take the expansion of 2^{n_i} and move the decimal point over n_i places:

$$5^{-n_i} = .00\cdots 0a_{i1}a\cdots a_{ij}.$$

Here $a_{i1}a\cdots a_{ij}$ is the expansion of 2^{n_i} so that $b_1\cdots b_l$ does not occur in the foregoing. We now assume the n_i to increase so rapidly that none of the blocks of a_{ik} overlap. Then $b_1\cdots b_l$ does not occur in the decimal expansion of ξ, and $S_{10}\xi$ cannot be dense. Thus Conjecture 2′ is a consequence of Conjecture 2.

To find other candidates for transversality we exhibit another pair of transformation semigroups for which hypothesis **B** is valid. Let r be an integer > 1. Denote by I_r the ring of r-adic integers. I_r is the completion of the integers **Z** in the non-Archimedean metric for which the distance between two numbers goes to 0 as their difference is divisible by high powers of r. Each element of I_r has a unique expansion as an infinite series: $x = \sum_0^\infty \omega_n r^n$ $\omega_n \in \{0, \ldots, r-1\}$. If p divides some power of r then the operation $x \to \left[\frac{x}{p}\right]$ ($[y]$ denoting the greatest integer in y) is uniformly continuous on the integers and extends to a transformation D_p on I_r. One verifies that $D_{pq} = D_p D_q$. We now take $X = I_r$. We shall see presently that as a sequence space I_r is endowed with a natural notion of dimension for which $\dim I_r = \log r$. We now have

THEOREM 3. *Let $r = pq$ where $p > 1$ and $q > 1$ are not powers of the same integer. Let S'_p and S'_q denote, respectively, the semigroups of transformations of I_r generated by D_p and D_q. If A is a proper closed subset of I_r invariant under both S'_p and S'_q, then* $\dim A = 0$.

Theorem 3 is a non-Archimedean analogue of Theorem 1 but it is also a consequence of the latter. We shall sketch a proof. Let Ω_r denote the space of all doubly infinite sequences $\omega = (\omega_n)$ with entries in $\{0, \ldots, r-1\}$. $T: \Omega_r \to \Omega_r$ denotes the shift operation. We have a map $\varphi^+: \Omega_r \to I_r$ with $\varphi^+(\omega) = \sum_0^\infty \omega_n r^n$ as well as $\varphi^-: \Omega_r \to K$ with $\varphi^-(\omega) = \sum_{-\infty}^{-1} \omega_n r^n$. Suppose now that A is a closed subset of I_r invariant under D_p and D_q. Then it is invariant under $D_{pq} = D_r$ which is simply the shift operation on one-sided sequences. Let $\tilde{A}$ be the set of sequences ω in Ω_r satisfying $\varphi^+(T^m\omega) \in A$ for each m. It can be shown that $\tilde{A}$ is infinite if dim $A > 0$. The operations D_p and D_q may be extended to Ω_r and one finds that they leave $\tilde{A}$ invariant. For

$$D_p\left(\sum_0^\infty \omega_n r^n\right) = \sum_0^\infty \omega_n' r^n$$

with

$$\omega_n' \equiv \left[\frac{\omega_n}{p}\right] + q^{\omega_{n+1}} \pmod{r}, \tag{3}$$

and (3) defines an operation on Ω_r. Now take $B = \varphi^-(\tilde{A})$. It follows from the foregoing and the shift invariance of $\tilde{A}$ that if $\sum_{-\infty}^{-1} \xi_n r^n \in B$ so is $\sum_{-\infty}^{-1} \xi_n' r^n$, where

$$\xi_n' \equiv \left[\frac{\xi_{n-1}}{p}\right] + q\xi_n \pmod{r}.$$

But this means that if $x \in B$, so is qx. Similarly B is invariant under $x \to px$. By Theorem 1, B is therefore either finite or all of K. One sees that this implies that $\tilde{A}$ is either finite or all of Ω_r. But the former implies that dim $A = 0$ and the latter implies that $A = I_r$.

In analogy with the Archimedean case we formulate the following:

Conjecture 3. The operations D_p and D_q are transverse on I_r.

For a final example we take $X = I_p$ where p is a prime. Let π be a p-adic integer divisible by p but not by p^2. In addition to the standard p-adic representation, each $x \in I_r$ has a representation $x = \sum_0^\infty \omega_n \pi^n$ with $\omega_n \in \{0, \ldots, p-1\}$. The shift operation on sequences now induces a transformation D_π on I_r with $D_\pi x = \pi^{-1}(x - h(x))$ where $h(x) \in \{0, \ldots, p-1\}$ and $h(x) \equiv x \pmod{p}$. Now let π_1 and π_2 be two such p-adic numbers.

Conjecture 4. D_{π_1} and D_{π_2} are transverse on I_p provided $\pi_2\pi_1^{-1}$ is not a root of unity.

Of course $\pi_2\pi_1^{-1}$ is, in any case, a p-adic unit. In case $\pi_1 = p$ and $\pi_2 = pq$ where q is relatively prime to p, then it may be shown that Conjecture 4 is a consequence of Conjecture 3. We omit the proof.

3. Strong transversality

The present investigation is devoted to a result which lends further plausibility to the conjectures of the preceding section. To formulate this we shall identify subsets of the circle K with subsets of the real line. A closed subset $A \subset [0, 1]$ will be called a *p-set* if $pA \subset A \cup (A + 1) \cup$

$(A + 2) \cup \cdots \cup (A + p - 1)$. This corresponds to an S_p-invariant subset of K. Now if A and B are subsets of the line, it is easy to see that

$$\dim A \cap B = \dim A \times B \cap l_0,$$

where l_0 represents the diagonal line $x = y$ of the plane. Similarly if l is an arbitrary line of the plane, $l: y = ux + t$, then $\dim (uA + t) \cap B = \dim A \times B \cap l$.

The remainder of this paper is devoted to a proof of the following property of p-sets.

THEOREM 4. *Let A be a p-set and B a q-set where p and q are not powers of the same integer, and let $C = A \times B$. Let δ be an arbitrary number. If there is some line l with positive, finite slope which intersects C in a set of dimension $> \delta$, then for almost every $u > 0$ (in the sense of Lebesgue measure) there is a line of slope u intersecting C in a set of dimension $> \delta$.*

We can reformulate this using a variant of transversality. We say that two closed subsets A, B of the line are *strongly transverse* if every translate $A + t$ of A is transverse to B. We remark that for arbitrary closed A, B, *almost* every translate of A is transverse to B. From Theorem 4 we deduce the following theorem:

THEOREM 5. *Let A and B be as in Theorem 4. Assume that for a set of u of positive Lebesgue measure the dilation uA of A is strongly transverse to B. Then A and B are strongly transverse.*

For, if not, there is a line with positive slope that intersects C in a set of dimension $> \dim A + \dim B - 1$ (resp. 0). Then for almost all directions this will be the case and $\dim (uA + t) \cap B > \dim A + \dim B - 1$ (resp. 0), and since $\dim A = \dim uA$, uA and B are not strongly transverse.

In this connection let us mention a further conjecture:

Conjecture 5. For arbitrary compact subsets A, B of the line, almost all dilations of A are strongly transverse to B.

The reason for expecting almost all dilations of a set to be well behaved with respect to another set is the following result (for which I am grateful to J.-P. Kahane):

THEOREM 6. *If A and B are arbitrary compact subsets of the line, for almost all dilations uA of A we have*

$$\dim (B + uA) = \begin{cases} \dim A + \dim B, & \text{if this is } \leqq 1, \\ 1, & \text{otherwise.} \end{cases}$$

Here $B + uA$ denotes the set of sums $ux + y$, $x \in A$, $y \in B$. The foregoing is a special case of a more general result asserting that if C is a compact set in the plane with $\dim C = \gamma$, $\gamma \leqq 1$, then in almost every direction C projects onto a linear set of dimension γ. This is easily proven using the characterisation of dimension in terms of capacities (see [2]).

Clearly, Conjecture 5, together with the theorem we are about to prove, imply the validity of Conjecture 1.

In the non-Archimedean case a similar result is true. Specifically consider subsets A, B of I_p. I_p is contained in the field of all p-adic numbers and we may speak of lines in the p-adic plane intersecting $A \times B$. We shall be interested in lines of the form $y = ux + t$ where the "slope" u is a p-adic unit.

THEOREM 7. *Let A be a closed subset of I_p invariant under D_{π_1}, B a closed subset invariant under D_{π_2}, and assume $\pi_2\pi_1^{-1}$ is not a root of unity. There is a subgroup U of finite index in the group of p-adic units such that if a single line $y = u_0x + t_0$ intersects $A \times B$ in a set of dimension $> \delta$, where $u_0 \in U$, then for almost every $u \in U$ (with respect to Haar measure on U) there is a line with slope u intersecting $A \times B$ in a set of dimension $> \delta$.*

The proofs of Theorems 4 and 7 are very similar and we shall confine our attention to the former.

4. Trees and dimension

Let Λ be a finite set with r elements. Denote by Ω_Λ the product $\Lambda \times \Lambda \times \Lambda \times \cdots$ endowed with the usual topology that renders it a compact Hausdorff space. We shall denote by Λ^* the free semigroup generated by Λ: $\Lambda^* = \bigcup_0^\infty \Lambda^n$ where Λ^0 consists of the empty word which we denote by 1, and where multiplication is by juxtaposition. If $\sigma \in \Lambda^n$ we shall write $l(\sigma) = n$. We shall use the term *factor* to denote *left factor*: σ is a factor of τ if $\tau = \sigma\rho$ for some ρ.

DEFINITION 1. A subset $\Delta \subset \Lambda^*$ is called a *tree* if

(i) $1 \in \Delta$.
(ii) If $\sigma \in \Delta$ then every factor of $\sigma \in \Delta$.
(iii) If $\sigma \in \Delta$ then some $\sigma a \in \Delta$ for $a \in \Lambda$.

The following example will justify the terminology:

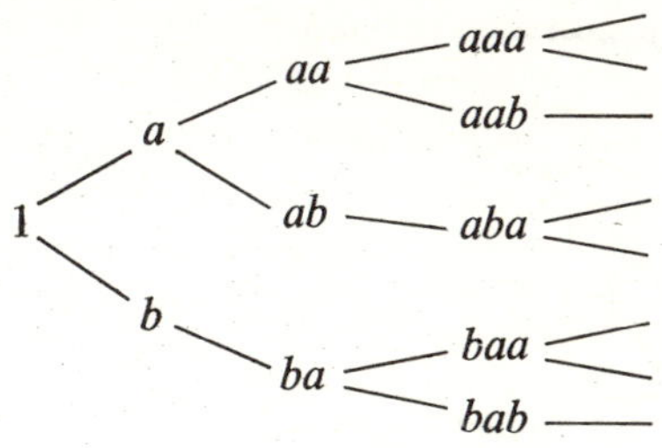

One can set up a correspondence between closed subsets of Ω_Λ and trees in Λ^*. If $A \subset \Omega_\Lambda$ let A^* denote the set of all initial segments (including 1) occurring in sequences of A. Clearly A^* is a tree. Conversely if Δ is a tree, we associate to it the set of all infinite sequences all of whose initial segments are words in Δ. It is easily seen that this is a one-to-one correspondence. (If, however, A is not required to be closed, then the set associated to A^* is the closure of A.)

DEFINITION 2. A *section* of a tree Δ is a finite subset $\Pi \subset \Delta$ satisfying the following conditions:

(i) With finitely many exceptions, each element of Δ has a factor in Π.
(ii) If ρ is a factor of σ and both $\rho, \sigma \in \Pi$ then $\rho = \sigma$.

If $a_1, a_1a_2, \ldots, a_1a_2, \ldots, a_n$ is an increasing sequence of elements of Δ, then one of these must eventually possess a factor in Π by (i), and by (ii) it follows that the foregoing sequence intersects Π in exactly one element. We see from this that a section of the tree A^* corresponds to an irredundant open covering of the compact set A.

If Π is a section we denote by $l(\Pi)$ the minimum of $l(\sigma)$, $\sigma \in \Pi$.

DEFINITION 3. The dimension of a tree Δ is defined as the g.l.b. of the set of λ with the property: $\exists$ sections $\Pi \subset \Delta$ with $l(\Pi)$ arbitrarily large and

$$\sum_{\sigma \in \Pi} e^{-\lambda l(\sigma)} < 1.$$

If A is a closed subset of Ω_Λ we set $\dim A = \dim A^*$.

The connection with Hausdorff dimension is given in the following:

LEMMA 1. *If $\Lambda = \{0, \ldots, r-1\}$ and A is a closed subset of Ω_Λ, set*

$$\bar{A} = \left\{\sum_1^\infty \omega_n r^{-n} : \omega = (\omega_n) \in A\right\} \subset [0, 1].$$

Then

$$\dim A = \log r \times \text{Hausdorff dim}\,(\bar{A}).$$

The proof is straightforward. One notices as a consequence that the dimension of a set depends only on the "geometry" of the associated tree and not upon how one labels its vertices. Thus if in Δ, each $\sigma \in \Delta$ is followed by exactly m successors σa_i, then $\dim \Delta = \log m$.

We introduce into the space of trees a compact Hausdorff topology by setting

$$D(\Delta, \Delta') = \frac{1}{m+1} \quad \text{if } \Delta \cap \Lambda^m = \Delta' \cap \Lambda^m \quad \text{but} \quad \Delta \cap \Lambda^{m+1} \neq \Delta' \cap \Lambda^m$$

If Δ is a tree and if $\rho \in \Delta$, then $\{\sigma | \rho\sigma \in \Delta\}$ is a tree which we denote Δ^ρ.

Definition 4. If Δ is a tree we denote by $\mathscr{D}(\Delta)$ the closure of the set $\{\Delta^\rho, \rho \in \Delta\}$ in the space of trees. The trees of $\mathscr{D}(\Delta)$ are called *derived trees* of Δ.

Just as trees correspond to sets in Ω_Λ we can define objects corresponding to measures on Ω_Λ as follows:

Definition 5. A real-valued function θ on Λ^* is a T-function if

(i) $0 \leqq \theta(\sigma) \leqq 1$,
(ii) $\theta(1) = 1$,
(iii) $\theta(\sigma) = \sum_{a \in \Lambda} \theta(\sigma a)$.

The set of T-functions will be denoted by T_Λ.

The support $|\theta|$ of a T-function is always a tree. Conversely every tree is the support of some T-function. If μ is a probability measure on Ω_Λ and we denote by $\rho\Omega_\Lambda$ the set of sequences that begin with ρ, then the function $\theta(\rho) = \mu(\rho\Omega_\Lambda)$ clearly represents a T-function. The space T_Λ is a compact metric space where we set

$$\mathscr{D}(\theta, \theta') = \sum_{\sigma \in \Lambda^*} 2^{-l(\sigma)}|\theta(\sigma) - \theta'(\sigma)|.$$

The convergence of this series follows from the fact that $\sum_{l(\sigma)=n} \theta(\sigma) = 1$.

Remark. Quite generally, if Π is a section of $|\theta|$ we have $\sum_{\sigma \in \Pi} \theta(\sigma) = 1$. This is a special case of a still more general assertion: if ρ is a factor of an element of Π then $\sum \theta(\sigma) = \theta(\rho)$, the sum being extended over those $\sigma \in \Pi$ of which ρ is a factor. To see this, note that it is obviously true if $\rho \in \Pi$. If $\rho \notin \Pi$ then all its successors ρa, $a \in \Lambda$ are factors of elements of Π and it suffices to establish the assertion for these. But then our assertion follows by induction on $\max_{\sigma \in \Pi} l(\sigma) - l(\rho)$.

If θ is a T-function, then for each $\rho \in |\theta|$ we may set

$$\theta^\rho(\sigma) = \frac{\theta(\rho\sigma)}{\theta(\rho)}.$$

θ^ρ is again a T-function and $|\theta^\rho| = |\theta|^\rho$.

Definition 6. If θ is a T-function we denote by $\mathscr{D}(\theta)$ the closure in T_Λ of the set $\{\theta^\sigma : \theta(\sigma) > 0\}$. The functions of $\mathscr{D}(\theta)$ are *derived* T-functions of θ.

We can define a notion of dimension for T-functions:

DEFINITION 7. If θ is a T-function we set

$$\dim \theta = \liminf_{\Pi \text{ a section of } |\theta|} \frac{\sum_{\sigma\in\Pi} \theta(\sigma) \log \frac{1}{\theta(\sigma)}}{\sum_{\sigma\in\Pi} \theta(\sigma) l(\sigma)}.$$

LEMMA 2. (i) $\dim \theta \leqq \dim |\theta|$.
(ii) *If* Δ *is a tree*, $\dim \Delta = \sup \{\dim \theta : |\theta| \subset \Delta\}$.

PROOF. If

$$\sum_{\sigma\in\Pi} e^{-\lambda l(\sigma)} < 1$$

then

$$\sum_{\sigma\in\Pi} \theta(\sigma) e^{-\lambda l(\sigma) + \log 1/\theta(\sigma)} < 1,$$

whence by Jensen's inequality (using $\sum \theta(\sigma) = 1$):

$$\sum_{\sigma\in\Pi} \theta(\sigma) \log (e^{-\lambda l(\sigma) + \log 1/\theta(\sigma)}) < 0$$

or

$$-\lambda \sum_{\sigma\in\Pi} \theta(\sigma) l(\sigma) + \sum_{\sigma\in\Pi} \theta(\sigma) \log \frac{1}{\theta(\sigma)} < 0,$$

so that $\lambda > \dim \theta$, and hence $\dim |\theta| \geqq \dim \theta$. This proves (i).

To prove (ii) we need only prove that for $\lambda < \dim \Delta$, $\exists\, \theta$ with $|\theta| \subset \Delta$ and $\dim \theta \geqq \lambda$. We use [2, Théorème II, p. 27], which asserts that if E is a compact subset of the line with Hausdorff dimension $> \beta$, then there exists a probability measure μ on E with $\mu([a, b] < C|b - a|^{\beta})$. If $\Delta = A^*$ then $A \subset \Omega_\Lambda$ corresponds to a subset of $[0, 1]$ obtained by viewing Ω_Λ as expansions of real numbers to the base r (= card. Λ). This set has Hausdorff dimension $> \lambda/\log r$. Choose $\beta = \lambda/\log r$ and let μ be a measure satisfying the foregoing inequality. We lift μ to a measure $\tilde{\mu}$ on A and we find that $\tilde{\mu}(\rho\Omega_\Lambda) = \mu(I)$ where I is an interval of length $r^{-l(\rho)}$. Thus if $\theta(\rho) = \tilde{\mu}(\rho\Omega_\Lambda)$, $\theta(\rho) \leq Cr^{-\beta l(\rho)}$, or

$$\log \frac{1}{\theta(\rho)} \geqq -\log C + \beta l(\rho) \log r = -\log C + \lambda l(\rho)$$

so that

$$\liminf_{l(\Pi)\to\infty} \frac{\sum_{\rho\in\Pi} \theta(\rho) \log \frac{1}{\theta(\rho)}}{\sum_{\rho\in\Pi} \theta(\rho) l(\rho)} \geqq \lambda.$$

Here we use the fact that

$$\sum_{\rho\in\Pi} \theta(\rho) l(\rho) \geqq l(\Pi).$$

This completes the proof of Lemma 2.

5. Markov processes on T_Λ

The space of T-functions, T_Λ, is endowed with a natural set of transition probabilities which induce a family of Markov processes on T_Λ. If $\theta \in T_\Lambda$, the numbers $\theta(a)$, $a \in \Lambda$ constitute a probability distribution on Λ, and we may define transition probabilities on T_Λ by assigning probability $\theta(a)$ to the transition $\theta \to \theta^a$. We call these the *canonical transition probabilities* on T_Λ. Fixing an initial distribution for a T_Λ-valued variable z_0, there is determined a unique process $\{z_n\}$ in accordance with these transition probabilities. Inasmuch as $\theta(a)$ is a continuous function of θ, it follows that the induced Markov operator transforms continuous functions to continuous functions. There will exist stationary measures; these constitute a compact convex set, and the extremals of this set determine ergodic stationary processes $\{z_n\}$.

DEFINITION 8. A *C-process* is a stationary Markov process on T_Λ with the canonical transition probabilities.

The main step in the proof of Theorem 4 is provided by the following theorem.

THEOREM 8. *If Δ is a tree with* $\dim \Delta > \delta$, *then there exists an ergodic C-process $\{z_n\}$ such that with probability* 1, *$|z_n|$ is contained in trees of $\mathscr{D}(\Delta)$, and such that with probability* 1, $\dim |z_n| > \delta$.

The purpose of this theorem is to exhibit an abundance of trees of dimension $> \delta$ once we have a single tree with dimension $> \delta$. The following shows how a single T-function θ_0 may generate a C-process:

DEFINITION 9. A measure μ on T_Λ is *compatible* with a T-function θ_0 if there exists a sequence $N_k \to \infty$ such that for each continuous function f on T_Λ,

$$\int_{T_\Lambda} f(\theta)\, d\mu(\theta) = \lim_{k\to\infty} \frac{1}{N_k + 1} \sum_{n=0}^{N_k} \sum_{\tau \in \Lambda^n \cap |\theta_0|} \theta_0(\tau) f(\theta_0^\tau). \tag{4}$$

LEMMA 3. *If μ is compatible with θ then μ has its support on $\mathscr{D}(\theta)$. In addition μ is a stationary measure for the canonical transition probabilities.*

PROOF. The first statement is obvious and the second involves a straightforward verification. If $p(\theta, \theta')$ denotes the canonical transition probabilities, then

$$p(\theta_0^\tau, \theta_0^{\tau a}) = \frac{\theta_0(\tau a)}{\theta_0(\tau)}.$$

To show that μ is stationary we must show that

$$\int \left(\sum p(\theta, \theta')f(\theta')\right) d\mu(\theta) = \int f(\theta)\, d\mu(\theta).$$

But

$$\lim \frac{1}{N_k + 1} \sum_{n=0}^{N_k} \sum_{\tau} \theta_0(\tau) \sum_{\alpha\in\Lambda} p(\theta_0^{\tau}, \theta_0^{\tau a})f(\theta_0^{\tau a})$$

$$= \lim \frac{1}{N_k + 1} \sum_{n=0}^{N_k} \sum_{\tau} \sum_{a} \theta_0(\tau a) f(\theta_0^{\tau a})$$

$$= \lim \frac{1}{N_k + 1} \sum_{n=1}^{N_k+1} \sum_{\tau} \theta_0(\tau) f(\theta_0^{\tau}) = \int f(\theta)\, d\mu(\theta).$$

DEFINITION 10. If μ is a stationary measure on T_Λ we set

$$E(\mu) = \int_{T_\Lambda} \sum_{\theta'} p(\theta, \theta') \log \frac{1}{p(\theta, \theta')}\, d\mu(\theta).$$

LEMMA 4. *If μ is compatible with θ_0 then $E(\mu) \geqq \dim \theta_0$.*

PROOF.

$$E(\mu) = \lim_{k\to\infty} \frac{1}{N_k + 1} \sum_{n} \sum_{\tau} \sum_{a} \theta_0(\tau) \cdot \frac{\theta_0(\tau a)}{\theta_0(\tau)} \log \frac{\theta_0(\tau)}{\theta_0(\tau a)}$$

$$= \lim_{k\to\infty} \frac{1}{N_k + 1} \sum_{n} \sum_{\tau} \sum_{a} \theta_0(\tau a)\{\log \theta_0(\tau) - \log \theta_0(\tau a)\}$$

$$= \lim_{k\to\infty} \frac{1}{N_k + 1} \sum_{l(\tau) = N_k + 1} \theta_0(\tau) \log \frac{1}{\theta_0(\tau)}.$$

Now let Π_k denote the section $|\theta_0| \cap \Lambda^{N_k+1}$. Then the last expression in the proof just given is the same as

$$\lim_{k\to\infty} \frac{\sum_{\tau\in\Pi_k} \theta_0(\tau) \log \frac{1}{\theta_0(\tau)}}{\sum_{\tau\in\Pi_k} \theta_0(\tau) l(\tau)} \geqq \dim \theta_0,$$

which proves the lemma.

LEMMA 5. *For any T-function θ_0 there exists an ergodic stationary measure μ for the canonical transition probabilities with support on $\mathscr{D}(\theta_0)$ and with $E(\mu) \geqq \dim \theta_0$.*

PROOF. To begin with, there exist measures μ' compatible with θ_0 since we can choose $\{N_k\}$ so that (4) converges for every function in $\mathscr{C}(T_\Lambda)$. μ' is supported on the derived set of θ_0. The set of stationary measures with support in $\mathscr{D}(\theta_0)$ forms a compact convex set, and it follows that μ' can be expressed as an integral of ergodic measures with support in $\mathscr{D}(\theta_0)$: $\mu' = \int \mu_\omega \, d\nu(\omega)$. It follows that

$$E(\mu') = \int E(\mu_\omega) \, d\nu(\omega);$$

therefore some $E(\mu_\omega) \geqq E(\mu') \geqq \dim \theta_0$.

LEMMA 6. *If μ is an ergodic stationary measure, then for almost all θ with respect to μ,* $\dim \theta \geqq E(\mu)$.

PROOF. Let $\{z_n\}$ denote the T_Λ-valued variables of the stationary Markov process determined by μ. Define the function H on T_Λ:

$$H(\theta) = \sum_{a \in \Lambda} \theta(a) \log \frac{1}{\theta(a)}.$$

$H(\theta)$ is a bounded continuous function. By the ergodic theorem we have, with probability 1,

$$\frac{H(z_0) + H(z_1) + \cdots + H(z_{n-1})}{n} \to \int H(\theta) \, d\mu(\theta) = E(\mu).$$

Now let λ be any number $< E(\mu)$, and denote by $S(n, \lambda)$ the subset of the underlying probability space for which some average

$$\frac{H(z_0) + H(z_1) + \cdots + H(z_{n-1})}{N} < \lambda$$

with $N \geqq n$. We have $P(S(n, \lambda)) \to 0$ as $n \to \infty$. Hence the conditional probability $P(S(n, \lambda)|z_0) \to 0$ as $n \to \infty$ for almost all z_0. Now given $z_0 = \theta$, the conditional probability that $z_1 = \theta^{a_1}, z_2 = \theta^{a_1 a_2}, \ldots, z_n = \theta^{a_1 a_2 \cdots a_n}$ is $\theta(a_1 a_2 \cdots a_n)$. Let Π be a section of $|\theta|$ with $l(\Pi) > n$. Given $z_0 = \theta$, we can assert that there will be a value of m for which $z_m = \theta^{a_1 \cdots a_m}$ and $\sigma = a_1 \cdots a_m \in \Pi$. We will then have

$$G(\sigma) = G(a_1 \cdots a_m) = \frac{H(\theta) + H(\theta^{a_1}) + \cdots + H(\theta^{a_1 \cdots a_{m-1}})}{m} \geqq \lambda$$

unless the sequence $\theta^{a_1} \cdots \theta^{a_1 \cdots a_m}$ corresponds to a point of $S(n, \lambda)$. Since the conditional probability of obtaining this sequence is $\theta(a_1 \cdots a_m)$, we conclude that the sum of the $\theta(\sigma)$ for which $\sigma \in \Pi$, and $G(\sigma) < \lambda$, is less than $P(S(n, \lambda)|z_0 = \theta)$. Hence

$$(5) \qquad \sum_{\Pi} \theta(\sigma) G(\sigma) l(\sigma) \geqq \lambda\{1 - P(S(n, \lambda)|z_0 = \theta)\} \sum \theta(\sigma) l(\sigma).$$

We proceed to evaluate the left side of (5). First a preliminary observation. Let $\rho|\sigma$ denote that ρ is a proper factor of σ; $\rho|\Pi$ denotes that ρ is a proper factor of an element in Π. By our remark following Definition 5 we have

$$\sum_{\rho|\sigma,\sigma\in\Pi} \theta(\sigma) = \theta(\rho). \tag{6}$$

Now by (6),

$$\sum_{\Pi} \theta(\sigma)G(\sigma)l(\sigma) = \sum_{\sigma\in\Pi}\sum_{\rho|\sigma} H(\theta^\rho)\theta(\sigma) = \sum_{\rho|\Pi} H(\theta^\rho)\theta(\rho).$$

Also

$$\begin{aligned}\sum_{\rho|\Pi} H(\theta^\rho)\theta(\rho) &= \sum_{\rho|\Pi}\sum_{a\in\Lambda} \theta(\rho a)\log\frac{\theta(\rho)}{\theta(\rho a)} \\ &= \sum_{\rho|\Pi}\left\{\sum_{a\in\Lambda}\theta(\rho a)\log\frac{1}{\theta(\rho a)} - \theta(\rho)\log\frac{1}{\theta(\rho)}\right\} \\ &= \sum_{\sigma\in\Pi}\theta(\sigma)\log\frac{1}{\theta(\sigma)}.\end{aligned}$$

Hence

$$\frac{\sum_{\Pi}\theta(\sigma)\log\dfrac{1}{\theta(\sigma)}}{\sum_{\Pi}\theta(\sigma)l(\sigma)} \geqq \lambda\{1 - P(S(n,\lambda)|z_0 = \theta)\} \to \lambda \quad \text{as } n\to\infty \text{ a.e.}$$

This proves the lemma.

Taking stock of what has occurred in the foregoing lemma we see that we have arrived at a proof of Theorem 8.

6. Proof of Theorem 4

Let p and q be two positive integers not both powers of the same integer and let $A \subset [0, 1]$, $B \subset [0, 1]$ be respectively a p-set and a q-set. We shall assume $q > p$. If we set $g(x) = [px]$ and $h(y) = [qx]$, then A is closed under $x \to px - g(x)$, and B is closed under $y \to qy - h(y)$. Let us define two transformations of the unit square:

$$\begin{aligned}\Phi_1(x, y) &= (px - g(x), y)\\ \Phi_2(x, y) &= (px - g(x), qy - h(y)).\end{aligned} \tag{7}$$

Then $A \times B$ is invariant under both transformations Φ_1 and Φ_2. Each Φ_i transforms a line into a finite number of line segments, and if l is a line with slope u, then each line of $\Phi_1(l)$ has slope u/p and each line of $\Phi_2(l)$ has slope qu/p.

Now suppose that l is a line that intersects $A \times B$ in a set of dimension $> \delta$. The same will be true of at least one of the lines of $\Phi_1(l)$ and of one of the lines of $\Phi_2(l)$. Proceeding in this manner we find infinitely many lines with the property that they intersect $A \times B$ in a set of dimension $> \delta$. If the first of these lines had slope u, we will find all the slopes $u' = q^m u/p^n$, $n > m$, represented. Note that the set of u' of this form is dense in $(0, \infty)$ precisely when p and q are not powers of the same integer. For that is equivalent to $\log q/\log p$ being irrational, which implies that

$$\log u' = \log p\left(m\frac{\log q}{\log p} - n + \frac{\log u}{\log p}\right)$$

is dense in the reals. Thus under the hypotheses of the theorem, we will obtain a dense set of slopes with the desired property. The machinery of the preceding section will be invoked to extend the conclusion to almost all u'. Note, however, that the present argument shows that it suffices to establish the assertion of the theorem for slopes in some finite interval. We shall do so for $1 \leqq u \leqq q$.

We now introduce a number of spaces that will play a role in proving Theorem 4.

We denote the subset $A \times B$ of the unit square in the plane by C. L will denote the set of lines $Y = uX + t$ with $1 \leqq u \leqq q$ and which intersect the unit square. We will speak interchangeably of the line l and the pair (u, t). We denote by W the set of ordered pairs: a point of C, a line of L through the point. Thus $W = \{(x, y, u, t):(x, y) \in C, y = ux + t\}$. We define a transformation Φ of W into itself as follows:

$$\Phi(x, y, u, t) = \begin{cases} \left(\Phi_2(x, y), \dfrac{q}{p}u, t'\right) & \text{if } 1 \leqq u < p \\ \left(\Phi_1(x, y), \dfrac{1}{p}u, t'\right) & \text{if } p \leqq u < q. \end{cases} \tag{8}$$

Here t' is uniquely determined by the condition that the resulting point belongs to W. As defined, the transformation Φ is discontinuous. This is already true for the functions $g(x)$, $h(x)$. The functions g and h become continuous if, instead of the unit interval, we take as their domains the totally disconnected sequence spaces representing the unit interval to the bases p and q, respectively. A similar modification of the u-interval renders the function Φ continuous at $u = p$. We shall suppose w to have been modified in this manner and we retain the same notation for the modified space.

L^∞ denotes the compact space $L \times L \times L \times \cdots$. We define a map $\gamma: W \to L^\infty$ by associating to $w \in W$ its orbit $\Phi^n(w)$, and setting $\gamma(w) = \{(u_0, t_0), (u_1, t_1), \ldots\}$ where $\Phi^n(w) = (x_n, y_n, u_n, t_n)$. We denote the range of this map by $\tilde{L}$.

Let $(x_0, y_0, l_0) \in W$, and form $\Phi^n(x_0, y_0, l_0) = (x_n, y_n, l_n)$. The numbers $g(x_n)$ are readily seen to constitute the successive digits of the expansion of x_0 to the base p. From (8) one sees that the same will not be true for $h(y_n)$ and the expansion of y to the base q. Inasmuch as certain successive y_n are identical, the numbers $h(y_n)$ represent with repetitions the digits of this expansion. Suppose that in the sequence $h(y_0), h(y_1), \ldots, h(y_n)$ the first n' digits occur. We wish to find the relationship between n and n'.

LEMMA 7. $|n' \log q - n \log p| \leqq \log q$.

PROOF. n' represents the number of values of $i \leqq n$ for which $u_n \leqq p$. We have

$$\log \frac{u_n}{u_0} = \sum_{i=0}^{n-1} \log \frac{u_{i+1}}{u_i} = n' \log \frac{q}{p} + (n - n') \log \frac{1}{p} = n' \log q - n \log p.$$

This proves the lemma.

Denote by U' the finite set of rationals $\frac{p_i}{q_j}$, $0 < i < p$, $0 < j < q$.

LEMMA 8. *If* $\lambda = (l_0, l_1, \ldots) \in \tilde{L}$ *with* $l_i = (u_i, t_i)$, *and if none of the* $u_i \in U'$, *then there is a unique point* $w \in W$ *with* $\gamma(w) = \lambda$.

PROOF. According to (8)

$$t_{n+1} = y_{n+1} - u_{n+1}x_{n+1} = \begin{cases} q^{t_n} - \left\{h(y_n) - \frac{q}{p}\, u_n g(x_n)\right\} & \text{if } u_n < p, \\ t_n + \frac{u_n}{p}\, g(x_n) & \text{if } u_n \geqq p. \end{cases}$$

Thus the u_n sequence and the t_n sequence determine the digits $g(x_n)$, $h(y_n)$, and hence the point (x_0, y_0) provided that the numbers $\mu - \frac{q}{p} u_n \xi$ uniquely determine $\xi \in \{0, \ldots, p - 1\}$ and $\mu \in \{0, \ldots, q - 1\}$. But this is the case provided $u_n \notin U'$.

Now let u_0 be a value for which no successive value of u_n enters U'.

LEMMA 9. *Let* $\lambda' = \gamma(x', y', l_0)$, $\lambda'' = \gamma(x'', y'', l_0)$ *be two points in* $\tilde{L}$. *If* $l_i(\lambda') = l_i(\lambda'')$ *for* $i \leqq n$, *then* $|x' - x''| < p^{-n}$.

PROOF. This is clear since the $g(x_n)$ represent the successive digits of x_0, and these are determined by the l_i, $i \leqq n$.

LEMMA 10. *There is a constant H such that for each n, if l_0 is a line of L and if J is a subinterval of l_0 of length $< p^{-n}$, then there are at most H distinct sequences $l_0(\lambda), \ldots, l_n(\lambda)$ that can occur for all the points $\lambda = \gamma(w)$ satisfying $w = (x, y, l_0)$ and $(x, y) \in J$.*

PROOF. The sequence $\{l_i(\lambda), i \leqq n\}$ is determined by the two sequences $\{g(x_i), i \leqq n\}$, $\{h(y_i), i \leqq n\}$, inasmuch as l_0 is kept fixed. Since the $g(x_i)$ represent the first n digits in the expansion of $x_0 = x$ and the possible values of x come from an interval of length $\leqq p^{-n}$, we see that there are at most two possibilities for the sequence $g(x_i)$, $i \leqq n$. The possible values of y lie in an interval of length $\leqq pq^{-n}$. Now

$$qp^{-n} = q^{1-n \log p/\log q}$$

and this is smaller than $cq^{-n'}$ according to Lemma 7. Here n' denotes the number of places in the expansion of y to the base q represented in $h(y_0), \ldots, h(y_n)$. But then the possible number of such expansions is bounded. This completes the proof of the lemma.

If we fix a line l_0 in L, we can consider on the one hand all the points of C lying on l_0. On the other hand we may consider the set of all $\lambda \in \tilde{L}$ which start with $l_0(\lambda) = l_0$. The foregoing lemmas show that these sets are closely related. We shall now see that the two sets have the same dimension.

DEFINITION 11. An *L-tree* is a subset $D \subset \bigcup_1^\infty L^n$ satisfying

(i) There is a unique element l_0 of L in D,
(ii) If $(l_0, \ldots, l_m, l_{m+1}) \in D$ so is $(l_0, \ldots, l_m)$,
(iii) If $(l_0, \ldots, l_m) \in D$ so is $(l_0, \ldots, l_m, l_{m+1})$ for some l_{m+1},
(iv) If $(l_0, \ldots, l_m) \in D$ then there exists $\lambda \in \tilde{L}$ with $l_i = l_i(\lambda)$.

As in the case of ordinary trees each closed subset of L corresponding to a fixed l_0 determines an L-tree, and conversely. Note that each element $(l_0, \ldots, l_m)$ of a tree can be followed by at most pq lines l_{m+1}. This follows from the fact that l_{m+1} is a portion of either $\Phi_1(l_m)$ or $\Phi_2(l_m)$. It follows that one may carry over to L-trees the theory of dimension developed earlier.

In particular, let $D(l_0)$ denote the maximal tree which starts with l_0. We then have

LEMMA 11. $\dim D(l_0) = \log p \times \dim (l_0 \cap C)$.

PROOF. We may assume that none of the successors of l_0 have slope in the exceptional set U'. For, in any case, this will be true if instead of l_0, we consider its successors l_m for a sufficiently large m. One then checks that the validity of the lemma is not affected by this change. Now assume that $\beta > \dim D(l_0)$, and let Π be a section of $D(l_0)$ with $\sum_{\sigma \in \Pi} e^{-\beta l(\sigma)} < 1$. If $(x, y) \in l_0 \cap C$ then $\gamma(x, y, l_0)$ begins with some sequence of Π. If (x', y'), (x'', y'') are two points corresponding to the same element σ of Π then, by

Lemma 9, $|x' - x''| < p^{-l(\sigma)}$. Hence Π determines a covering of $l_0 \cap C$ by intervals of respective length $p^{-l(\sigma)}$. But we have

$$\sum_{\sigma} \{p^{-l(\sigma)}\}^{\beta/\log p} < 1,$$

so that $\beta \geqq \log p \times \dim (l_0 \cap C)$. It follows that $\dim D(l_0) \geqq \log p \times \dim (l_0 \cap C)$.

Conversely suppose $\alpha > \dim (l_0 \cap C)$ and that $\{J_i\}$ is a covering of $l_0 \cap C$ with $\sum |J_i|^\alpha < H^{-1}p^{-\alpha}$. Here $|J|$ denotes the length of the interval J. Define n_i by $p^{-n_i-1} \leqq |J_i| < p^{-n_i}$. Next take the set of all segments in $D(l_0)$ of length n which occur in $\gamma(x, y, l_0)$ where (x, y) is a point of J_i. Denote this set Π_i. By Lemma 10, there are at most H elements in Π_i. Now $\bigcup \Pi_i = \Pi'$ is a finite set which clearly contains a section of $D(l_0)$. We have

$$\sum_{\sigma \in \Pi'} e^{-\alpha \log pl(\sigma)} \leqq H \sum e^{-\alpha \log pn_i} = H \sum p^{-\alpha n_i} \leqq Hp^\alpha \sum |J_i|^\alpha < 1,$$

so that $\alpha \log p \geqq \dim D(l_0)$. This proves the lemma.

One can also introduce the notion of T-functions in the context of L-trees.

Definition 12. A T_L-function is a function θ on $\bigcup_1^\infty L^n$ satisfying

(i) θ has its support in some $D(l_0)$,
(ii) $0 \leqq \theta \leqq 1$,
(iii) $\theta(l_0) = 1$,
(iv) $\theta(l_0, l_1, \ldots, l_m) = \sum_{l'} \theta(l_0, l_1, \ldots, l_m, l')$.

We denote the compact space of all T_L-functions by T_L. On this space we have a system of canonical transition probabilities which may be described as follows. If θ has support in $D(l_0)$ consider the set of l_1^i with $l_0 l_1^i$ in $D(l_0)$ for which $\theta(l_0 l_1^i) > 0$. This is a finite set. We then set

$$\theta^i(l_1, l_2, \ldots, l_m) = \begin{cases} 0, & \text{if } l_1 \neq l_1^i \\ \dfrac{\theta(l_0, l_1, l_2, \ldots, l_m)}{\theta(l_0, l_1^i)}, & \text{if } l_1 = l_1^i. \end{cases}$$

The transitions for the canonical transition probabilities are from θ to θ^i with probability $\theta(l_0, l_1^i)$. Note that here l_0 plays the role of the empty word in the case of T-functions. But for this modification the theories are perfectly analogous. In analogy with Theorem 8 one has the following theorem:

Theorem 9. *If D is an L-tree with* $\dim D > \delta'$, *then there exists an ergodic stationary Markov process $\{z_n\}$ with state space T_L, with the canonical transition probabilities, and with* $\dim |z_n| > \delta$ (*with probability* 1).

We are now in a position to prove Theorem 4. Suppose dim $(l_0 \cap C) > \delta$. By Lemma 11, dim $D(l_0) > \delta/\log p$. Now apply Theorem 9 and let $\{z_n\}$ denote the stationary process provided by the theorem. For any T-function, denote by $l(\theta)$ the line determined by $|\theta| \subset D(l(\theta))$. Consider now the stationary, line-valued process $l(z_n)$. Since with probability 1, dim $|z_n| > \delta/\log p$, *a fortiori*, dim $D(l(z_n)) > \delta/\log p$. By Lemma 11, we have, with probability 1, dim $(l(z_n) \cap C) > \delta$. We now determine the distribution of the slopes of $l(z_n)$. From the form of the canonical transition probabilities we see that if u_n is the slope of $l(z_n)$, the slope of $l(z_{n+1})$ is given by

$$u_{n+1} = \begin{cases} \dfrac{q}{p} u_n & \text{if } u_n < p, \\[2ex] \dfrac{1}{p} u_n & \text{if } u_n > p. \end{cases}$$

Let ν denote the distribution of $\dfrac{\log u}{\log q}\cdot$ Then ν is a measure on the unit interval invariant under the transformation $t \to t - \dfrac{\log p}{\log q}$ (modulo 1). But this is a version of the irrational rotation of the circle and Lebesgue measure is the unique invariant measure. It follows that the distribution of u is a measure equivalent to Lebesgue measure. The assertion of Theorem 4 is now a consequence of the Fubini theorem.

HEBREW UNIVERSITY
JERUSALEM

REFERENCES

1. FURSTENBERG, H., "Disjointness in ergodic theory, minimal sets, and a problem in diophantine approximation," *Mathematical Systems Theory, 1* (1967), 1–49.
2. KAHANE, J.-P., and R. SALEM, *Ensembles Parfaits et Series Trigonometriques.* Paris: Hermann & Cie., 1963.

Kählersche Mannigfaltigkeiten mit hyper-q-konvexem Rand

HANS GRAUERT UND
OSWALD RIEMENSCHNEIDER

Einleitung

1953 zeigte K. Kodaira [3], daß die Kohomologiegruppen $H^s(X, \underline{F})$ für $s < n$ verschwinden, wenn X eine n-dimensionale kompakte kählersche Mannigfaltigkeit und F ein negatives komplex-analytisches Geradenbündel auf X ist. Hierbei bezeichnet $\underline{F}$ wie üblich die Garbe der Keime von lokalen holomorphen Schnitten in F. Kodaira verwandte beim Beweis dieses Ergebnisses neben der harmonischen Analysis wesentlich eine Ungleichung, die zuerst von S. Bochner angegeben wurde (man vgl. [2]). Bis heute ist es nicht gelungen, diesen Satz für vollständige projektiv-algebraische Mannigfaltigkeiten auf algebraischem Wege herzuleiten.

Man gelangte jedoch zu mehreren Verschärfungen und Verallgemeinerungen. Von Y. Akizuki und S. Nakano wurde bewiesen [1], daß in der obigen Situation sogar $H^s(X, \underline{F} \otimes \Omega^r) = 0$ ist für $r + s < n$, wenn Ω^r die Garbe der Keime von lokalen holomorphen r-Formen bezeichnet. Nakano [6] gewann danach eine Aussage über negative Vektorraumbündel V auf kompakten kählerschen Mannigfaltigkeiten X: Es gilt stets $H^s(X, \underline{V}) = 0$ für $s < n$. Schließlich übertrug E. Vesentini [8] 1959 einige Resultate auf semi-negative Geradenbündel. Da kompakte komplexe Mannigfaltigkeiten mit solchen Geradenbündeln nicht mehr notwendig kählersch sind, stellte sich die Frage, ob die "Vanishing"-Aussagen wesentlich an der Kähler-Struktur hängen. Diese Frage wurde, ebenfalls von Vesentini, im positiven Sinne beantwortet.

Ein Geradenbündel F heißt negativ, wenn es eine offene, relativ-kompakt in F gelegene Umgebung der Nullschnittfläche gibt, die in jedem Randpunkt streng pseudokonvex ist. Ein Vektorraumbündel mit der entsprechenden Eigenschaft nennen wir schwach negativ. Es hat sich gezeigt, daß negative Vektorraumbündel (im Sinne der Definition von Nakano) auch schwach negativ sind. Das Umgekehrte gilt jedoch nicht. Es gibt sogar, wie wir zeigen werden, schwach negative Vektorraumbündel V über dem $\mathbf{P}^2$ mit $H^1(\mathbf{P}^2, \underline{V}) \neq 0$. Wir zeigen dies auch für beliebige projektiv-algebraische Mannigfaltigkeiten.

In der vorliegenden Arbeit sollen die Sätze von Kodaira, Akizuki und Nakano weiter verallgemeinert werden. Zu diesem Zwecke müssen Ergebnisse von J. Kohn [4] wesentlich herangezogen werden. Wir betrachten eine n-dimensionale komplexe Mannigfaltigkeit X, die relativ-kompakter offener Teilbereich einer komplexen Mannigfaltigkeit $\hat{X}$ ist. Auf $\hat{X}$ sei eine kählersche Metrik gegeben. Ferner sei der Rand ∂X von X glatt und beliebig oft differenzierbar. Ist ∂X streng pseudokonvex (wir sagen auch 1-konvex), so zeigen wir, daß $H_c^s(X, \mathcal{O})$ für $s < n$ und $H^s(X, \Omega^n)$ für $s > 0$ verschwindet. Dabei ist $\mathcal{O} = \Omega^0$ die Strukturgarbe von X und $H_c^s(X, \mathcal{O})$ die s-te Kohomologiegruppe mit kompaktem Träger. Ersetzt man die 1-Konvexität durch die schwächere q-Konvexität, $q > 1$, so sind im allgemeinen $H_c^s(X, \mathcal{O})$ für $s \leqq n - q$ und $H^s(X, \Omega^n)$ für $s \geqq q$ von Null verschieden, obgleich man zunächst das Gegenteil vermuten möchte. Ein Beispiel hierzu kann aus einem schwach negativen Vektorraumbündel über dem $\mathbf{P}^2$ gewonnen werden, für das nicht das Vanishing-Theorem von Nakano gilt. Wir müssen daher die q-Konvexität für $q > 1$ durch eine stärkere Forderung ersetzen, die wir als Hyper-q-Konvexität bezeichnen.

In Teil 1 der Arbeit werden bekannte Ergebnisse zusammengestellt, die später benötigt werden. Der Teil 2 enthält den Beweis der Hauptaussage, und in Teil 3 wird gezeigt, wie dieser Beweis in Beziehung steht zu den Vanishing-Theoremen von Kodaira, Akizuki und Nakano. Ferner werden dort Gegenbeispiele gebracht.

1. Ergebnisse von J. Kohn

A. Wir stellen zunächst einige bekannte Tatsachen aus der Theorie der kählerschen Mannigfaltigkeiten zusammen. Es sei X eine überall n-dimensionale komplexe Mannigfaltigkeit; es bezeichne A^l den $\mathbf{C}$-Vektorraum der (beliebig oft) differenzierbaren komplexen Differentialformen vom Grade l auf X und $A^{r,s}$ den Vektorraum der Formen vom Typ (r, s). Es gilt also $A^l = \bigoplus_{r+s=l} A^{r,s}$. Die totale Ableitung $d: A^l \to A^{l+1}$ läßt sich in der Form $d = d' + d''$ schreiben, wobei $d': A^{r,s} \to A^{r+1,s}$ die totale Ableitung nach den Variablen $z_1, \ldots, z_n$ und $d'': A^{r,s} \to A^{r,s+1}$ die totale Ableitung nach den $\bar{z}_1, \ldots, \bar{z}_n$ bezeichnet. Es gilt: $d^2 = d'^2 = d''^2 = 0$, $d'd'' + d''d' = 0$, $d\bar{\varphi} = \overline{d\varphi}$, $d'\bar{\varphi} = \overline{d''\varphi}$ und $d(\varphi \wedge \psi) = d\varphi \wedge \psi + (-1)^l \varphi \wedge d\psi$, wenn $\varphi \in A^l$ ist (die letzte Formel gilt analog auch für d' und d'').

B. Es sei auf X eine (positiv-definite) hermitesche Metrik $ds^2 = \sum g_{\nu\bar{\mu}}\, dz_\nu\, d\bar{z}_\mu$ gegeben. Dann ist der Metrik ds^2 ein Anti-Isomorphismus $\bar{*}\colon A^{r,s} \to A^{n-r,n-s}$, d.h. eine $\mathbf{R}$-lineare bijektive Abbildung mit $\bar{*}c\varphi = \bar{c}\bar{*}\varphi$, $c \in \mathbf{C}$, zugeordnet, so daß $\bar{*}^2\varphi = (-1)^l\varphi$ für $\varphi \in A^l$ gilt. Eine positiv-definite hermitesche Metrik auf X heißt *kählersch*, wenn die zugeordnete (reelle) Form $\omega := \frac{i}{2}\sum g_{\nu\bar{\mu}}\, dz_\nu \wedge d\bar{z}_\mu$ geschlossen ist, d.h. wenn $d\omega = 0$ gilt.

DEFINITION 1. Es sei $ds^2 = \sum g_{\nu\bar{\mu}}\, dz_\nu\, d\bar{z}_\mu$ eine hermitesche Metrik in X und $x_0 \in X$ ein Punkt. Unter einem geodätischen Koordinatensystem zu x_0 und ds^2 versteht man ein System von lokalen komplexen Koordinaten $z_1, \ldots, z_n$ in einer Umgebung $U(x_0)$, so daß x_0 die Koordinaten $z_\nu(x_0) = 0$, $\nu = 1, \ldots, n$, besitzt und für alle $\nu, \mu = 1, \ldots, n$ gilt: $g_{\nu\bar{\mu}}(x_0) = \delta_{\nu\mu}$, $dg_{\nu\bar{\mu}}(x_0) = 0$.

Geodätische Koordinaten heißen in der angelsächsischen Literatur auch "normal coordinates". Es gilt:

Eine hermitesche Metrik ist genau dann kählersch, wenn es zu jedem Punkt $x_0 \in X$ ein geodätisches Koordinatensystem gibt.

Hat man in x_0 ein geodätisches Koordinatensystem, so gilt für $\varphi = \sum a_{\nu_1\cdots\nu_r,\bar{\mu}_1\cdots\bar{\mu}_s}\, dz_{\nu_1} \wedge \cdots \wedge dz_{\nu_r} \wedge d\bar{z}_{\mu_1} \wedge \cdots \wedge d\bar{z}_{\mu_s} = \sum a_{\nu\bar{\mu}}\, d\mathfrak{z}_\nu \wedge d\bar{\mathfrak{z}}_\mu \in A^{r,s}$ im Punkte x_0:

$$\text{(1)} \qquad \bar{*}\varphi = b(r, s, n) \sum \operatorname{sign}(\nu, *\nu) \operatorname{sign}(\mu, *\mu) \bar{a}_{\nu\bar{\mu}}\, d\mathfrak{z}_{*\nu} \wedge d\bar{\mathfrak{z}}_{*\mu},$$
$$b(r, s, n) = (-1)^{n(n-1)/2 + s(n-r)} 2^{r+s-n} i^n,$$

wobei $*\nu$ ein $(n - r)$-tupel bezeichnet, in dem genau alle natürlichen Zahlen von 1 bis n stehen, die nicht in $\nu = (\nu_1, \ldots, \nu_r)$ vorkommen.

Wir setzen wie üblich $\delta = -\bar{*}d\bar{*}$, $\delta' = -\bar{*}d'\bar{*}$, $\delta'' = -\bar{*}d''\bar{*}$. Man hat $\delta = \delta' + \delta''$, $\delta^2 = \delta'^2 = \delta''^2 = 0$, $\delta'\delta'' + \delta''\delta' = 0$, $\delta\bar{\varphi} = \overline{\delta\varphi}$, $\delta'\bar{\varphi} = \overline{\delta''\varphi}$. δ' und δ'' sind homogene Operatoren vom Typ $(-1, 0)$ bzw. $(0, -1)$, d.h. $\delta'\colon A^{r,s} \to A^{r-1,s}$, $\delta''\colon A^{r,s} \to A^{r,s-1}$. Wir definieren nun den reellen Laplace-Operator Δ und den komplexen Laplace–Beltrami-Operator $\square$ durch $\Delta := d\delta + \delta d$ und $\square := d''\delta'' + \delta''d''$.

Ist $\omega = \frac{i}{2}\sum g_{\nu\bar{\mu}}\, dz_\nu \wedge d\bar{z}_\mu$ die der hermiteschen Metrik ds^2 zugeordnete (1, 1)-Form, so definiert man schließlich noch Operatoren $L\colon A^l \to A^{l+2}$ und $\Lambda\colon A^l \to A^{l-2}$ durch $L\varphi := \varphi \wedge \omega$ und $\Lambda := (-1)^l\bar{*}L\bar{*}$. Eine Form φ heißt primitiv, wenn $\Lambda\varphi = 0$ ist. Es gilt für primitive (r, s)-Formen die Gleichung

$$\text{(2)} \qquad \bar{*}L^e\varphi = (-1)^{l(l+1)/2} \frac{e!}{(n - l - e)!}\, i^{s-r} L^{n-l-e}\bar{\varphi}$$

für $e = 0, 1, \ldots, n - l$ und $l = r + s$. Ist $\varphi \in A^l$ mit $l \leqq n$, und gilt $L^{n-l}\varphi = 0$, so folgt $\varphi = 0$.

Falls die gegebene Metrik kählersch ist, gilt $\Box = \frac{1}{2}\Delta = d'\delta' + \delta'd'$; insbesondere ist dann $\Box$ ein reeller Operator: $\Box\bar{\varphi} = \overline{\Box\varphi}$. Außerdem hat man in diesem Fall die folgenden Relationen:

$$d'L = Ld', \quad d''L = Ld'', \quad \delta'L - L\delta' = -id'', \quad \delta''L - L\delta'' = id',$$
$$\delta'\Lambda = \Lambda\delta', \quad \delta''\Lambda = \Lambda\delta'', \quad d'\Lambda - \Lambda d' = -i\delta'', \quad d''\Lambda - \Lambda d'' = i\delta',$$
$$d'\delta'' + \delta''d' = 0, \quad d''\delta' + \delta'd'' = 0,$$

und $\Box$ ist vertauschbar mit den Operatoren L, Λ, d', d'', δ', und δ''.

C. Im folgenden sei X stets mit einer kählerschen Metrik $ds^2 = \sum g_{\nu\bar{\mu}}\, dz_\nu\, d\bar{z}_\mu$ versehen. Es sei weiter $V \xrightarrow{\pi} X$ ein komplex-analytisches Vektorraumbündel vom Range m über X. Dann gibt es eine Überdeckung $\mathfrak{U} = \{U_\iota : \iota \in I\}$ und Trivialisierungen $\tau_\iota: U_\iota \times \mathbf{C}^m \to V|U_\iota$, so daß die durch $\tau_\kappa^{-1} \circ \tau_\iota(z, w_\iota) = (z, e_{\iota\kappa} \circ w_\iota)$ definierten Abbildungen $e_{\iota\kappa}: U_\iota \cap U_\kappa \to GL(m, \mathbf{C})$ einen holomorphen Kozyklus bilden (hierbei bezeichnet $w_\iota = (w_1^{(\iota)}, \ldots, w_m^{(\iota)})$ das m-tupel der Faserkoordinaten über U_ι).

Das zu V duale Bündel V^* wird bezüglich derselben Überdeckung $\mathfrak{U}$ durch den Kozyklus $(e_{\iota\kappa}^t)^{-1}$ definiert (dabei bezeichnet $e_{\iota\kappa}^t$ das Transponierte von $e_{\iota\kappa}$).

Wir bezeichnen den $\mathbf{C}$-Vektorraum der Formen mit Koeffizienten—wir sagen auch: mit Werten—in V durch $A^{r,s}(V)$ und setzen $A^l(V) := \bigoplus_{r+s=l} A^{r,s}(V)$. Lokal wird $\varphi \in A^{r,s}(V)$ auf U_ι gegeben durch einen Vektor $\varphi_\iota = \varphi|U_\iota = (\varphi_1^{(\iota)}, \ldots, \varphi_m^{(\iota)})$, dessen Komponenten gewöhnliche Differentialformen vom Typ (r, s) auf U_ι sind, so daß auf $U_\iota \cap U_\kappa$ gilt: $\varphi_\iota = e_{\iota\kappa} \circ \varphi_\kappa$. Da $\{e_{\iota\kappa}\}$ ein holomorpher Kozyklus ist, wird durch $(d''\varphi)_\iota := (d''\varphi_1^{(\iota)}, \ldots, d''\varphi_m^{(\iota)})$ eindeutig ein Operator

$$d'': A^{r,s}(V) \to A^{r,s+1}(V)$$

definiert. Ebenso ist $L: A^l(V) \to A^{l+2}(V)$ durch $(L\varphi)_\iota = (L\varphi_1^{(\iota)}, \ldots, L\varphi_m^{(\iota)})$ erklärt.

Wir setzen weiter voraus, daß auf den Fasern von V eine hermitesche Form gegeben sei; d.h. auf jedem U_ι ist eine (von $z \in U_\iota$ differenzierbar abhängende) positiv-definite hermitesche Matrix $h_\iota = (h_{ik}^{(\iota)})$ gegeben, so daß die Form von der Gestalt $\sum h_{ik} w_k \bar{w}_i$ ist. Auf $U_\iota \cap U_\kappa$ gilt dann: $h_\iota = \bar{e}_{\kappa\iota}^t h_\kappa e_{\kappa\iota}$. Man kann eine solche Metrik stets einführen, sofern X parakompakt ist. Wir wollen zunächst zeigen, daß man dann die Trivialisierungen in spezieller Weise wählen kann.

Definition 2. Eine Trivialisierung $\tau_\iota: U_\iota \times \mathbf{C}^m \to V|U_\iota$ heißt normal im Punkte $x_0 \in U_\iota$, wenn $(h_\iota(x_0))$ die Einheitsmatrix und $dh_\iota(x_0) = 0$ ist.

SATZ 1. *Zu jedem $x_0 \in X$ gibt es eine Umgebung $U = U(x_0)$ und eine in x_0 normale Trivialisierung $\tau^*: U \times \mathbf{C}^m(w_1^*, \ldots, w_m^*) \to V|U$. Ist $\tau: U \times \mathbf{C}^m(w_1, \ldots, w_m) \to V|U$ eine weitere normale Trivialisierung, so ist $\tau^{-1} \circ \tau^*$ in x_0 eine unitäre Transformation, und es ist dort $df_{ik} = 0$, wenn $f_{ik} = f_{ik}(x)$ die Komponentenfunktionen der Abbildungsmatrix sind.*

BEWEIS. Es sei $h_{ik}(x_0) = \delta_{ik}$ und $\tau^{-1} \circ \tau^*$ von der Form: $w_i = \sum f_{ik} w_k^* = w_i^* + \sum_{k,\nu} a_{ik\nu} w_k^* z_\nu$ + Glieder höherer Ordnung. Wir setzen weiter voraus, daß die Koordinaten $z_1, \ldots, z_n$ von x_0 Null sind. Dann folgt:

$$\begin{aligned}\sum_{i,k} h_{ik}^* w_k^* \bar{w}_i^* &= \sum_{i,k} h_{ik} w_k \bar{w}_i \\ &= \sum_{i,k} h_{ik}\Big(w_k^* + \sum_{j,\nu} a_{kj\nu} w_j^* z_\nu + \cdots\Big)\Big(\bar{w}_i^* + \sum_{j,\nu} \bar{a}_{ij\nu} \bar{w}_j^* \bar{z}_\nu + \cdots\Big) \\ &= \sum_{i,k} \Big(h_{ik} + \sum_\nu (a_{ik\nu} z_\nu + \bar{a}_{ki\nu} \bar{z}_\nu) + \cdots\Big) w_k^* \bar{w}_i^*,\end{aligned}$$

d.h.

$$h_{ik}^*(z) = h_{ik}(z) + \sum_\nu (a_{ik\nu} z_\nu + \bar{a}_{ki\nu} \bar{z}_\nu) + \cdots.$$

Es gilt somit $dh_{ik}^*(x_0) = 0$ genau dann, wenn $h_{ik,\nu}(x_0) := h_{ik,z_\nu}(x_0) = -a_{ik\nu}$ ist. Ist τ schon eine normale Trivialisierung, so muß also $df_{ik} = 0$ gelten. Das ist auch richtig, wenn $f_{ik}(x_0)$ nicht gleich δ_{ik} ist; denn die Transformation $(f_{ik}(x_0))^{-1} \circ \tau^{-1} \circ \tau^*$ führt ebenfalls eine normale Trivialisierung in eine normale über und hat die vorausgesetzte Form.

Um die Existenz zu zeigen, dürfen wir annehmen, daß schon $h_{ik}(x_0) = \delta_{ik}$ gilt; denn das kann man ja durch eine lineare Transformation in $\mathbf{C}^m(w_1, \ldots, w_m)$ erreichen. Es ist dann $\gamma: w_i = w_i^* - \sum_{k,\nu} h_{ik,\nu}(x_0) w_k^* \cdot z_\nu$ für hinreichend kleines $U = U(x_0)$ eine fasertreue biholomorphe Abbildung von $U \times \mathbf{C}^m(w_1^*, \ldots, w_m^*)$ nach $U \times \mathbf{C}^m(w_1, \ldots, w_m)$, so daß $\tau^* = \tau \circ \gamma$ eine Trivialisierung mit den gewünschten Eigenschaften ist.

Auf Grund der Eindeutigkeitsaussage von Satz 1 sind Differentialoperatoren von höchstens erster Ordnung auf $A^{r,s}(V)$ schon dann eindeutig definiert, wenn sie in jedem Punkte $x_0 \in X$ in Bezug auf eine normale Trivialisierung invariant gegenüber unitären Transformationen im $\mathbf{C}^m$ angegeben sind. Sei also $\tau: U \times \mathbf{C}^m \to V|U$ eine in $x_0 \in U$ normale Trivialisierung, so setzen wir für $\varphi \in A^{r,s}(V)$ und $\varphi|U = (\varphi_1, \ldots, \varphi_m)$ in x_0: $\bar{*}\varphi := (\bar{*}\varphi_1, \ldots, \bar{*}\varphi_m)$, $d'\varphi := (d'\varphi_1, \ldots, d'\varphi_m)$ und erhalten somit globale Operatoren $\bar{*}: A^{r,s}(V) \to A^{n-r,n-s}(V^*)$ und $d': A^{r,s}(V) \to A^{r+1,s}(V)$. Wir definieren weiter analog zu den $\mathbf{C}$-wertigen Formen: $\Lambda\varphi = (-1)^l \bar{*} L \bar{*} \varphi$, $\varphi \in A^l(V)$, $\delta'' = -\bar{*} d'' \bar{*}$ und $\Box = d''\delta'' + \delta'' d''$.

Ist $\tau: U_\iota \times \mathbf{C}^m \to V|U_\iota$ eine beliebige Trivialisierung, $\varphi \in A^{r,s}(V)$ und $\varphi_\iota = \varphi|U_\iota = (\varphi_1, \ldots, \varphi_m)$, so sei $d_0'\varphi_\iota := (d'\varphi_1^{(\iota)}, \ldots, d'\varphi_m^{(\iota)})$. Ist $\Theta_\iota = h_\iota^{-1} d_0' h_\iota$, so wird durch $(\partial\varphi)_\iota := d_0'\varphi_\iota + \Theta_\iota \wedge \varphi_\iota$ ein globaler Operator $\partial: A^{r,s}(V) \to A^{r+1,s}(V)$ definiert (vgl. Nakano [6]), der in normalen Trivialisierungen mit d' übereinstimmt. Da weiter in normaler Trivialisierung $\delta''L - L\delta'' = id'$ gilt, so folgt allgemein

$$d' = \partial = i(L\delta'' - \delta''L).$$

Im Falle des trivialen Bündels $V = X \times \mathbf{C}$ hat man als hermitesche Metrik die euklidische Metrik auf den Fasern und die Gleichheit $A^{r,s}(V) = A^{r,s}$. Es stimmen alle in diesem Abschnitt definierten Operatoren mit den entsprechenden Abbildungen aus Abschnitt B überein.

D. Wir bezeichnen mit $\mathcal{O}$ die Strukturgarbe von X und mit Ω^r die Garbe der Keime von lokalen holomorphen r-Formen $\varphi = \sum a_{\nu_1 \cdots \nu_r}\, dz_{\nu_1} \wedge \cdots \wedge dz_{\nu_r}$. Ferner bezeichne, falls $V \to X$ ein komplex-analytisches Vektorraumbündel über X ist, $\underline{V}$ die Garbe der Keime von lokalen holomorphen Schnitten in V und $\underline{A}^l(V)$ bzw. $\underline{A}^{r,s}(V)$ die (feine) Garbe der Keime von lokalen differenzierbaren Formen mit Werten in V vom Grade l bzw. vom Typ (r, s). Wegen des Lemmas von Dolbeault ist die Sequenz

$$0 \to \underline{V} \otimes \Omega^r \to \underline{A}^{r,0}(V) \xrightarrow{d''} \underline{A}^{r,1}(V) \to \cdots \to \underline{A}^{r,n}(V) \to 0$$

für alle $r = 0, \ldots, n$ exakt. Setzt man $Z^{r,s}(V) := \{\varphi \in A^{r,s}(V): d''\varphi = 0\}$ und $B^{r,s}(V) := d''A^{r,s-1}(V)$, so sind die de Rhamschen Gruppen definiert durch $H^{r,s}(V) := Z^{r,s}(V)/B^{r,s}(V)$, und wegen der obigen exakten Sequenz von feinen Garben gilt:

$$H^{r,s}(V) \cong H^s(X, \underline{V} \otimes \Omega^r).$$

Setzt man $A_c^{r,s}(V) = \{\varphi \in A^{r,s}(V): \operatorname{Tr} \varphi \text{ ist kompakt in } X\}$, $Z_c^{r,s}(V) = \{\varphi \in A_c^{r,s}(V): d''\varphi = 0\}$, $B_c^{r,s}(V) = d''A_c^{r,s-1}(V)$ und $H_c^{r,s}(V) = Z_c^{r,s}(V)/B_c^{r,s}(V)$, so gilt entsprechend

$$H_c^{r,s}(V) \cong H_c^s(X, \underline{V} \otimes \Omega^r),$$

wobei $H_c^s(X, \underline{V} \otimes \Omega^r)$ die s-te Kohomologiegruppe von X mit kompaktem Träger und Werten in $\underline{V} \otimes \Omega^r$ bezeichnet.

Ist $V = X \times \mathbf{C}$ das triviale Bündel, so schreiben wir $H^{r,s}$ und $H_c^{r,s}$ anstelle von $H^{r,s}(X \times \mathbf{C})$ und $H_c^{r,s}(X \times \mathbf{C})$. Man hat in diesem Fall

$$H^{r,s} \cong H^s(X, \Omega^r), \qquad H_c^{r,s} \cong H_c^s(X, \Omega^r).$$

E. Im folgenden sei $\hat{X}$ eine n-dimensionale komplexe kählersche Mannigfaltigkeit und $V \xrightarrow{\pi} \hat{X}$ ein komplex-analytisches Vektorraumbündel über $\hat{X}$. Ferner sei $X \subset \hat{X}$ ein relativ-kompakter Teilbereich; der Rand

∂X von X sei glatt und beliebig oft differenzierbar. Darüber hinaus gebe es zu jedem Punkt $x_0 \in \partial X$ eine Umgebung $U = U(x_0)$ und eine in U beliebig oft differenzierbare streng q-konvexe Funktion $p(x)$ mit $dp(x) \neq 0$ für alle $x \in U$, so daß $X \cap U = \{x \in X : p(x) < 0\}$. Wir setzen X also als streng q-konvex voraus. Jede andere Funktion $p_1(x)$ mit diesen Eigenschaften unterscheidet sich von $p(x)$ in einer Umgebung von x_0 nur durch einen positiven, beliebig oft differenzierbaren Faktor.

Ein komplexer Tangentialvektor im Punkte $x_0 \in X$ an ∂X ist ein kontravarianter Vektor $\xi = \sum a_\nu \dfrac{\partial}{\partial z_\nu} + \sum \bar{a}_\nu \dfrac{\partial}{\partial \bar{z}_\nu}$, so daß für alle $a\xi :=$ $\sum a a_\nu \dfrac{\partial}{\partial z_\nu} + \sum \bar{a}\bar{a}_\nu \dfrac{\partial}{\partial \bar{z}_\nu}$ im Punkte x_0 gilt: $(a\xi)(p) = 0$. Diese Definition ist unabhängig von der speziellen Auswahl von p. Die Menge $T_{x_0}(\partial X, \mathbf{C})$ der komplexen Tangentialvektoren im Punkte $x_0 \in X$ an ∂X ist ein komplex-$(n-1)$-dimensionaler Vektorraum, der (als reeller Untervektorraum) in dem reell-$(2n-1)$-dimensionalen Vektorraum $T_{x_0}(\partial X)$ der reellen Tangentialvektoren im Punkte $x_0 \in X$ an ∂X enthalten ist. Es sei schließlich $T_{x_0} = T_{x_0}(\hat{X})$ der reelle Vektorraum der Tangentialvektoren in x_0 an $\hat{X}$. Dann ist eine Differentialform φ vom Grade l auf $\hat{X}$ im Punkte x_0 eine alternierende l-Form auf T_{x_0}.

Wir bezeichnen nun mit $\dot{A}^{r,s}(V)$ den Untervektorraum derjenigen (r, s)-Formen auf $\bar{X}$ mit Werten in $V|\bar{X}$, die noch auf dem Rand ∂X von $\bar{X}$ beliebig oft differenzierbar sind, d.h. es ist $\dot{A}^{r,s}(V) = \{\hat{\varphi}|\bar{X} : \hat{\varphi}$ eine (r, s)-Form auf $\hat{X}$ mit Werten in $V\}$. Ist $\varphi \in \dot{A}^{r,s}(V)$, so ist—da die Einbettung $id \colon \partial X \hookrightarrow X$ beliebig oft differenzierbar ist—$id^*\varphi$ eine C^∞-Form vom Typ (r, s) mit Werten in $V|\partial X$. Wir setzen $\varphi|\partial X := id^*\varphi$. Im Falle des trivialen Bündels schreiben wir wieder $\dot{A}^{r,s}$ für $\dot{A}^{r,s}(\hat{X} \times \mathbf{C})$.

Wir sagen, daß die Beschränkung einer Form $\varphi = \sum dz_{\nu_1} \wedge \cdots \wedge dz_{\nu_r} \wedge \varphi_{\nu_1 \cdots \nu_r} \in \dot{A}^{r,s}(V)$ mit $\varphi_{\nu_1 \cdots \nu_r} \in \dot{A}^{0,s}(V)$ auf dem komplexen Rand von X verschwindet, wenn für alle $x_0 \in \partial X$ und alle $\xi_1, \ldots, \xi_s \in T_{x_0}(\partial X, \mathbf{C})$ der Wert $\varphi_{\nu_1 \cdots \nu_r}(x_0)(\xi_1, \ldots, \xi_s) = 0$ ist, und schreiben dafür auch $\varphi|\partial_a X = 0$. Es gilt für 0-Formen $\varphi|\partial_a X = 0$ genau dann, wenn $\varphi|\partial X = 0$ ist.

Die eben definierte Eigenschaft stimmt mit dem Begriff "complex normal" in Kohn–Rossi [5] überein. Es gilt nämlich

SATZ 2. *$\varphi \in \dot{A}^{r,s}(V)$ verschwindet genau dann auf dem analytischen Rand von X, wenn es zu jedem Punkt $x_0 \in \partial X$ eine Umgebung $U = U(x_0) \subset \hat{X}$, eine reelle C^∞-Funktion $p(x)$ auf U mit $dp(x) \neq 0$ für alle $x \in U$ und $X \cap U = \{x \in U : p(x) < 0\}$ und Formen ψ_1 und ψ_2 auf U vom Typ $(r, s-1)$ bzw. (r, s) gibt, so daß $\varphi|U = d''p \wedge \psi_1 + p\psi_2$ ist.*

BEWEIS. Es sei $x_0 \in \partial X$. Nach Voraussetzung gibt es eine Umgebung $U = U(x_0)$ und eine reelle C^∞-Funktion $p(x)$ in U mit $X \cap U = \{p(x) < 0\}$.

Wir können U so klein wählen, daß man in U komplexe Koordinaten $z_1, \ldots, z_n$ mit den folgenden Eigenschaften einführen kann: x_0 hat die Koordinaten $z_1 = \cdots = z_n = 0$, und $p(z)$ ist (eventuell nach Multiplikation mit einer positiven Konstanten) von der Form $p(z) = z_1 + \bar{z}_1 +$ Glieder höherer Ordnung. In x_0 gilt $\varphi | \partial_a X = 0$ für $\varphi \in \dot{A}^{r,s}(V)$ genau dann, wenn alle Komponenten von $\varphi(x_0)$ den Faktor $d\bar{z}_1$ enthalten, da die Ebene $\{z_1 = 0\}$ die komplexe Tangente an ∂X in x_0 bildet. Da in x_0 außerdem $d''p = d\bar{z}_1$ gilt, kann man in x_0 aus φ den Faktor $d''p$ eindeutig ausklammern. Auf diese Weise bestimmt man ein ψ_1 auf U vom Typ (r, s), so daß auf $U \cap \partial X$ die Gleichheit $\varphi = d''p \wedge \psi_1$ besteht. Mit $\psi_2 = (1/p) \times (\varphi - d''p \wedge \psi_1)$ folgt die Behauptung.

Gilt umgekehrt $\varphi = d''p \wedge \psi_1 + p\psi_2$, so ist in x_0: $\varphi = d\bar{z}_1 \wedge \psi_1$, d.h. $\varphi | \partial_a X = 0$.

Man beweist weiter leicht: *Für* $\varphi \in \dot{A}^{r,s}(V)$ *gilt* $\varphi | \partial_a X = 0$ *genau dann, wenn für alle* $\alpha \in \dot{A}^{n-r, n-s-1}(V^*)$ *das Integral*

$$\int_{\partial X} \varphi \wedge \alpha$$

verschwindet. Hierbei bedeutet $\wedge$ das Produkt $A^{r,s}(V) \times A^{r*,s*}(V^*) \to A^{r+r*,s+s*}$, das wegen der Paarung $V \times V^* \to \mathbf{C}$ wie bei $\mathbf{C}$-wertigen Formen definiert werden kann.

F. Aus dem letzten Kriterium folgt unmittelbar, daß die Räume

$$\mathfrak{B}^{r,s}(V) := \{\varphi \in \dot{A}^{r,s}(V) : \varphi = d''\psi, \psi \in \dot{A}^{r,s-1}(V)\}$$
$$\mathfrak{D}^{r,s}(V) := \{\varphi \in \dot{A}^{r,s}(V) : \varphi = \delta''\psi, \psi \in \dot{A}^{r,s+1}(V), \bar{*}\psi | \partial_a X = 0\}$$
$$\mathfrak{H}^{r,s}(V) := \{\varphi \in \dot{A}^{r,s}(V) : d''\varphi = \delta''\varphi = 0, \bar{*}\varphi | \partial_a X = 0\}$$

in Bezug auf das hermitesche Skalarprodukt

$$(\varphi, \psi) := \int_X \varphi \wedge \bar{*}\psi, \qquad \varphi, \psi \in \dot{A}^{r,s}(V)$$

paarweise orthogonal sind. Wir setzen weiter

$$\mathring{A}^{r,s}(V) := \{\varphi \in \dot{A}^{r,s}(V) : \bar{*}\varphi | \partial_a X = \bar{*}d''\varphi | \partial_a X = 0\}.$$

Da X streng q-konvex ist, gibt es nach Kohn [4] für $s \geqq q$ einen beschränkten Operator N: $\mathring{A}^{r,s}(V) \to \mathring{A}^{r,s}(V)$, so daß für $\varphi \in \mathring{A}^{r,s}(V)$ gilt: $h := \varphi - \square N\varphi \in \mathfrak{H}^{r,s}(V)$. Setzen wir $\varphi_1 := \delta'' N\varphi$ und $\varphi_2 := d'' N\varphi$, so folgt $\varphi = d''\varphi_1 + d''\varphi_2 + h$ mit $d''\varphi_1 \in \mathfrak{B}^{r,s}(V)$ und $\delta''\varphi_2 \in \mathfrak{D}^{r,s}(V)$. Man hat also das Zerlegungstheorem

$$\mathring{A}^{r,s}(V) = \mathfrak{B}^{r,s}(V) \oplus \mathfrak{D}^{r,s}(V) \oplus \mathfrak{H}^{r,s}(V) \quad \text{für } s \geqq q.$$

Ist nun $\varphi \in \mathring{A}^{r,s}(V)$ mit $d''\varphi = 0$ und $s \geqq q$, so gestattet φ die Zerlegung $\varphi = \varphi_0 + h$ mit $\varphi_0 \in \mathfrak{D}^{r,s}(V)$ und $h \in \mathfrak{H}^{r,s}(V)$. Die harmonische Form h

hängt nur von der de Rhamschen Klasse von φ ab. Man erhält so einen Isomorphismus

$$H^{r,s}(V) \cong \mathfrak{H}^{r,s}(V), \qquad s \geqq q.$$

Zur Untersuchung der Kohomologie mit kompaktem Träger setzen wir

$$\mathfrak{H}_c^{r,s}(V) := \{\varphi \in \mathring{A}^{r,s}(V) : d''\varphi = \delta''\varphi = 0, \varphi|\partial_a X = 0\}.$$

Der $\bar{*}$-Operator gibt einen Isomorphismus $\mathfrak{H}_c^{r,s}(V) \to \mathfrak{H}^{n-r,n-s}(V^*)$. Außerdem hat man nach Serre [7] für q-konvexe Räume einen Isomorphismus $H^{n-s}(X, \underline{V}^* \otimes \Omega^{n-r}) \to H_c^s(X, \underline{V} \otimes \Omega^r)$ für $s \leqq n - q$. Also gilt auch

$$H_c^{r,s}(V) \cong \mathfrak{H}_c^{r,s}(V), \qquad s \leqq n - q.$$

Im nächsten Teil werden wir zeigen, daß

$$\mathfrak{H}_c^{0,s}(V) = 0$$

für $s \leqq n - q$ gilt, sofern X hyper-q-konvex und V semi-negativ ist. Insbesondere ist dann gezeigt:

$$H_c^s(X, \mathcal{O}) = 0, \qquad s \leqq n - q,$$
$$H^s(X, \Omega^n) = 0, \qquad s \geqq q.$$

2. Beweis des Hauptresultates

A. Es sei $\hat{X}$ im folgenden eine n-dimensionale komplexe kählersche Mannigfaltigkeit, $V \xrightarrow{\pi} \hat{X}$ ein komplex-analytisches Vektorraumbündel über $\hat{X}$ vom Range m, und auf den Fasern von V sei eine hermitesche Form gegeben. Ferner sei $X \subset \hat{X}$ ein offener relativ-kompakter Teilbereich von $\hat{X}$ mit glattem, beliebig oft differenzierbarem Rand ∂X. Wir wollen untersuchen, unter welchen Voraussetzungen an V und ∂X die Räume $\mathfrak{H}^{r,s}(V)$ verschwinden. Dazu leiten wir zunächst eine wichtige Gleichung her.

Es sei $\varphi \in \mathfrak{H}^{r,s}(V)$; dann folgt $d'\varphi = i(L\delta'' - \delta''L)\varphi = -i\delta''L\varphi$, und wegen $\bar{*}\delta''L\varphi = (-1)^l d''\bar{*}L\varphi$, $l = r + s$, $d(d'\varphi \wedge \bar{*}L\varphi) = d''(d'\varphi \wedge \bar{*}L\varphi) = d''d'\varphi \wedge \bar{*}L\varphi + (-1)^{l+1} d'\varphi \wedge d''\bar{*}L\varphi$ ergibt sich mit Hilfe des Stokes'schen Satzes

$$(d'\varphi, d'\varphi) = i(d'\varphi, \delta''L\varphi) = i\int_X d'\varphi \wedge \bar{*}\delta''L\varphi$$
$$= -i\int_{\partial X} d'\varphi \wedge \bar{*}L\varphi + i\int_X d''d'\varphi \wedge \bar{*}L\varphi.$$

Definiert man noch wie üblich $\chi := \frac{1}{2\pi i} d''\Theta$, so ergibt sich $d''d'\varphi = d''(d_0'\varphi + \Theta \wedge \varphi) = d''\Theta \wedge \varphi = 2\pi i\chi \wedge \varphi$ und somit unsere Hauptgleichung

$$-(d'\varphi, d'\varphi) = 2\pi(\chi \wedge \varphi, L\varphi) + i \int_{\partial X} d'\varphi \wedge \bar{*}L\varphi. \tag{3}$$

Es gilt $(\chi \wedge \varphi, L\varphi) \geqq 0$ für $\varphi \in \dot{A}^{0,s}(V)$, $s < n$, wie wir im nächsten Abschnitt zeigen werden, wenn $V|X$ ein semi-negatives Vektorraumbündel im Sinne von Nakano ist. Im Abschnitt C untersuchen wir die Bedingungen an ∂X, unter denen auch das Randintegral $i \int_{\partial X} d'\varphi \wedge \bar{*}L\varphi \geqq 0$ wird. Da stets $(d'\varphi, d'\varphi) \geqq 0$ gilt, muß dann bei semi-negativen Bündeln sogar $\int_{\partial X} d'\varphi \wedge \bar{*}L\varphi = 0$ gelten. Daraus wird dann in Abschnitt D unter den gegebenen Voraussetzungen $\mathfrak{H}_c^{r,s}(V) = 0$ folgen.

B. Es sei $\varphi \in \dot{A}^{0,s}(V)$, $s < n$, und $\psi := (\chi \wedge \varphi) \wedge \bar{*}L\varphi \in \dot{A}^{n,n}$. Ist $x_0 \in X$ ein Punkt, so können wir in einer Umgebung U von x_0 eine Trivialisierung von V finden, die in x_0 normal ist. Gilt bezüglich dieser Trivialisierung $\varphi = (\varphi_1, \ldots, \varphi_m)$, $\chi = (\chi_{ij})$, so folgt in x_0:

$$\psi = \sum_i (\chi \wedge \varphi)_i \wedge \bar{*}L\varphi_i = \sum_{i,j} (\chi_{ij} \wedge \varphi_j) \wedge \bar{*}L\varphi_i,$$

und daraus wegen $\bar{*}L\varphi_i = (-1)^{s(s+1)/2} . \frac{i^s}{(n-s-1)!} L^{n-s-1}\bar{\varphi}_i$:

$$\psi = (-1)^{s(s+1)/2} . \frac{i^s}{(n-s-1)!} \sum_{i,j} \chi_{ij} \wedge \varphi_j \wedge \bar{\varphi}_i \wedge \omega^{n-s-1}.$$

Wir können weiter in U komplexe Koordinaten einführen, die in x_0 geodätisch sind. Es ist dann

$$\omega = \frac{i}{2} \sum_{\nu=1}^{n} dz_\nu \wedge d\bar{z}_\nu.$$

Setzt man $\varphi_i = \sum_{1 \leq \nu_1 < \cdots < \nu_s \leq n} {}_i a_{\nu_1 \cdots \nu_s} d\bar{z}_{\nu_1} \wedge \cdots \wedge d\bar{z}_{\nu_s}$, $i = 1, \ldots, m$, und für $1 \leqq \nu_1 < \cdots < \nu_{s-1} \leqq n$:

$${}_i a^{(\nu)}_{\nu_1 \cdots \nu_{s-1}} = \begin{cases} 0, & \text{falls } \nu \in \{\nu_1, \ldots, \nu_{s-1}\} \\ \text{sign } (\nu, \nu_1, \ldots, \nu_{s-1}) \cdot {}_i a_{\mu_1 \cdots \mu_s}, & \text{falls } \nu \notin \{\nu_1, \ldots, \nu_{s-1}\}, \text{ wobei} \\ \mu_1, \ldots, \mu_s \text{ die natürliche Anordnung von } \nu, \nu_1, \ldots, \nu_{s-1} \text{ ist,} & \end{cases}$$

so folgt wegen

$$\chi_{ij} = \frac{i}{2\pi} \sum h_{ij,\nu,\bar{\mu}} \, dz_\nu \wedge d\bar{z}_\mu:$$

$$\begin{aligned} &2\pi(\chi \wedge \varphi, L\varphi) \\ &= 2^{s+1} \sum_{1 \leqq \nu_1 < \cdots < \nu_{s-1} \leqq n} \left(\sum_{i,j=1}^{m} \sum_{\nu,\mu=1}^{n} h_{ij,\nu,\bar{\mu}} \, {}_j a^{(\nu)}_{\nu_1 \cdots \nu_{s-1}} \, {}_i \bar{a}^{(\mu)}_{\nu_1 \cdots \nu_{s-1}} \right) do, \end{aligned} \tag{4}$$

wobei $d\mathfrak{v} = dx_1 \wedge dy_1 \wedge \cdots \wedge dx_n \wedge dy_n$, $z_\nu = x_\nu + iy_\nu$, $\nu = 1, \ldots, n$, das Volumenelement von X im Punkte x_0 ist.

DEFINITION 3 (Nakano). Ein komplex-analytisches Vektorraumbündel $V \xrightarrow{\pi} X$ heißt positiv [semi-positiv, (semi-)negativ], wenn es eine hermitesche Form $\sum h_{ij} w_j \bar{w}_i$ auf den Fasern von V gibt, so daß in jedem $x_0 \in X$ bezüglich aller normalen Trivialisierungen die hermitesche Form $\sum h_{ij,\nu,\bar{\mu}} \xi_{j\nu} \bar{\xi}_{i\bar{\mu}}$ negativ definit [negativ semi-definit, positiv (semi-)definit] ist.

Aus dieser Definition folgt unmittelbar:

SATZ 3. *Ist $V|X$ semi-negativ und $\varphi \in \dot{A}^{0,s}(V)$ mit $s < n$, so gilt* $(\chi \wedge \varphi, L\varphi) \geqq 0$.

C. Als nächstes untersuchen wir des Randintegral $i \int_{\partial X} d'\varphi \wedge \bar{*} L\varphi$ für Formen $\varphi \in \dot{A}^{r,s}(V)$ mit $r + s < n$, $\varphi|\partial_a X = 0$ und $\Lambda\varphi = 0$; wir setzen φ also als primitiv voraus. Wählt man in x_0 eine normale Trivialisierung von V, so ist in x_0 der Integrand gleich $\sum_i d'\varphi_i \wedge \bar{*} L\varphi_i$, $\varphi = (\varphi_1, \ldots, \varphi_m)$, und es gilt dort: $\Lambda\varphi_i = 0$ und $\varphi_i|\partial_a X = 0$ für $i = 1, \ldots, m$. Es genügt also, das folgende Problem zu behandeln:

Es sei $\varphi \in \dot{A}^{r,s}$, $l = r + s < n$, $x_0 \in \partial X$, und in x_0 gelte $\Lambda\varphi = 0$ und $\varphi|\partial_a X = 0$. Unter welchen Voraussetzungen an ∂X gilt dann $\gamma := i \cdot d'\varphi \wedge \bar{*} L\varphi|\partial X \geqq 0$ in x_0?

Es sei p eine reelle differenzierbare Funktion in einer Umgebung U von x_0 mit $dp \neq 0$ und $U \cap X = \{x \in U : p(x) < 0\}$. Dann gilt $U \cap \partial X = \{x \in U : p(x) = 0\}$, und es ist $dp|\partial X = d'p|\partial X + d''p|\partial X = 0$. Da φ auf dem analytischen Rand von X verschwindet, gibt es nach Satz 2 in einer Umbegung von x_0 eine Darstellung $\varphi = d''p \wedge \psi + p\psi_0$. Damit folgt $d'\varphi \wedge \bar{\varphi} = (d'd''p \wedge \psi - d''p \wedge d'\psi + d'p \wedge \psi_0 + pd'\psi_0) \wedge (d'p \wedge \bar{\psi} + p\bar{\psi}_0)$, das heißt: $d'\varphi \wedge \bar{\varphi}|\partial X = d'd''p \wedge \psi \wedge d'p \wedge \bar{\psi}|\partial X = (-1)^{l+1} d'p \wedge d'd''p \wedge \psi \wedge \bar{\psi}|\partial X$, und wegen Formel (2) hat man in x_0:

$$
\begin{aligned}
\gamma = id'\varphi \wedge \bar{*} L\varphi &= (-1)^{l(l+1)/2} \cdot \frac{i^{s-r+1}}{(n-l-1)!} d'\varphi \wedge \bar{\varphi} \wedge \omega^{n-l-1} \\
&= -(-1)^{l(l-1)/2} \cdot \frac{i^{s-r+1}}{(n-l-1)!} d'p \wedge d'd''p \wedge \psi \wedge \bar{\psi} \wedge \omega^{n-l-1},
\end{aligned}
\tag{5}
$$

wobei auf der rechten Seite stets die Beschränkung auf ∂X zu nehmen ist.

Wir führen nun in x_0 geodätische Koordinaten ein, so daß dort p von der Gestalt ist: $p(z) = x_1 +$ Glieder höherer Ordnung. Es gilt in x_0: $\varphi = \frac{1}{2} d\bar{z}_1 \wedge \psi$; wir können deshalb annehmen, daß ψ kein $d\bar{z}_1$ enthält. Setzen wir

$$
\varphi = \sum_{\substack{1 \leqq \nu_1 < \cdots < \nu_r \leqq n \\ 1 \leqq \mu_1 < \cdots < \mu_s \leqq n}} a_{\nu_1 \cdots \nu_r, \bar{\mu}_1 \cdots \bar{\mu}_s} dz_{\nu_1} \wedge \cdots \wedge dz_{\nu_r} \wedge d\bar{z}_{\mu_1} \wedge \cdots \wedge d\bar{z}_{\mu_s},
$$

und definieren wir für $1 \leqq \rho_1 < \cdots < \rho_{r-1} \leqq n, 1 \leqq \sigma_1 < \cdots < \sigma_{s-1} \leqq n$:

$$a^{(\nu)}_{\rho_1 \cdots \rho_{r-1}, \bar{\sigma}_1 \cdots \bar{\sigma}_{s-1}} = \begin{cases} 0, & \text{falls } \nu \in \{\rho_1, \ldots, \rho_{r-1}, \sigma_1, \ldots, \sigma_{s-1}\} \\ \operatorname{sign}(\nu, \rho_1, \ldots, \rho_{r-1}) \operatorname{sign}(\nu, \sigma_1, \ldots, \sigma_{s-1}) \cdot & \\ \qquad a_{\nu_1 \cdots \nu_r, \bar{\mu}_1 \cdots \bar{\mu}_s}, & \end{cases}$$

falls $\nu_1, \ldots, \nu_r$ die natürliche Anordnung von $\nu, \rho_1, \ldots, \rho_{r-1}$ und $\mu_1, \ldots, \mu_s$ die natürliche Anordnung von $\nu, \sigma_1, \ldots, \sigma_{s-1}$ bedeutet, so ist $\Lambda\varphi = 0$ äquivalent zu

$$\sum_{\nu=1}^{n} a^{(\nu)}_{\rho_1 \cdots \rho_{r-1}, \bar{\sigma}_1 \cdots \bar{\sigma}_{s-1}} = 0 \tag{6}$$

für alle $1 \leqq \rho_1 < \cdots < \rho_{r-1} \leqq n$, $1 \leqq \sigma_1 < \cdots < \sigma_{s-1} \leqq n$. Nehmen wir einen Koeffizienten der Form $a_{1\rho_1 \cdots \rho_{r-1}, \bar{1}\bar{\sigma}_1 \cdots \bar{\sigma}_{s-1}}$ her, so gilt

$$a_{1\rho_1 \cdots \rho_{r-1}, \bar{1}\bar{\sigma}_1 \cdots \bar{\sigma}_{s-1}} = \sum_{\nu=1}^{n} a^{(\nu)}_{\rho_1 \cdots \rho_{r-1}, \bar{\sigma}_1 \cdots \bar{\sigma}_{s-1}} = 0,$$

da in jedem Summanden von φ der Term $d\bar{z}_1$ auftritt. Das heißt aber: In φ und damit auch in ψ kommt kein dz_1 vor. Wir können also ψ im Punkte x_0 als Form auf der komplexen Tangentialebene $\{z_1 = 0\}$ von ∂X in x_0 ansehen. Kennzeichnen wir Operatoren auf dieser Ebene mit einem Strich, so folgt aus (6) unmittelbar:

$$\Lambda'\psi = 0.$$

Folglich gilt für $L'\psi = \psi \wedge \omega' := \psi \wedge \left(\frac{i}{2} \sum_{\nu=2}^{n} dz_\nu \wedge d\bar{z}_\nu\right)$:

$$\bar{*}'L'\psi = (-1)^{l(l-1)/2} \cdot \frac{i^{s-r-1}}{(n-l-1)!} \bar{\psi} \wedge \omega'^{n-l-1},$$

und hieraus ergibt sich zusammen mit (5) und $d'p = \frac{1}{2} dz_1 = \frac{i}{2} dy_1$:

$$\gamma = dy_1 \wedge \left(\frac{i}{2} d'd''p \wedge \psi\right) \wedge \bar{*}'(L'\psi).$$

Schließlich setzen wir noch $\omega_0' = \frac{i}{2} d'd''p | \partial_a X$, $L_0'\psi = \omega_0' \wedge \psi$, und erhalten

$$\gamma = dy_1 \wedge (L_0'\psi \wedge \bar{*}'L'\psi). \tag{7}$$

Definition 4. Wir sagen, X hat im Punkte $x_0 \in \partial X$ die Eigenschaft $B(r, s)$, wenn es eine Umgebung $U = U(x_0)$ von x_0 in $\hat{X}$, eine in U beliebig oft differenzierbare reelle Funktion $p(x)$ mit $dp(x) \neq 0$ und $U \cap X = \{x : p(x) < 0\}$ und ein geodätisches Koordinatensystem $z_1, \ldots, z_n$ in x_0 gibt, so daß folgendes gilt:

(i) $\{z_1 = 0\}$ ist die komplexe Tangentialebene $\partial_a X$ an X in x_0,
(ii) $d'd''p|\partial_a X$ hat Diagonalgestralt, d.h.

$$d'd''p|\partial_a X = \sum_{\nu=2}^{n} e_\nu \, dz_\nu \wedge d\bar{z}_\nu, \qquad e_\nu = \frac{\partial^2 p}{\partial z_\nu \, \partial \bar{z}_\nu},$$

(iii) Für alle primitiven Formen $\psi \neq 0$ auf $\partial_a X$ vom Typ $(r, s-1)$ gilt: $L_0'\psi \wedge \bar{*}'L'\psi > 0$.

Da für eine positive differenzierbare Funktion α die Gleichheit $d'd''(\alpha p)|\partial_a X = \alpha d'd''p|\partial_a X$ besteht, ist die obige Definition unabhängig von der speziellen Wahl von p.

Aus (7) folgt unmittelbar

SATZ 4. *Hat X die Eigenschaft $B(r, s)$ (d.h. gilt $B(r, s)$ in jedem Punkt $x_0 \in \partial X$), so ist für jede primitive Form $\varphi \in \dot{A}^{r,s}(V)$ mit $r + s < n$ und $\varphi|\partial_a X = 0$, deren Koeffizienten nicht sämtlich auf ∂X verschwinden:*

$$i \int_{\partial X} d'\rho \wedge \bar{*}L\varphi > 0.$$

Wir wollen im folgenden die Eigenschaft (iii) aus der Definition 4 näher untersuchen. Es sei

$$\psi = \sum_{\substack{2 \leqq \nu_1 < \cdots < \nu_r \leqq n \\ 2 \leqq \mu_1 < \cdots < \mu_s \leqq n}} a_{\nu_1 \cdots \nu_r, \bar{\mu}_1 \cdots \bar{\mu}_s} \, dz_{\nu_1} \wedge \cdots \wedge dz_{\nu_r} \wedge d\bar{z}_{\mu_1} \wedge \cdots \wedge d\bar{z}_{\mu_s}$$

mit $\Lambda'\psi = 0$. Wir setzen für $2 \leqq \rho_1 < \cdots < \rho_{r+1} \leqq n$, $2 \leqq \sigma_1 < \cdots < \sigma_{s+1} \leqq n$:

$$a^{(\bar{\nu})}_{\rho_1 \cdots \rho_{r+1}, \bar{\sigma}_1 \cdots \bar{\sigma}_{s+1}} = \begin{cases} 0, & \text{falls } \nu \notin \{\rho_1, \ldots, \rho_{r+1}\} \cap \{\sigma_1, \ldots, \sigma_{s+1}\} \\ \operatorname{sign}(\nu, \nu_1, \ldots, \nu_r) \operatorname{sign}(\nu, \mu_1, \ldots, \mu_s) \cdot a_{\nu_1 \cdots \nu_r, \bar{\mu}_1 \cdots \bar{\mu}_s}, \end{cases}$$

falls $\{\nu, \nu_1, \ldots, \nu_r\} = \{\rho_1, \ldots, \rho_{r+1}\}$, $2 \leqq \nu_1 < \cdots < \nu_r \leqq n$, $\{\nu, \mu_1, \ldots, \mu_s\} = \{\sigma_1, \ldots, \sigma_{s+1}\}$ und $2 \leqq \mu_1 < \cdots < \mu_s \leqq n$. Dann gilt

$$L_0'\psi = \frac{i}{2} \sum_{\substack{2 \leqq \rho_1 < \cdots < \rho_{r+1} \leqq n \\ 2 \leqq \sigma_1 < \cdots < \sigma_{s+1} \leqq n}} \left(\sum_{\nu=2}^{n} a^{(\bar{\nu})}_{\rho_1 \cdots \rho_{r+1}, \bar{\sigma}_1 \cdots \bar{\sigma}_{s+1}} e_\nu \right) \cdot dz_{\rho_1} \wedge \cdots \wedge dz_{\rho_{r+1}} \wedge d\bar{z}_{\sigma_1} \wedge \cdots \wedge d\bar{z}_{\sigma_{s+1}},$$

$$L'\psi = \frac{i}{2} \sum \left(\sum_{\nu=2}^{n} a^{(\bar{\nu})}_{\rho, \bar{\sigma}} \right) d\mathfrak{z}_\rho \wedge d\bar{\mathfrak{z}}_\sigma,$$

und folglich

$$L_0'\psi \wedge \bar{*}'L'\psi = 2^{r+s} \sum_{\substack{2 \leqq \rho_1 < \cdots < \rho_{r+1} \leqq n \\ 2 \leqq \sigma_1 < \cdots < \sigma_{s+1} \leqq n}} \left(\sum_{\nu=2}^{n} e_\nu a^{(\vartheta)}_{\rho_1 \cdots \rho_{r+1}, \bar{\sigma}_1 \cdots \bar{\sigma}_{s+1}} \right) \cdot \left(\sum_{\nu=2}^{n} \bar{a}^{(\vartheta)}_{\rho_1 \cdots \rho_{r+1}, \bar{\sigma}_1 \cdots \bar{\sigma}_{s+1}} \right) d\mathfrak{o}'$$

mit $d\mathfrak{o}' = dx_2 \wedge dy_2 \wedge \cdots \wedge dx_n \wedge dy_n$. Äquivalent zu (iii) [im Falle $(r, s+1)$] ist also die Aussage

$$\sum_{\rho,\bar{\sigma}} \left(\sum_{\nu=2}^{n} a^{(\vartheta)}_{\rho,\bar{\sigma}} e_\nu \right) \left(\sum_{\nu=2}^{n} \bar{a}^{(\vartheta)}_{\rho,\bar{\sigma}} \right) > 0 \tag{8}$$

für alle $(a_{\nu_1 \cdots \nu_r, \bar{\mu}_1 \cdots \bar{\mu}_s}) \neq 0$ mit

$$\sum_{\iota=2}^{n} a^{(\iota)}_{\lambda_1 \cdots \lambda_{r-1}, \bar{\kappa}_1 \cdots \bar{\kappa}_{s-1}} = 0, \qquad 2 \leqq \lambda_1 < \cdots < \lambda_{r-1} \leqq n, \quad 2 \leqq \kappa_1 < \cdots < \kappa_{s-1} \leqq n. \tag{9}$$

(8) ist offenbar genau dann erfüllt, wenn für alle $t = 1, \ldots, \min(r+1, s+1)$ gilt:

$$\sum_{\rho,\bar{\sigma}}^{t} \left(\sum_{\nu=2}^{n} a^{(\vartheta)}_{\rho,\bar{\sigma}} e_\nu \right) \left(\sum_{\nu=2}^{n} \bar{a}^{(\vartheta)}_{\rho,\bar{\sigma}} \right) > 0, \tag{10}$$

wobei $\sum\limits_{\rho,\bar{\sigma}}^{t}$ die Summation über alle $\rho_1, \ldots, \rho_{r+1}$ und $\sigma_1, \ldots, \sigma_{s+1}$ mit t-elementigem Durchschnitt $\{\rho_1, \ldots, \rho_{r+1}\} \cap \{\sigma_1, \ldots, \sigma_{s+1}\} = \{\tau_1, \ldots, \tau_t\}$ bedeutet. Man sieht leicht, daß ferner (10) genau dann erfüllt ist, wenn für alle Systeme $\{b_{\tau_1 \cdots \tau_t}\} \neq 0$ gilt:

$$\sum_{2 \leqq \tau_1 < \cdots < \tau_t \leqq n} \left(\sum_{\nu=2}^{n} b^{(\vartheta)}_{\tau} e_\nu \right) \left(\sum_{\nu-2}^{n} \bar{b}^{(\vartheta)}_{\tau} \right) > 0, \tag{11}$$

wobei $b^{(\vartheta)}_{\tau}$ analog zu den $a^{(\vartheta)}_{\rho,\bar{\sigma}}$ definiert wird.

Man sieht unmittelbar:

Es gilt $B(r, s)$ für alle r und s, wenn $e_2 = \cdots = e_n > 0$.

Diese Bedingung ist jedoch nicht notwendig, wie wir jetzt am Fall $(r, s) = (0, q)$ zeigen wollen. Da jede $(0, q)$-Form primitiv ist, sind die Zusatzbedingungen leer, und wir erhalten

$$\begin{aligned} 0 &< \sum_{\substack{2 \leqq \rho \leqq n \\ 2 \leqq \sigma_1 < \cdots < \sigma_q \leqq n}} e_\rho |a^{(\hat{\rho})}_{\rho, \bar{\sigma}_1 \cdots \bar{\sigma}_q}|^2 \\ &= \sum_{2 \leqq \tau_1 < \cdots < \tau_{q-1} \leqq n} (e_{\rho_1} + \cdots + e_{\rho_{n-q}}) |a_{\bar{\tau}_1 \cdots \bar{\tau}_{q-1}}|^2, \end{aligned}$$

wobei $\{\rho_1, \ldots, \rho_{n-q}\} \cup \{\tau_1, \ldots, \tau_{q-1}\} = \{2, \ldots, n\}$ ist. Man sieht sofort:

SATZ 5. *X besitzt die Eigenschaft $B(0, q)$ im Punkte $x_0 \in \partial X$ genau dann, wenn neben (i) und (ii) in der Definition 4 gilt:*

$$\text{(iii)}' \qquad e_{\rho_1} + \cdots + e_{\rho_{n-q}} > 0, \qquad 2 \leqq \rho_1 < \cdots < \rho_{n-q} \leqq n.$$

DEFINITION 5. X heißt hyper-q-konvex, wenn X die Eigenschaft $B(0, n - q)$ hat.

Insbesondere besitzt dann $d'd''p|\partial_a X$ mindestens $(n - q)$ positive Eigenwerte, d.h.:

Eine hyper-q-konvexe Mannigfaltigkeit ist q-konvex. Im Falle $q = 1$ stimmen beide Begriffe überein. Da $B(0, q)$ die Eigenschaft $B(0, q - 1)$ impliziert, ist jede hyper-q-konvexe Mannigfaltigkeit auch hyper-$(q + 1)$-konvex.

Wir erhalten aus Satz 4 das

KOROLLAR. *Ist X hyper-q-konvex, so ist für jede Form $\varphi \in \dot{A}^{0,s}(V)$ mit $s \leqq n - q$ und $\varphi|\partial_a X = 0$, deren Koeffizienten nicht alle auf ∂X verschwinden:*

$$i \int_{\partial X} d'\rho \wedge \bar{*}L\varphi > 0.$$

D. Es sei nun $\varphi \in \mathfrak{H}_c^{r,s}(V)$, $l = r + s < n$, φ sei primitiv, und X besitze die Eigenschaft $B(r, s)$. Im Falle $r = 0$ sei V als semi-negativ vorausgesetzt, im Falle $r > 0$ sei $V = X \times \mathbf{C}$ das triviale Bündel. Dann folgt aus unserer Hauptgleichung und den Abschnitten B und C:

$$0 \leqq (d'\varphi, d'\varphi) = -2\pi (\chi \wedge \varphi, L\varphi) - i \int_{\partial X} d'\varphi \wedge \bar{*}L\varphi \leqq 0,$$

d.h. $(d'\varphi, d'\varphi) = 0$ und $i \int_{\partial X} d'\varphi \wedge \bar{*}L\varphi = 0$. Wir erhalten also

$$(12) \qquad d'\varphi = 0,$$

und wegen Satz 4 gilt mit $\varphi_i = \sum {}_i a_{\nu,\bar{\mu}} d\mathfrak{z}_\nu \wedge d\bar{\mathfrak{z}}_\mu$:

$$(13) \qquad {}_i a_{\nu_1 \cdots \nu_r, \bar{\mu}_1 \cdots \bar{\mu}_s} | \partial X = 0$$

für alle $1 \leqq \nu_1 < \cdots < \nu_r \leqq n$, $1 \leqq \mu_1 < \cdots < \mu_s \leqq n$ und $1 \leqq i \leqq m$. Schließlich folgt wegen $d''\varphi = \Lambda\varphi = 0$ und $\delta' = i(d''\Lambda - \Lambda d'')$ noch

$$(14) \qquad \delta'\varphi = 0.$$

Wir wollen zeigen, daß dann schon $\varphi = 0$ gelten muß. Dazu führen wir in x_0 geodätische Koordinaten $z_1, \ldots, z_n$ ein, und es gelte wieder $p(z) = x_1 +$ Glieder höherer Ordnung. Wegen (13) sind dann alle Ableitungen

$$(15) \qquad \begin{aligned} &\frac{\partial}{\partial x_j} {}_i a_{\nu_1 \cdots \nu_r, \bar{\mu}_1 \cdots \bar{\mu}_s}, \qquad j = 2, \ldots, n, \\ &\frac{\partial}{\partial y_j} {}_i a_{\nu_1 \cdots \nu_r, \bar{\mu}_1 \cdots \bar{\mu}_s}, \qquad j = 1, \ldots, n, \end{aligned}$$

in x_0 gleich Null. Definiert man entsprechend wie im Abschnitt B die Ausdrücke ${}_ia^{(\bar{\mu})}_{\nu_1\cdots\nu_r,\bar{\mu}_2\cdots\bar{\mu}_s}$, so folgt wegen $\delta''\varphi = 0$ in x_0:

$$\frac{\partial}{\partial z_1}\,{}_ia_{\nu_1\cdots\nu_r,\bar{1}\bar{\mu}_2\cdots\bar{\mu}_s} = \sum_{\mu=1}^{n}\frac{\partial}{\partial z_\mu}\,{}_ia^{(\bar{\mu})}_{\nu_1\cdots\nu_r,\bar{\mu}_2\cdots\bar{\mu}_s} = 0$$

d.h. für alle ${}_ia_{\nu_1\cdots\nu_r,\bar{1}\bar{\mu}_2\cdots\bar{\mu}_s}$, $2 \leqq \mu_2 < \cdots < \mu_s \leqq n$, $1 \leqq \nu_1 < \cdots < \nu_r \leqq n$, $i = 1, \ldots, m$, verschwinden sämtliche partiellen Ableitungen in x_0. Daraus schließt man, da auch $d''\varphi = 0$ in x_0 gilt, für $2 \leqq \mu_1 < \cdots < \mu_s \leqq n$:

$$\frac{\partial}{\partial \bar{z}_1}\,{}_ia_{\nu_1\cdots\nu_r,\bar{\mu}_1\cdots\bar{\mu}_s} = \sum_{\mu=1}^{n}\frac{\partial}{\partial \bar{z}_\mu}\,{}_ia^{(\hat{\mu})}_{\nu_1\cdots\nu_r,\bar{1}\bar{\mu}_1\cdots\bar{\mu}_s} = 0.$$

Also verschwinden neben den Koeffizienten ${}_ia_{\nu,\bar{\mu}}$ auch alle ersten partiellen Ableitungen dieser Koeffizienten auf ∂X. Da das System ${}_ia_{\nu,\bar{\mu}}$ einem elliptischen Operator zweiter Ordnung genügt, nämlich $\Box$, müssen alle ${}_ia_{\nu,\bar{\mu}} = 0$ sein, d.h. φ verschwindet identisch auf X.

Wir haben also den folgenden Satz bewiesen:

SATZ 6. *Es sei $\varphi \in \mathfrak{H}_c^{r,s}(V)$, $l = r + s < n$, eine primitive Form, X besitze die Eigenschaft $B(r, s)$. Ferner sei*

(i) *V semi-negativ, falls $r = 0$,*
(ii) *$V = X \times \mathbf{C}$, falls $r > 0$.*

Dann gilt $\varphi = 0$.

Ist insbesondere X hyper-q-konvex, so folgt für alle $\varphi \in \mathfrak{H}_c^{0,s}(V)$ mit $s \leqq n - q{:}\varphi = 0$. Zusammen mit den Ausführungen in Teil 1 erhält man das in der Einleitung angegebene Vanishing-Theorem:

SATZ 7. *Es sei $\hat{X}$ eine kählersche Mannigfaltigkeit und X ein relativ-kompakter offener Teilbereich mit hyper-q-konvexem Rand, $q \geqq 1$. Ferner sei V ein semi-negatives komplex-analytisches Vektorraumbündel über $\hat{X}$. Dann gilt*

$$H_c^s(X, \underline{V}) = 0, \qquad s \leqq n - q,$$
$$H^s(X, \underline{V}^* \otimes \Omega^n) = 0, \qquad s \geqq q.$$

Insbesondere erhält man $H_c^s(X, \mathcal{O}) = 0$ für $s \leqq n - q$ und $H^s(X, \Omega^n) = 0$ für $s \geqq q$.

E. Wir wollen Satz 6 noch etwas verallgemeinern:

SATZ 8. *Es sei $\varphi \in \mathfrak{H}_c^{r,s}$, und die Bedingungen $B(e - \lambda, s - \lambda)$, $\lambda =$ $0, 1, \ldots,$ min (r, s), seien für X erfüllt. Ferner gelte für alle λ: $\Lambda^\lambda|\partial_a X = 0$. Dann ist $\varphi = 0$.*

BEWEIS. Wegen $\delta''\varphi = 0$ ist auch $\delta''\Lambda^\lambda \varphi = \Lambda^\lambda\delta''\varphi = 0$. Außerdem folgt mit $\Box\varphi = 0: \delta'' d''\Lambda^\lambda\varphi = \Box\Lambda^\lambda\varphi = \Lambda^\lambda\Box\varphi = 0$, und hieraus ergibt sich $d(\Lambda^\lambda\varphi \wedge \bar{*}d''\Lambda^\lambda\varphi) = d''\Lambda^\lambda\varphi \wedge \bar{*}d''\Lambda^\lambda\varphi + (-1)^l\Lambda^\lambda\varphi \wedge d''\bar{*}d''\Lambda^\lambda\varphi$, $l = r + s$. Da der zweite Ausdruck verschwindet, erhält man durch Integration

$$(d''\Lambda^\lambda\varphi, d''\Lambda^\lambda\varphi) = \int_{\partial X} \Lambda^\lambda\varphi \wedge \bar{*}d''\Lambda\varphi.$$

Nun gilt $\Lambda^\lambda\varphi|\partial_a X = 0$. Folglich ist das Randintegral wegen des Kriteriums aus Abschnitt D von Teil 1 gleich Null, und es ist deshalb $d''\Lambda^\lambda\varphi = 0$, d.h. $\Lambda^\lambda\varphi \in \mathfrak{H}_c^{r-\lambda,s-\lambda}$ für alle λ.

Wir führen nun vollständige Induktion nach l. Für $l = 0$ und 1 ist φ primitiv und der Satz schon bewiesen. Ist nun $l > 1$ und der Satz für Formen vom Grade $\leqq l - 1$ schon hergeleitet, so ist wegen $\Lambda\varphi \in \mathfrak{H}_c^{r-1,s-1}$ die Form φ primitiv, und die Behauptung folgt aus Satz 6.

Es ist den Verfassern nicht bekannt, ob Satz 8 anwendbar ist.

3. Die Sätze von Akizuki und Nakano. Beispiele

A. Es sei $X = \hat{X}$ eine kompakte kählersche Mannigfaltigkeit und V ein negatives komplex-analytisches Vektorraumbündel über X. Aus unserer Hauptgleichung folgt dann

$$-(\chi \wedge \varphi, L\varphi) \geqq 0$$

für alle $\varphi \in \mathfrak{H}^{r,s}(V) = \{\varphi \in A^{r,s}(V): d''\varphi = \delta''\varphi = 0\}$. Andererseits wurde in Abschnitt B von Teil 2 gezeigt, daß $(\chi \wedge \varphi, L\varphi) \geqq 0$ ist für $\varphi \in A^{0,s}(V)$, $s < n$. Da auf Grund der Gleichung (4) $(\chi \wedge \varphi, L\varphi) = 0$ nur dann gilt, wenn $\varphi = 0$ ist, erhalten wir das Vanishing-Theorem von Nakano:

$$H^s(X, \underline{V}) = \mathfrak{H}^{0,s}(V) = 0, \qquad s < n.$$

B. Ist F ein negatives Geradenbündel über der kompakten kählerschen Mannigfaltigkeit $X = \hat{X}$, so ist χ eine positiv-definite globale (1, 1)-Form auf X, die geschlossen ist. Man kann daher χ als die einer kählerschen Metrik auf X zugeordnete Form betrachten und erhält mit $L\varphi := \chi \wedge \varphi$ stets $(\chi \wedge \varphi, L\varphi) \geqq 0$. Andererseits ist wegen der Hauptgleichung $-(\chi \wedge \varphi, L\varphi) \geqq 0$ für Formen $\varphi \in \mathfrak{H}^{r,s}(F)$, d.h. $L\varphi = 0$. Daraus folgt im Falle $r + s < n$ schon $\varphi = 0$. Damit ist auch das Vanishing-Theorem von Akizuki und Nakano bewiesen:

$$H^s(X, \underline{F} \otimes \Omega^r) = \mathfrak{H}^{r,s}(F) = 0, \qquad r + s < n.$$

C. Wir geben in diesem Abschnitt das in der Einleitung angekündigte Beispiel eines komplex-analytischen Vektorraumbündels an, das schwach negativ, jedoch nicht negativ ist.

Es sei Y eine n-dimensionale komplexe projektiv-algebraische Mannigfaltigkeit, $n \geqq 2$. F bezeichne das zu einem Hyperebenenschnitt gehörende komplex-analytische Geradenbündel; F ist somit positiv. Es gibt dann in der l-ten Tensorpotenz F^l von F, $l \geqq l_0$, schon $n + 1$ Schnittflächen $s_0, s_1, \ldots, s_n$, die keine gemeinsame Nullstelle in Y haben. Man erhält deshalb einen Epimorphismus $(n + 1)\mathcal{O} \to \underline{F}^l \to 0$, wobei $\mathcal{O}$ die Strukturgarbe von Y bezeichnet. Wir tensorieren mit $\underline{F}^{-1}$ und bekommen so einen Epimorphismus ϵ: $(n + 1)\underline{F}^{-1} \to \underline{F}^{l-1}$. Es sei $\underline{V}$ der Kern von ϵ. Die Garbe $\underline{V}$ ist lokal frei und wird von einem n-rangigen Untervektorraumbündel V von $(n + 1)F^{-1}$ geliefert. Da F^{-1} und mithin auch $(n + 1)F^{-1}$ negativ ist, muß V schwach negativ sein.

Mit $H^0(Y, (n + 1)\underline{F}^{-1}) = H^1(Y, (n + 1)\underline{F}^{-1}) = 0$ folgt aus der exakten Garbensequenz $0 \to \underline{V} \to (n + 1)\underline{F}^{-1} \to \underline{F}^{l-1} \to 0$:

$$H^1(Y, \underline{V}) \cong H^0(Y, \underline{F}^{l-1}), \qquad l \geqq l_0.$$

Da schließlich $H^0(Y, \underline{F}^{l-1})$ von Null verschieden ist für hinreichend großes, l_0 erhält man

$$H^1(Y, \underline{V}) \neq 0.$$

Nach dem Satz von Nakano müßte jedoch $H^1(Y, \underline{V}) = 0$ gelten, wenn V negativ wäre.

D. Zum Schluß entwickeln wir aus dem vorigen Abschnitt ein Beispiel dafür, daß unser Vanishing-Theorem nicht mehr richtig bleibt, wenn man die Hyper-q-Konvexität im Falle $q > 1$ durch die schwächere q-Konvexität ersetzt.

Es seien Y und V wie in Abschnitt C gewählt. Es gibt eine hermitesche Form $\sum h_{ik} w_k \bar{w}_i$ auf den Fasern von V, so daß durch $\{\sum h_{ik} w_k \bar{w}_i < 1\}$ eine relativ-kompakte streng pseudokonvexe Umgebung der Nullschnittfläche definiert wird. Auf den Fasern des dualen Bündels V^* ist dann ebenfalls eine hermitesche Form $\sum k^{ik} w_k^* \bar{w}_i^*$ gegeben, wobei $(h^{ik}) = (h^t)^{-1}$, $h = (h_{ik})$, ist. Obwohl in normaler Trivialisierung beide Formen übereinstimmen, ist die Nullumgebung $\{\sum h^{ik} w_k^* \bar{w}_i^* < 1\}$ nicht mehr 1-konvex. Schließt man jedoch die Fasern von V^* zu komplex-projektiven Räumen ab und bezeichnet man mit $\bar{V}^*$ das auf diese Weise erhaltene projektive Bündel, so ist das Komplement X der oben definierten Umgebung der Nullschnittfläche bezüglich $\bar{V}^*$ eine n-konvexe Umgebung der unendlich fernen Punkte $V_\infty = \bar{V}^* - V^*$.

Wir zeigen, daß $H_c^2(X, \mathcal{O}) \neq 0$ ist. Es ist dann das gewünschte Beispiel in X gefunden.

Es sei $\mathfrak{U} = \{U_\iota : \iota \in \mathrm{I}\}$ eine endliche Steinsche Überdeckung von Y. Nach Konstruktion gibt es einen Kozyklus $\xi \in Z^1(\mathfrak{U}, \underline{V})$, der nicht kohomolog zu Null ist. Da die Elemente von V_y, $y \in Y$, Linearformen auf V_y^* sind, erhält man durch ξ einen Kozyklus $\hat{\xi} \in Z^1(\hat{\mathfrak{U}}, \mathcal{O})$, wobei $\hat{\mathfrak{U}}$ das Urbild der Überdeckung $\mathfrak{U}$ bezüglich der Projektion $\bar{V}^* \to Y$ bezeichnet. Ist $\hat{\xi} = \{\hat{\xi}_{\iota_1\iota_2}\}$, so sind die Funktionen $\hat{\xi}_{\iota_1\iota_2}$ auf den Durchschnitten $\hat{U}_{\iota_1\iota_2}$ holomorph und entlang der Fasern von $\bar{V}^*$ linear; sie besitzen deshalb im Unendlichfernen eine Polstelle erster Ordnung.

Es sei $W_R = \{\sum h^{ik} w_k^* \bar{w}_i^* > R\} \subset\subset X$ eine Umgebung von V_∞. Die Kohomologieklasse von $\hat{\xi} | V^* - \overline{W}_R$ läßt sich dann nicht nach $\bar{V}^*$ fortsetzen. Wäre dies nämlich doch der Fall, so gibt es wegen $\hat{U}_\iota = U_\iota \times \mathbf{P}^n$ und $H^1(\hat{U}_\iota, \mathcal{O}) = 0$ ein $\gamma \in Z^1(\hat{\mathfrak{U}}, \mathcal{O})$ und ein $\eta \in C^0(\hat{\mathfrak{U}} | \bar{V}^* - \overline{W}_R, \mathcal{O})$, so daß $\gamma = \hat{\xi} - \delta\eta$ in $\bar{V}^* - \overline{W}_R$ ist. Entwickelt man dann $\hat{\xi}$, γ und η in eine Potenzreihe nach den Koordinaten auf den Fasern und bezeichnet man mit $\hat{\xi}^{(1)}$, $\gamma^{(1)}$ und $\eta^{(1)}$ den linearen Anteil, so gilt $\hat{\xi}^{(1)} = \xi$, $\gamma^{(1)} = 0$ und mithin $\xi = \delta\eta^{(1)}$ im Gegensatz zu der Voraussetzung.

Aus dem Vorhergehenden folgt weiter, daß auch $\hat{\xi} | X - \overline{W}_R$ nicht nach X hinein fortgesetzt werden kann. Wir stellen nun die Kohomologieklasse von $\hat{\xi} | X - V_\infty$ bezüglich der Dolbeault-Isomorphie durch eine (0, 1)-Form φ dar. Es sei ϵ eine beliebig oft differenzierbare positive Funktion auf X, die in $X - \overline{W}_R$ identisch 1 ist und in einer Umgebung von V_∞ identisch verschwindet. Offenbar hat $\psi = d''(\epsilon\varphi)$ kompakten Träger. Gäbe es eine (0, 1)-Form α in X mit kompaktem Träger und $d''\alpha = \psi$, so hätte man $d''(\epsilon\varphi - \alpha) = 0$. Ist $R > 1$ hinreichend klein, so besteht jedoch in $X - \overline{W}_R$ die Gleichheit $\varphi = \epsilon\varphi - \alpha$. φ wäre also nach X hinein fortsetzbar. Da dies nicht der Fall ist, wie eben gezeigt wurde, repräsentiert $d''(\epsilon\varphi)$ nicht die Nullklasse von $H_c^2(X, \mathcal{O})$.

UNIVERSITÄT GÖTTINGEN

LITERATUR

1. AKIZUKI, Y., and S. NAKANO, "Note on Kodaira–Spencer's proof of Lefschetz theorems," *Proc. Japan. Acad.*, *30* (1954), 266–272.
2. BOCHNER, S., "Curvature and Betti numbers I, II," *Ann. of Math.*, *49* (1948), 379–390; *50* (1949), 77–93.
3. KODAIRA, K., "On a differential geometric method in the theory of analytic stacks," *Proc. Nat. Acad. Sci.*, *39* (1953), 1268–1273.
4. KOHN, J. J., "Harmonic integrals on strongly pseudo-convex manifolds, I and II," *Ann. of Math.*, *78* (1963), 112–148; *79* (1964), 450–472.
5. KOHN, J. J., and H. ROSSI, "On the extension of holomorphic functions from the boundary of a complex manifold," *Ann. of Math.*, *81* (1965), 451–472.
6. NAKANO, S., "On complex analytic vector bundles," *J. Math. Soc. Japan*, *7* (1955), 1–12.
7. SERRE, J.-P., "Un théorème de dualité," *Comment. Math. Helv.*, *29* (1955), 9–26.
8. VESENTINI, E., "Osservazioni sulle strutture fibrate analitiche sopra una varietà kähleriana compatta, I, II," *Atti Accad. Naz. Lincei Rend. Cl. Sci. Fis. Mat. Natur.* (8), *23* (1957), 231–241; *24* (1958), 505–512.

Iteration of Analytic Functions of Several Variables

SAMUEL KARLIN AND JAMES McGREGOR[1]

Professor Salomon Bochner has contributed notably in many areas of mathematics including the theory of functions of several complex variables and the theory of stochastic processes.

In line with these interests, this paper reports results on iteration of holomorphic functions of several complex variables motivated by investigations pertaining to multi-type branching Markoff processes. Apart from its intrinsic importance and independent interest, iteration of holomorphic mappings plays a fundamental role in celestial mechanics, population genetics, numerical analysis, and other areas.

Consider a vector-valued mapping

$$(1)\qquad \begin{aligned} f(z) &= (f_1(z), f_2(z), \ldots, f_p(z)),\\ z &= (z_1, z_2, \ldots, z_p) \in Z_p, \end{aligned}$$

where Z_p denotes the space of p-tuples of complex numbers. Suppose the mapping has a fixed point in its domain of definition and that at the fixed point the Jacobian matrix is nonsingular and all its eigenvalues are of modulus less than 1. One of the main objectives is to determine a canonical representation for the iterates

$$(2)\qquad \begin{aligned} f(z; n) &= f(f(z; n-1), 1), \qquad n = 1, 2, \ldots\\ f(z; 0) &= z \end{aligned}$$

from which the complete structure and the asymptotic behavior of iterates of high order is easily ascertained.

The case $p = 1$ has a long history and an extensive literature and has been the subject of considerable recent research, e.g., see Jabotinsky [6], Baker [1], Karlin and McGregor [8], [9] and Szekeres [17]; the last named

[1] Research supported in part at Stanford University, Stanford, California, under contract N0014-67-A-0112-0015.

work contains a substantial bibliography of earlier writings on this topic. Consider a complex-valued function of one complex variable $f(z)$ analytic in the neighborhood of the origin and satisfying $f(0) = 0$ so that the origin is a fixed point. It was already known in the nineteenth century that if $c = f'(0)$ satisfies $0 < |c| < 1$, then the limit function

$$A(z) = \lim_{n\to\infty} \frac{f(z;n)}{c^n} \tag{3}$$

exists, is holomorphic at $z = 0$ and satisfies

$$A(f(z)) = cA(z). \tag{4}$$

The function $A(z)$ may be characterized as the unique solution of the functional equation (4) holomorphic at 0 with prescribed initial conditions $A(0) = 0$, $A'(0) = 1$. If the mapping inverse to $w = A(z)$ is $z = B(w)$, then, from

$$A(f(z, n)) = c^n A(z), \qquad n = 0, \pm 1, \pm 2, \ldots,$$

we obtain the representation of the iterates

$$f(z, n) = B(c^n A(z)). \tag{5}$$

The formula

$$f(z; t) = B(e^{t \log c} A(z)), \qquad -\infty < t < \infty \tag{6}$$

defines a continuous extension of $f(z; n)$ where each $f(z; t)$ is analytic at 0 and keeps this point fixed. It is not difficult to show that (6), for the various determinations of $\log c$, gives the only continuous embeddings of the iterates of f in a one-parameter group (see [8]).

The expression (6) provides a canonical representation of the group of mappings $f(z; t)$ with the nature of the dependence on z and t clearly displayed.

DEFINITION 1.1. If $z \to \psi(z)$ is a holomorphic mapping defined in a neighborhood of the fixed point of $f(z)$ with the same fixed point, then the mapping $z \to g(z) = \psi^{-1}(f(\psi(z)))$ is called *conjugate* to $f(z)$.

It is clear that for g conjugate to f then

$$g(z; n) = \psi^{-1}(f(\psi(z); n)),$$

and the structure of the iterates of g is readily discernible from those of f and conversely.

The representation (5) *asserts that if* $f(z)$ *is holomorphic at* 0, $f(0) = 0$, *and* $f'(0) = c$ $(0 < |c| < 1)$, *then* $f(z)$ *is conjugate to the linear function* $g(z) = cz$.

In this article we outline the analogues of (5) and (6) in the case $p \geqq 2$ and indicate some applications of the representation formula. In order to state the principal result we require some additional notation. There is given a mapping $z \to f(z) = (f_1(z), \ldots, f_p(z))$ in accordance with (1) with each component function holomorphic in a neighborhood of the origin. Suppose for definiteness that 0 is a fixed point of $f(z)$. Thus each $f_\nu(z)$ admits a power series expansion in the p variables $z_1, z_2, \ldots, z_p$ without constant term, of the form

$$f_\nu(z) = \sum_{j=\Delta} a_\nu(j) z^j$$

where Δ denotes the set of all $j = (j_1, j_2, \ldots, j_p)$, j_ν non-negative integers not all zero and $z^j = z_1^{j_1} \cdot z_2^{j_2} \cdots z_p^{j_p}$. The length of the vector $z = (z_1, \ldots, z_p)$ is denoted by $|z| = \max(|z_1|, |z_2|, \ldots, |z_p|)$.

The gradient matrix of the mapping at the fixed point is

$$C = f'(0) = \|C_{\alpha\beta}\|, \tag{7}$$

where

$$C_{\alpha\beta} = \left.\frac{\partial f_\alpha}{\partial z_\beta}\right|_{z=0} \qquad \alpha, \beta = 1, 2, \ldots, p.$$

The gradient plays the role, for general p, of the multiplying factor $f'(0) = c$ occurring in the case $p = 1$.

Our *main assumption* is that every eigenvalue λ of C satisfies

$$0 < |\lambda| < 1. \tag{8}$$

The first inequality, $0 < |\lambda|$, is simply equivalent to the assertion that C is nonsingular, or to the assertion that the mapping has an inverse holomorphic at the fixed point. The second inequality, $|\lambda| < 1$, is a restriction essential for our methods. If, contrary to (8), all eigenvalues satisfy $|\lambda| > 1$, then the inverse mapping satisfies (8) and our methods and results remain applicable.

The eigenvalues of C are listed as $\lambda_1, \lambda_2, \ldots, \lambda_p$ and are arranged such that

$$1 > |\lambda_1| \geqq |\lambda_2| \geqq \cdots \geqq |\lambda_p| > 0.$$

(Repeated λ's are to appear consecutively.)

Algebraic complications may develop which do not arise in the case $p = 1$. These are of two kinds: (i) The canonical form must necessarily be recast whenever C possesses nonsimple elementary divisors. This situation can occur only when $p \geqq 2$. (ii) A more substantial modification is required whenever algebraic relations among the eigenvalues $\{\lambda_1, \lambda_2, \ldots, \lambda_p\}$ exist. *An algebraic relation is an identity of the form*

$$\lambda_1^{r_1} \lambda_2^{r_2} \cdots \lambda_p^{r_p} = \lambda_\nu \tag{9}$$

where $r_1, r_2, \ldots, r_p$ are positive integers, and $\sum_{\mu=1}^{p} r_\mu \geqq 2$. When algebraic relations are not present and C admits no elementary divisors, a simple diagonal canonical form prevails. Even under these circumstances a further technical obstacle pointed out by Bellman [2] and other authors is that the most straightforward generalization of the limit formula (3) is no longer forthcoming. It is correct that the first component function $A_1(z)$ can, in fact, be constructed in the spirit of (3), but the other component functions $A_\nu(z)$, $\nu = 2, \ldots, p$, cannot be so simply determined.

The form of the general canonical representation is the content of the following basic theorem $\left(\text{we write } |j| = \sum_{\nu=1}^{p} |j_\nu| \text{ for } j \in \Delta\right)$.

THEOREM 1.[2] *Let the assumption* (8) *hold. There exists a mapping* $A(z) = (A_1(z), A_2(z), \ldots, A_p(z))$ *holomorphic at the origin such that*

$$A(0) = 0, \qquad \left.\frac{\partial A_\nu}{\partial z_\mu}\right|_{z=0} = \delta_{\mu\nu} \tag{10}$$

and $A_\nu(z)$ *satisfies a functional relation*

$$A_\nu(f(z;n)) = \lambda_\nu^n \Big\{ A_\nu(z) + \sum_{\substack{\beta<\nu \\ \lambda_\beta=\lambda_\nu}} q_{\nu,\beta}(n) A_\beta(z) + \sum_{|j|\geqq 2,\, \lambda^j=\lambda_\nu} R_{\nu,j}(n)[A(z)]^j \Big\}, \tag{11}$$

$$n = 0, \pm 1, \pm 2, \ldots, \quad (\nu = 1, 2, \ldots, p),$$

where $q_{\nu,\beta}(n)$ *and* $R_{\nu,j}(n)$ *are polynomials in the variable* n, $[A(z)]^j$ *for* $j = (j_1, j_2, \ldots, j_p)$ *is an abbreviation of*

$$[A(z)]^j = [A_1(z)]^{j_1}[A_2(z)]^{j_2} \cdots [A_p(z)]^{j_p},$$

and λ^j *signifies* $\lambda^j = \lambda_1^{j_1}\lambda_2^{j_2} \cdots \lambda_p^{j_p}$.

The equations (11) can be equivalently written for $f(z; n)$ by applying the inverse transformation $B = A^{-1}$. Bounds on the degrees of the polynomials $q_{\nu,\beta}$ and $R_{\nu,j}$ are also available.

The first sum on the right of (11) appears if C possesses elementary divisors. The second sum generally occurs in the presence of algebraic relations.

The significance of Theorem 1 can be succinctly stated in the following terms. *The mapping* $f(z)$ *satisfying* (8) *is conjugate to a polynomial mapping*

$$h(z) = (h_1(z), \ldots, h_p(z)) \tag{12}$$

[2] See the last paragraph of this paper.

where

$$(13)\quad h_\nu(z) = \lambda_\nu\left\{z_\nu + \sum_{\beta=1}^{\nu-1} q_{\nu,\beta}(1)z_\beta + \sum_{\substack{j\in\Delta\\ |j|\geqq 2,\lambda^j=\lambda_\nu}} R_{\nu,j}(1)z_1^{j_1}z_2^{j_2}\cdots z_p^{j_p}\right\}$$

and thus $f(z)$ *can be written in the form* $f(z) = A^{-1}(h(A(z)))$.

It is not difficult to prove that the map $\zeta \to h(z)$ is a univalent holomorphic map of Z_p *onto* Z_p. If no algebraic relations are present then $h(z)$ is linear and $f(z)$ is conjugate to a linear transformation and even more specifically $f(z)$ is conjugate to the linear mapping $\tilde{h}(z) = Cz$.

An important corollary of Theorem 1 is that the iterates of the mapping (2) can be embedded in a continuous one-parameter group $f(z; t)$ of holomorphic mappings satisfying

$$f(z; t + \tau) = f(f(z; t), \tau),$$
$$f(z; 1) = f(z).$$

The parameter t assumes all complex values and $f(z; t)$ is a holomorphic function of the $p + 1$ variables $z_1, \ldots, z_p, t$ and for each t is a function of z vanishing at $z = 0$ and holomorphic in a 0-neighborhood. The embedding is not unique, but in many cases it can be made unique by imposing simple and natural auxiliary conditions. For example, in many applications the power series $f_\nu(z)$ of the given mapping have real (or non-negative) coefficients and the eigenvalues λ_ν are non-negative. The embedding is then unique if it be required that the corresponding power series for $f(z; t)$ have real coefficients for $-\infty < t < \infty$ (non-negative for $0 \leqq t < \infty$).

The main ideas in our method of proof will be sketched very briefly. After a preliminary linear change of variables we can assume the matrix (7) is in Jordan canonical form, and then it follows that $|f(z)| < |z|$ if $|z_\nu| < \rho_\nu$, $\nu = 1, \ldots, p$ where ρ_ν are small positive numbers in a suitable geometric progression. Then after a further linear change of variables we can assume $f(z)$ maps the closed unit polydisc $\{z; |z| \leqq 1\}$ into its own interior (contraction property).

Let $\mathscr{H}$ be the Hilbert space consisting of all power series

$$\xi(z) = \sum_{j\in\Delta} \hat{\xi}(j)z^j$$

with

$$\|\xi\|^2 = \sum_{j\in\Delta} |\hat{\xi}(j)|^2 < \infty.$$

Note that $\xi(0) = 0$. In $\mathscr{H}$ a linear transformation T is defined by $(T\xi)(z) = \xi(f(z))$. One can readily show that, because of the contraction property, T is completely continuous.

The proof now proceeds by applying the well known theory of completely continuous linear operators. The subspace of $\mathscr{H}$ consisting of all power series beginning with homogeneous terms of degree $\geqq r$ is invariant under T and dually the subspace of polynomials of degree $\leqq r$ is invariant under the dual operator T'. One readily verifies that in the subspace consisting of polynomials of first degree, T' acts like the matrix $C = f'(0)$. Hence there are p linearly independent linear polynomials $v_\nu(z)$, $\nu = 1, \ldots, p$ such that $(T' - \lambda_\nu)^n v_\nu = 0$ for suitable positive integers $n = n(\nu)$. It follows from the theory of completely continuous operators that each λ_ν is an eigenvalue of T. Moreover, if $\mathscr{E}(\lambda_\nu)$ denotes the null space of $(T - \lambda_\nu)^n$ for large n and $\mathscr{E}$ is the (direct) sum of $\mathscr{E}(\lambda_\nu)$, $\nu = 1, \ldots, p$ then $\mathscr{E}$ is of finite dimension $N \geqq p$. A basis $u_1, u_2, \ldots, u_N$ can be so chosen in $\mathscr{E}$ that

$$(u_\alpha, v_\beta) = \delta_{\alpha,\beta}, \qquad \alpha, \beta = 1, 2, \ldots, p,$$

with $u_\alpha \in \mathscr{E}(\lambda_\alpha)$. Since the v_β are linear polynomials it follows from the definition of the inner product in $\mathscr{H}$ that the linear parts of the holomorphic functions $u_\nu(z)$, $\nu = 1, \ldots, p$ are linearly independent. Consequently any basis in $\mathscr{E}$ has p vectors whose linear parts are linearly independent.

We can choose a basis $w_1, w_2, \ldots, w_N$ in $\mathscr{E}$ such that

$$Tw_\nu = \gamma_\nu(w_\nu + \delta_\nu w_{\nu-1}), \qquad \nu = 1, \ldots, N$$

where each γ_ν is one of the λ_α and each δ_ν is either zero or one, and is zero if $\gamma_\nu \neq \gamma_{\nu-1}$. From the contraction property it is easy to show by an inductive argument that each w_ν is holomorphic in a polydisc $|z| < 1 + \epsilon$. A subsequence $A_1, \ldots, A_p$ of $w_1, \ldots, w_N$ is now chosen to satisfy the following condition. $A_\nu = w_{\alpha(\nu)}$ where $\alpha(\nu)$ is the smallest integer such that the linear parts of $A_1, A_2, \ldots, A_\nu$ are linearly independent.

Since each $A_\nu(z)$ is holomorphic in $|z| < 1 + \epsilon$, the functions

$$A^j(z) = A_1^{j_1}(z) A_2^{j_2}(z) \cdots A_p^{j_p}(z), \qquad j \in \Delta$$

are all in $\mathscr{H}$. These functions are evidently linearly independent and span $\mathscr{H}$.

LEMMA. *If a, b are holomorphic in $|z| < 1 + \epsilon$ and there are complex numbers α, β and positive integers m, n such that $(T - \alpha)^m a = 0$, $(T - \beta)^n b = 0$, then*

$$(T - \alpha\beta)^{m+n-1} ab = 0.$$

By repeated application of this lemma one shows that for sufficiently large $n = n(j)$

$$(T - \lambda^j)^n A^j = 0, \qquad j \in \Delta.$$

Thus the basis $A^j, j \in \Delta$, provides a complete decomposition of the space $\mathscr{H}$ into eigenmanifolds of T. The eigenvalues of T are the numbers

$$\lambda^j = \lambda_1^{j_1} \cdots \lambda_p^{j_p}, \qquad j \in \Delta,$$

each listed with proper multiplicity (trivially, zero is not an eigenvalue of T).

In particular, each of the functions w_α is a linear combination of the finitely many A^j such that $\lambda^j = \gamma_\alpha$. Because of the way the λ_ν have been ordered, in any nontrivial relation $\lambda_\nu = \lambda^j$, $j \in \Delta$, $|j| \geqq 2$, only j_β with $\beta < \nu$ can be different from zero. Consequently, when the right member of

$$TA_\nu = Tw_{\alpha(\nu)} = \lambda_\nu(A_\nu + \gamma_{\alpha(\nu)} w_{\alpha(\nu)-1})$$

is expressed as a linear combination of the function A^j, the form (11) for $n = 1$ is obtained. The result for general n follows readily.

We also see that the dimension of $\mathscr{E}$ is

$$N = p + R$$

where R is the number of relations.

We pointed out, following the statement of Theorem 1, that when algebraic relations exist $f(z)$ is not generally conjugate to a linear mapping. However, it is possible to represent the mapping $f(z)$ in a higher dimensional space such that the conjugacy expression corresponds to that of a linear mapping. We illustrate the embedding first with an example in the case $p = 2$. Here, the eigenvalues λ_1, λ_2 of C fulfill exactly one relation, an identity $\lambda_2 = \lambda_1^n$ where n is an integer, $n \geqq 2$. For the example $f(z) = (\lambda z_1, \lambda^2 z_2)$, $0 < \lambda < 1$, we have a relation $\lambda_1 = \lambda$, $\lambda_2 = \lambda_1^2$ but nevertheless the map $h(\zeta)$ is linear; in fact we can take $A_\nu(z) = z_\nu$, $\nu = 1, 2$, and then $h(\zeta) = f(\zeta)$. On the other hand, for the example

$$f(z) = (\lambda z_1, \lambda^2 z_2 + z_1^2), \qquad (0 < \lambda < 1), \tag{14}$$

we again have a relation $\lambda_1^2 = \lambda_2 = \lambda^2$, but a simple computation reveals that $f(z)$ is not conjugate to Cz.

With the two dimensional mapping in (14) we shall associate a linear mapping $g(s)$ in the space of three complex variables $s = (s_1, s_2, s_3)$, viz.

$$g(s) = (\lambda s_1, \lambda^2 s_2, \lambda^2 s_3 + s_2). \tag{15}$$

It is easy to see that the algebraic surface

$$\mathscr{M} = \{s : s_2 = s_1^2\}$$

is mapped into itself by $g(s)$. The surface $\mathscr{M}$ is parametrized by the coordinates s_1, s_3. In fact, the map U defined by

$$U\begin{pmatrix} z_1 \\ z_2 \end{pmatrix} = \begin{pmatrix} z_1 \\ z_1^2 \\ z_2 \end{pmatrix}$$

provides a holomorphic and globally univalent map of the space of two complex variables onto $\mathscr{M}$. The inverse mapping is simply the restriction to $\mathscr{M}$ of the projection V,

$$V\begin{pmatrix} s_1 \\ s_2 \\ s_3 \end{pmatrix} = \begin{pmatrix} s_1 \\ s_3 \end{pmatrix}$$

It is apparent that

$$f(z) = V(g(U(z))).$$

Thus the mapping (15), although not conjugate to a linear transformation, is conjugate to the restriction, to an invariant algebraic manifold, of a linear transformation in a space of higher dimension. This situation is general. Suppose there are exactly R relations. Define

$$N = p + R.$$

Then we have

THEOREM 2. *There exists a univalent polynomial mapping U of Z_p onto an algebraic surface $\mathscr{M}$ in Z_N, and a nonsingular linear mapping g of Z_N onto Z_N such that*

(i) *If $\mathscr{L}$ is a linear subspace of Z_N and $\mathscr{L} \supseteq \mathscr{M}$ then $\mathscr{L} = Z_N$.*
(ii) *$Uh = gU$.*

Two remarks are worth appending:

(i) The inverse mapping U^{-1} of $\mathscr{M}$ onto Z_p is implemented by a *linear* map of Z_N *onto* Z_P.
(ii) The relation between h and g extends at once to the iterates h^n and g^n. That is,

$$Uh^n = g^n U, \qquad n = 0, \pm 1, \pm 2, \ldots.$$

Applications

MAPPINGS WHICH COMMUTE WITH $f(z)$. Let $e(z)$ be a mapping of Z_p into Z_p which is holomorphic in a neighborhood of the origin, leaves the origin fixed, and commutes with $f(z)$; that is,

$$e(f(z)) = f(e(z))$$

for all z in a neighborhood of the origin. The following Theorem serves to characterize such mappings.

THEOREM 3. *Using the notation of Theorems* 1 *and* 2, *the formula*

$$E = U(A(e(A^{-1}(U^{-1}))))$$

sets up a 1-1 *correspondence between the set of all mappings* $e(z)$ *of* Z_p, *holomorphic in a neighborhood of the origin, leaving the origin fixed and commuting with* f, *and the set of all linear mappings* E *of* Z_N *which map* $\mathcal{M}$ *into itself and commute with* g.

The important special case where no algebraic relations are present leads to the corollary.

COROLLARY. *Under the conditions of Theorem* 3, *if no relations exist, then* $e(z)$ *commutes with* $f(z)$ *if and only if* e *has the form*

$$e(z) = A^{-1}(E(A(z)))$$

where $E(z)$ *is a linear mapping of* Z_p *onto itself commuting with the mapping* $h(z)$.

SUPERCRITICAL MULTI-TYPE BRANCHING PROCESSES. Consider a multi-type branching Markov process of p-types with probability generating function (p.g.f) $f_\nu(z_1, \ldots, z_p) = \sum_j a_\nu(j)z^j$ for a single individual of the νth type, $\nu = 1, 2, \ldots, p$; i.e., the coefficient $a_\nu(j)$ is interpreted as the probability that a single parent of the νth type produces $|j|$ offspring consisting of j_1 individuals of type 1, j_2 of type 2, ..., j_p of type p where $j = (j_1, j_2, \ldots, j_p)$. Obviously $a_\nu(j) \geqq 0$ and $\sum_j a_\nu(j) = 1$ for each ν. Let $X(n) = (X_1(n), X_2(n), \ldots, X_p(n))$, $n = 0, 1, 2, \ldots$ be the vector random variable depicting the population structure at the nth generation; i.e., $X_\nu(n)$ denotes the number of individuals of the νth type in the nth generation.

The temporally homogeneous transition probability matrix governing the fluctuations of population size over successive generations is defined implicitly by the generating function relation

$$\sum_j Pr\{X(n+1) = j \mid X(n) = i\}z^j = [f(z)]^i \tag{16}$$

(we employ the previous notation here). The identity (16) characterizes Markov branching processes in that each individual bears offspring independently of the other existing individuals. Thus, if the current population makeup consists of i_r individuals of type r ($r = 1, 2, \ldots, p$), then the p.g.f. describing the next generation is given by

$$[f_1(z)]^{i_1}[f_2(z)]^{i_2}\cdots[f_p(z)]^{i_p}$$

and thus (16) obtains.

A standard result in the theory of Markov branching processes is that

$$\sum_j Pr\{X(n) = j \mid X(0) = i\} z^j = [f(z; n)]^i$$

is valid for $|z| \leqq 1$, where $f(z; n)$ denotes the nth iterate of $f(z; 1) = f(z)$.

By the nature of the stochastic process at hand, each of the functions $f_\nu(z)$ is defined in the polydisc $|z| \leqq 1$ and the vector $\mathbf{1} = (1, 1, \ldots, 1)$ is a fixed point of the mapping $f(z)$. We further postulate that $(\partial f_\nu/\partial z_\mu)(1) < \infty$ for all $\mu, \nu = 1, \ldots, p$. The branching process is said to be *supercritical* if the largest eigenvalue $\rho = \rho(D)$ of the matrix $D = \|(\partial f_\nu/\partial z_\mu)(1)\|$ exceeds 1. To avoid trivial technical alterations in the statement of the theorems we also assume $(\partial f_\nu/\partial z_\mu)(1) > 0$ for all ν, μ. In this circumstance the existence of another invariant point $\pi = (\pi_1, \pi_2, \ldots, \pi_p)$ with $0 \leqq \pi_i < 1$ $(i = 1, 2, \ldots, p)$ is assured; i.e., $f(\pi) = \pi$. Moreover, the iterates $f(z; n)$ converge to π as $n \to \infty$ for all $|z| \leqq 1$, $z \neq \mathbf{1}$ (see Harris [4], or Karlin [7], Chap. 11). If we assume $f(0) \neq 0$, then it follows that $\pi_i > 0$ for all i. We can now prove

PROPOSITION 1. *Under the conditions stated above, the eigenvalues of the matrix* $C = \|(\partial f_\nu/\partial z_\mu)(\pi)\|$ *satisfy* $0 \leqq |\lambda_i| < 1$ $(i = 1, 2, \ldots, p)$.

PROOF. Introduce the vector $\sqrt{\pi} = (\sqrt{\pi_1}, \sqrt{\pi_2}, \ldots, \sqrt{\pi_p})$. An application of the Schwartz inequality yields

$$\begin{aligned} f_\nu(\sqrt{\pi}) &= \sum_i a_\nu(i)(\sqrt{\pi})^i < \left(\sum_i a_\nu(i)\pi^i\right)^{1/2}\left(\sum_i a_\nu(i)\right)^{1/2} \\ &= \sqrt{f_\nu(\pi)} = \sqrt{\pi_\nu}. \end{aligned} \tag{17}$$

Strict inequality prevails because of the stipulations $0 < f_\nu(0)$, $(\partial f_\nu/\partial z_\mu)(1) > 0$ $(\nu, \mu = 1, 2, \ldots, p)$ and since $\rho(D) > 1$. Because each $f_\nu(z)$ is convex on the domain $\mathscr{D} = \{z : 0 \leqq z_\nu \leqq 1\}$, $\nu = 1, 2, \ldots, p$, we secure the vector inequality

$$f(z) - \pi = f(z) - f(\pi) \geqq C\cdot(z - \pi) \quad \text{for } z \in \mathscr{D},$$

where the notation signifies that the inequality holds componentwise. Substituting $z = \sqrt{\pi}$ and referring to (17) we find for the positive vector $u = \sqrt{\pi} - \pi$ that

$$u > Cu$$

with strict inequality in each component. This inequality implies, by virtue of the well known Frobenius theorem pertaining to matrices with positive entries, that all eigenvalues of C are in magnitude less than 1. The proof of Proposition 1 is complete.

If C is nonsingular, which is the case except if $\{f_\nu(z)\}_{\nu=1}^{p}$ are "essentially" functionally dependent, then Theorems 1 and 2 are relevant for the study of the iterates $f(z; n)$ in the neighborhood of the fixed point π. In particular, utilizing the canonical representation implicit in (11), it is not difficult to establish the strong ratio theorem. Specifically, it can be proved for any prescribed nonzero integer vectors i, j, k, l

$$\lim_{n\to\infty} \frac{Pr\{X(n) = j \mid X(0) = i\}}{Pr\{X(n) = k \mid X(0) = l\}}$$

exists. It is also possible to develop a spectral representation of the process (see Karlin and McGregor [8], [9]) and with its aid to uncover several right and left invariant functions of probabilistic importance. The elaboration of these applications as well as the proofs of Theorems 1 to 3 and related matters will be exposed in further publications.

We learned at the symposium whose proceedings are summarized in this volume that Sternberg [13] gave a proof of Theorem 1 using an elegant iterative scheme. He focused attention on the case where C admits no nonsimple elementary divisors. Undoubtedly his method, which is sketchy in places, can be adapted to handle this case as well. See also Sternberg [14], [15], [16], Hartman [5, Chapter 9], Chen [3] and Nelson [11] for discussion of the corresponding problem with mappings of class C^k for suitable k.

Numerous other authors have recently worked on the problem of Theorem 1 and succeeded in establishing partial cases of Theorem 1; e.g., see Schubert [12] and references given there. They are apparently unaware of Sternberg's important contribution on this subject.

Our method of proof of Theorem 1 differs from Sternberg's and may be of some interest.

STANFORD UNIVERSITY

REFERENCES

1. BAKER, I. N., "Fractional iteration near a fixed point of multiplier, 1," *J. Austral. Math. Soc.*, *4* (1964), 143–148.
2. BELLMAN, R., "The iteration of power series in two variables," *Duke Math. J.*, *19* (1952), 339–347.
3. CHEN, K. T., "Normal Forms of Local Diffeomorphisms on the Real Line," *Duke Math J. 35* (1968), 549–555.
4. HARRIS, T. E., *Theory of Branching Processes*. Berlin: Springer-Verlag, 1967.
5. HARTMAN, P., *Ordinary Differential Equations*. New York: John Wiley & Sons, Inc., 1964.
6. JABOTINSKY, E., "Analytic iteration," *Trans. Amer. Math. Soc.*, *108* (1963) 457–477.

7. Karlin, S., *A First Course in Stochastic Processes*. New York: Academic Press, 1966.
8. Karlin, S., and J. McGregor, "Embeddability of discrete time simple branching processes into continuous time branching processes," *Trans. Amer. Math. Soc.*, *132*, No. 1 (1968), 115–136.
9. ———, "Embedding iterates of analytic functions with two fixed points into continuous groups," *Trans. Amer. Math. Soc.*, *132*, No. 1 (1968), 137–145.
10. Karlin, S., and J. McGregor, "Spectral theory of branching processes, I and II," *Z. Wahrscheinlichkeitstheorie*, *5* (1966), 6–33 and 34–54.
11. Nelson, E., *The Local Structure of Vector Fields*, part of lecture notes at Princeton University, Section 3, 1969.
12. Schubert, C. F., "Solution of a generalized Schroeder equation in two variables," *J. Aust. Math. Soc.*, *4* (1964), 410–417.
13. Sternberg, S., "Local contractions and a theorem of Poincaré," *Amer. J. Math.*, *79* (1957), 809–824.
14. ———, "On the structure of local homeomorphisms," *Amer. J. Math.*, *80* (1958), 623–631.
15. ———, "The structure of local homeomorphisms, III," *Amer. J. Math.*, *81* (1959), 578–604.
16. ———, "Infinite Lie groups and the formal aspects of dynamical systems," *J. Math. Mech.*, *10* (1961), 451–474.
17. Szekeres, G., "Regular iteration of real and complex functions," *Acta Math.*, *100* (1958), 203–258.
18. ———, "Fractional iteration of entire and rational functions," *J. Aust. Math. Soc.*, *4* (1964), 129–142.

A Class of Positive-Definite Functions

J. F. C. KINGMAN

Among the many distinguished results associated with the name of Salomon Bochner, one of the more widely known is his characterization of the positive-definite functions, those continuous functions f of a real variable for which any choice of the values $t_1, t_2, \ldots, t_n$ makes

$$(f(t_\alpha - t_\beta);\ \alpha, \beta = 1, 2, \ldots, n) \tag{1}$$

a non-negative-definite Hermitian matrix. These are exactly the functions expressible in the form

$$f(t) = \int e^{i\omega t}\phi(d\omega), \tag{2}$$

for totally finite (positive) measures ϕ on the Borel subsets of the real line. In the case of a real-valued function f (necessarily even), (2) takes the form

$$f(t) = \int \cos \omega t\ \phi(d\omega), \tag{3}$$

and ϕ may then be regarded as concentrated on the interval $[0, \infty)$.

In the theory of probability these functions arise most commonly as the autocovariances of stationary stochastic processes. Thus let $Z = (Z_t; -\infty < t < \infty)$ be a stationary process, with $\mathbf{E}(Z_t^2)$ finite, and let $f(t)$ be the covariance of Z_s and Z_{s+t} (which is necessarily independent of s). Then f is a positive-definite function, and ϕ is the "spectral measure" of the process.

A particular case arises in the theory of Markov chains (for which we adopt the notation and terminology of Chung [1]). Suppose that $X = (X_t; -\infty < t < \infty)$ is a stationary Markov chain on a countable state space S, with

$$\begin{aligned} \pi_i &= \mathbf{P}(X_s = i), \\ p_{ij}(t) &= \mathbf{P}(X_{s+t} = j \mid X_s = i), \end{aligned} \tag{4}$$

for $t > 0$, $i, j \in S$. That is to say, X is a stochastic process such that, for $i_1, i_2, \ldots, i_n \in S$ and $t_1 < t_2 < \cdots < t_n$,

$$\mathbf{P}\{X_{t_\alpha} = i_\alpha\ (\alpha = 1, 2, \ldots, n)\} = \pi_{i_1} \prod_{\alpha=2}^{n} p_{i_{\alpha-1}, i_\alpha}(t_\alpha - t_{\alpha-1}). \tag{5}$$

As is usual, the chain is assumed to be *standard* in the sense that, for each $i \in S$,

$$\lim_{t \to 0} p_{ii}(t) = 1. \tag{6}$$

Let a be a particular state in S (with, to avoid triviality, $\pi_a > 0$), and define a stationary process Z by

$$\begin{aligned} Z_t &= 1 \quad \text{if } X_t = a, \quad \text{and} \\ &= 0 \quad \text{if } X_t \neq a. \end{aligned} \tag{7}$$

The autocovariance function f of Z is clearly given by

$$f(t) = \pi_a\{p_{aa}(t) - \pi_a\}. \tag{8}$$

Substituting this into (3), we obtain

$$p_{aa}(t) = \pi_a + \int \cos \omega t \lambda(d\omega), \tag{9}$$

with $\lambda = \pi_a^{-1}\phi$. Hence the function p_{aa}, which is a typical diagonal element of the transition matrix

$$P_t = (p_{ij}(t); i, j \in S), \tag{10}$$

is positive-definite.

Not every Markov transition matrix P_t can arise from a stationary Markov chain; this will be possible if and only if (for some a),

$$\lim_{t \to \infty} p_{aa}(t) > 0. \tag{11}$$

However, Kendall [6] has shown that conclusion (9) holds quite generally even without this condition and, furthermore, that the measure λ is absolutely continuous. Kendall's argument uses sophisticated Hilbert space techniques, but simpler arguments exist (cf. [15], [10]).

The diagonal elements p_{ii} of Markov chain transition matrices therefore form a subclass (denoted by $\mathscr{PM}$) of the class of real, positive-definite functions, and the problem arises of characterizing $\mathscr{PM}$. In a sense the answer is given by Markov chain theory, for a function p belongs to $\mathscr{PM}$

if and only if there exists a family $(p_{ij};\, i, j = 0, 1, 2, \ldots)$ of functions on $(0, \infty)$, satisfying

$$p_{ij}(t) \geqq 0, \qquad \sum_{j=0}^{\infty} p_{ij}(t) \leqq 1,$$

(12)

$$p_{ij}(s + t) = \sum_{k=0}^{\infty} p_{ik}(s)p_{kj}(t), \quad \text{and} \quad \lim_{t \to 0} p_{ii}(t) = 1,$$

and such that $p = p_{00}$. But this, although it (apparently) removes the problem from the province of the theory of probability, cannot be said to provide more than a "solution in principle."

To proceed further, observe that the function p_{aa} satisfies functional inequalities similar in form, but in addition to those defining the positive-definite functions, which can be written as

(13) $$\det\,(f(|t_\alpha - t_\beta|)) \geqq 0.$$

For example, if $s, t > 0$,

$$\begin{aligned} 0 &\leqq \mathbf{P}(X_s \neq a,\, X_{s+t} = a) \\ &= \mathbf{P}(X_{s+t} = a) - \mathbf{P}(X_s = X_{s+t} = a) \\ &= p_{aa}(s + t) - p_{aa}(s)p_{aa}(t), \end{aligned}$$

and

$$\begin{aligned} 0 &\leqq \mathbf{P}(X_s \neq a,\, X_{s+t} \neq a) \\ &= \mathbf{P}(X_s \neq a) - \mathbf{P}(X_s \neq a,\, X_{s+t} = a) \\ &= 1 - p_{aa}(s) - p_{aa}(s + t) + p_{aa}(s)p_{aa}(t). \end{aligned}$$

Thus every $p \in \mathscr{PM}$ satisfies

(14) $$p(s)p(t) \leqq p(s + t) \leqq p(s)p(t) + 1 - p(s).$$

These inequalities, when combined with the regularity condition

(15) $$\lim_{t \to 0} p(t) = 1,$$

which corresponds to (6), already imply a great deal about the function p. It follows quite easily [see 10] from (14) and (15) that p is strictly positive and uniformly continuous and that the (possibly infinite) limit

(16) $$q = -p'(0) = \lim_{t \to \infty} t^{-1}\{1 - p(t)\}$$

exists. It may be noted that not all functions of the form (3) with $f(0) = 1$ satisfy (14), which is not therefore a consequence of positive-definiteness.

The inequalities (14), which have often been used in Markov chain theory, are only the first of an infinite family of inequalities. For $t_1 < t_2 < \cdots < t_n$, the probability

$$\mathbf{P}\{X_{t_\alpha} \neq a\; (\alpha = 1, 2, \ldots, n - 1),\; X_{t_\alpha} = a\}$$

may be expressed as a polynomial

$$F(t_1, t_2, \ldots, t_n; p_{aa})$$

in the values of p_{aa} at the points $t_\alpha - t_\beta$ $(\alpha > \beta)$, the exact form of which is

$$(17)\quad F(t_1, t_2, \ldots, t_n; p) = \sum_{k=0}^{n-1} (-1)^k \sum_{0=\nu_0<\nu_1<\cdots<\nu_k=n} \prod_{j=0}^{k} p(t_{\nu_{j+1}} - t_{\nu_j}).$$

Moreover,

$$\mathbf{P}\{X_{t_\alpha} \neq a\ (\alpha = 1, 2, \ldots, n)\} = 1 - \sum_{r=1}^{n} F(t_1, t_2, \ldots, t_r; p_{aa})$$

Since these probabilities must be non-negative, it follows that every $p \in \mathscr{PM}$ satisfies the inequalities

$$(18)\qquad F(t_1, t_2, \ldots, t_n; p) \geqq 0, \qquad \sum_{r=1}^{n} F(t_1, t_2, \ldots, t_r; p) \leqq 1.$$

It should be noted that these inequalities are consequences of the fact that there exists a process Z, defined by (7), taking the values 0 and 1, and having finite-dimensional distributions determined by

$$\mathbf{P}\{Z_{t_\alpha} = 1\ (\alpha = 1, 2, \ldots, n) | Z_0 = 1\} = \prod_{\alpha=1}^{n} p_{aa}(t_\alpha - t_{\alpha-1}),$$

for $0 = t_0 < t_1 < \cdots < t_n$. They comprise, moreover, a sufficient as well as a necessary set of conditions, in the sense of the following theorem [see 10].

THEOREM 1. *In order that there should exist a stochastic process* $(Z_t; t > 0)$ *taking the values* 0 *and* 1 *and satisfying*

$$(19)\qquad \mathbf{P}\{Z_{t_\alpha} = 1\ (\alpha = 1, 2, \ldots, n)\} = \prod_{\alpha=1}^{n} p(t_\alpha - t_{\alpha-1})$$

for all $0 = t_0 < t_1 < \cdots < t_n$, *it is necessary and sufficient that the function* p *should satisfy the inequalities* (18).

It therefore becomes important to study the functions satisfying the inequalities (18), which, for want of a better name, are called *p-functions*. If the set of all functions from $(0, \infty)$ into $[0, 1]$ is given its product topology (compact by Tychonov's theorem), then (18) defines a closed subspace, so that the set of p-functions has a natural compact Hausdorff topology.

A p-function satisfying (15) is said to be *standard*, and the set of standard p-functions is denoted by $\mathscr{P}$. It follows from the derivation of (18) that, in any (standard) Markov chain, the function p_{aa} is a standard p-function, so that

$$\mathscr{P}\mathscr{M} \subseteq \mathscr{P}. \tag{20}$$

It will become clear that $\mathscr{P}\mathscr{M}$ is a proper, but dense, subset of $\mathscr{P}$.

For any $p \in \mathscr{P}$ and any $h > 0$, write

$$f_n = f_n(h) = F(h, 2h, \ldots, nh; p), \tag{21}$$

so that, from (18),

$$f_n \geqq 0 \quad (n \geqq 1), \qquad \sum_{n=1}^{\infty} f_n \leqq 1. \tag{22}$$

It follows after a little algebra from (17) that, for $n \geqq 1$,

$$p(nh) = \sum_{r=1}^{n} f_r(h) p(nh - rh), \tag{23}$$

a relation equivalent to the power series identity

$$\sum_{n=0}^{\infty} p(nh)z^n = \left\{1 - \sum_{n=1}^{\infty} f_n(h)z^n\right\}^{-1}, \tag{24}$$

if $|z| < 1$ and $p(0) = 1$.

For any sequence (f_n) satisfying (22), the sequence (u_n) defined recursively by

$$u_0 = 1, \qquad u_n = \sum_{r=1}^{n} f_r u_{n-r} \quad (n \geqq 1), \tag{25}$$

is called the *renewal sequence generated by* (f_n), and the class $\mathscr{R}$ of all renewal sequences is the discrete analogue of $\mathscr{P}$. Indeed, the preceding argument has, almost in its entirety, a discrete-time analogue in Feller's theory of recurrent events (see [4] and [5]) and in its application to discrete-time Markov chains. From that theory we draw one result due to Chung to the effect that, for any renewal sequence (u_n), there exists a Markov chain $(X_n; n = 0, 1, 2, \ldots)$ and a state a such that, for all $n \geqq 0$,

$$u_n = \mathbf{P}(X_n = a | X_0 = a). \tag{26}$$

The corresponding assertion in continuous time is false.

Comparing (23) with (25), we see that any function $p \in \mathscr{P}$ has the property that the sequence $(p(nh))$ belongs to $\mathscr{R}$ for all $h > 0$. Conversely, it is possible to show that a function continuous in $[0, \infty]$ with $(p(nh)) \in \mathscr{R}$ for arbitrarily small $h > 0$ belongs to $\mathscr{P}$. This means that some properties of functions in $\mathscr{P}$, notably those referring to behavior for large t, can be deduced from the extensive theory of renewal sequences. For example, a celebrated result of Erdös, Feller, and Pollard [3] states that, if (u_n) is any renewal sequence for which the set $\{n \geqq 1; u_n > 0\}$ has 1 as its greatest common divisor, then

$$\lim_{n \to \infty} u_n$$

exists. Hence, if $p \in \mathscr{P}$,

$$\lim_{n \to \infty} p(nh)$$

exists for all $h > 0$. Together with the uniform continuity of p, this implies that

$$p(\infty) = \lim_{t \to \infty} p(t) \tag{27}$$

exists. This is a straightforward deduction, but a systematic technique for deriving asymptotic properties by the "method of skeletons" has been elaborated in [9].

If in (24) we set $z = e^{-\theta h}$, multiply by h, and let $h \to 0$ (keeping $\theta > 0$ fixed), the left-hand side converges to the Laplace transform

$$r(\theta) = \int_0^\infty p(t)e^{-\theta t}\,dt \tag{28}$$

of p. The limiting behavior of the right-hand side of (24) may be examined with the aid of Helly's compactness principle, and the result is the following fundamental characterization of $\mathscr{P}$:

THEOREM 2. *If p belongs to $\mathscr{P}$, there exists a unique (positive) measure μ on $(0, \infty]$ such that*

$$\int_{(0,\infty]} (1 - e^{-x})\mu(dx) < \infty \tag{29}$$

and such that, for all $\theta > 0$,

$$\int_0^\infty p(t)e^{-\theta t}\,dt = \left\{\theta + \int_{(0,\infty]} (1 - e^{-\theta x})\mu(dx)\right\}^{-1}. \tag{30}$$

Conversely, if μ is any measure satisfying (29), *there exists exactly one continuous function p satisfying* (30), *and that function belongs to $\mathscr{P}$.*

In other words, (30) sets up a one-to-one correspondence between the class $\mathscr{P}$ and the set of measures satisfying (29). It is important to note that (29) does not imply that μ is totally finite, although it must have $\mu(\epsilon, \infty) < \infty$ for all $\epsilon > 0$. Indeed, an easy Abelian argument from (30) identifies the total mass of μ with the limit q in (16):

$$q = \mu(0, \infty]. \tag{31}$$

Letting $\theta \to 0$ in (30) gives

$$\int_0^\infty p(t)\,dt = [\mu\{\infty\}]^{-1}, \tag{32}$$

so that $\mu\{\infty\} > 0$ implies $p(\infty) = 0$. On the other hand, if $\mu\{\infty\} = 0$, it may be shown that

$$p(\infty) = \left\{1 + \int x\mu(dx)\right\}^{-1}. \tag{33}$$

In case q is finite, (30) may be written

$$\begin{aligned} r(\theta) &= \left\{\theta + q - \int e^{-\theta x}\mu(dx)\right)^{-1} \\ &= \sum_{n=0}^{\infty} (\theta + q)^{-n-1}\left\{\int e^{-\theta x}\mu(dx)\right\}^n \\ &= \sum_{n=0}^{\infty} (\theta + q)^{-n-1} \int e^{-\theta x}\mu_n(dx), \end{aligned}$$

where μ_n is the n-fold Stieltjes convolution of μ with itself. This inverts to give

$$p(t) = \sum_{n=0}^{\infty} \int_0^t \pi_n\{q(t - x)\}q^{-n}\mu_n(dx), \tag{34}$$

where π_n denotes the Poisson probability

$$\pi_n(\alpha) = e^{-\alpha}\alpha^n/n!. \tag{35}$$

This formula may be given a probabilistic interpretation as follows. Let $0 = T_0 < T_1 < T_2 < \cdots$ be random variables whose differences $\tau_n = T_{n+1} - T_n$ are independent, with distributions given by

$$\begin{aligned} \mathbf{P}(\tau_n \leqq x) &= 1 - e^{-qx} \qquad (n \text{ even}), \\ &= q^{-1}\mu(0, x] \qquad (n \text{ odd}), \end{aligned}$$

for $x > 0$. Define a process $(Z_t;\, t > 0)$ to be equal to 1 on the intervals $[T_{2m}, T_{2m+1}]$ and 0 elsewhere. Then routine calculations show that Z (which is a special type of alternating renewal process) has finite-dimensional distributions given by (19), where

$$p(t) = \sum_{m=0}^{\infty} \mathbf{P}(T_{2m} \leqq t \leqq T_{2m+1}). \tag{36}$$

Moreover,

$$\mathbf{P}(T_{2m} \leqq t \leqq T_{2m+1}) = \int_0^t \pi_m\{q(t - x)\}q^{-m}\mu_m(dx),$$

so that (34) and (36) coincide.

The question then arises of giving a similar construction for the case $q = \mu(0, \infty] = \infty$. This is a more complex problem, but the key to its solution lies in the observation that expressions similar to (30) arise in the theory of additive processes (processes with stationary independent increments). Indeed, for any μ satisfying (29), there exists an additive process $(\eta_t; t \geqq 0)$ (nondecreasing, and increasing only in jumps) such that

$$\mathbf{E}(e^{-\theta\eta_t}) = e^{-t\psi(\theta)}, \tag{37}$$

where

$$\psi(\theta) = \int (1 - e^{-\theta x})\mu(dx). \tag{38}$$

Adding a deterministic drift, we obtain a process

$$\xi_t = t + \eta_t,$$

with

$$\mathbf{E}(e^{-\theta\xi_t}) = e^{-t\{\theta + \psi(\theta)\}} = e^{-t/r(\theta)}. \tag{39}$$

It has been observed by Kendall (the proof being easiest if a strong Markov version of η is used) that the process defined by

$$\begin{aligned} Z_t &= 1 \quad \text{if } \xi_s = t \text{ for some } s, \\ &= 0 \quad \text{otherwise,} \end{aligned} \tag{40}$$

satisfies (19), where p is the function corresponding to μ in (30).

Our starting point was the positive-definiteness of functions in $\mathscr{PM}$, and it is therefore pertinent to remark that, as implied by the title of this article, the functions in $\mathscr{P}$ are also positive-definite. Indeed, it is proved in [10] that, if $p \in \mathscr{P}$, there exists a non-negative integrable function f such that

$$p(t) = p(\infty) + \int_0^\infty f(\omega) \cos \omega t \, d\omega. \tag{41}$$

No very simple proof of this result appears to be known. Although $\mathscr{P}$ is therefore a subset of the class of real positive-definite functions, it differs from that class in failing to be convex. This fact is important as the source of many of the difficulties encountered by the explorer of $\mathscr{P}$.

It will perhaps help to consider a few examples of functions in $\mathscr{P}$. Suppose for instance that the measure μ is concentrated in an atom of mass q at a single point a. Then (34) shows that

$$p(t) = \sum_{n=0}^{[t/a]} \pi_n\{q(t - na)\}. \tag{42}$$

This is an oscillating function, converging to the limit

$$p(\infty) = (1 + qa)^{-1}.$$

It is differentiable except at the point a, where it has left- and right-derivatives:

$$D_- p(a) = -qe^{-qa}, \qquad D_+ p(a) = q - qe^{-qa}.$$

According to a famous theorem of Austin and Ornstein (see [1], Section II.12) the functions p_{ij} in a Markov chain are all continuously differentiable in $(0, \infty)$. It therefore follows that every function in $\mathscr{PM}$ is so differentiable, and therefore that (42) cannot define a member of $\mathscr{PM}$. Thus $\mathscr{PM}$ is a *proper* subset of $\mathscr{P}$.

The local behavior of the particular p-function (42) is quite typical. For any μ satisfying (29) the function

$$m(t) = \mu(t, \infty] \tag{43}$$

is nonincreasing, right-continuous, and integrable over $(0, T)$ for any finite T. Equation (30) may be thrown into the form

$$r(\theta) = \theta^{-1}\left\{1 + \int_0^\infty m(t)e^{-\theta t}\, dt\right\}^{-1}. \tag{44}$$

Expanding this formally, we obtain

$$p(t) = 1 - \int_0^t b(s)\, ds, \tag{45}$$

where

$$b(s) = \sum_{n=1}^\infty (-1)^{n-1} m_n(s), \tag{46}$$

and m_n is the n-fold convolution of m with itself. It is shown in [12] that this formal expression is always valid, that the series (46) is uniformly absolutely convergent in every compact subinterval of $(0, \infty)$, and that m_n $(n \geqq 2)$ is continuous. Since m is continuous except for jump discontinuities at the atoms of μ, it follows that every function in $\mathscr{P}$ has left- and right-derivatives in $(0, \infty)$, and that

$$D_+ p(t) - D_- p(t) = \mu\{t\} \tag{47}$$

for all $t > 0$. In particular, p is continuously differentiable in any interval free from atoms of μ.

Reversing the logic, the Austin–Ornstein theorem on the continuous differentiability of members of $\mathscr{PM}$ is seen to be equivalent to the assertion that, if $p \in \mathscr{PM}$, the corresponding measure μ is nonatomic in $(0, \infty)$. In fact more is true, for a result of Lévy (see [13], I) implies that μ is absolutely continuous (except perhaps for an atom at ∞). We shall return later to the problem of describing the measures μ which correspond to members of $\mathscr{PM}$.

For another kind of example, consider any renewal sequence (u_n). By the theorem of Chung quoted previously, there exists a discrete-time Markov chain satisfying (26). Denote its transition matrix by P. If c is any positive number, and I denotes the identity matrix, then

$$Q = c(P - I) \tag{48}$$

defines the infinitesimal generator of a q-bounded Markov chain, with transition matrix

$$\begin{aligned} P_t &= \exp\{c(P - I)t\} \\ &= e^{-ct} \exp(ctP) \\ &= e^{-ct} \sum_{n=0}^{\infty} \frac{(ct)^n}{n!} P^n \\ &= \sum_{n=0}^{\infty} \pi_n(ct) P^n, \end{aligned}$$

where π_n is given by (35). Hence, from (26)

$$p_{aa}(t) = \sum_{n=0}^{\infty} \pi_n(ct) u_n.$$

Thus, for any $(u_n) \in \mathscr{R}$ and any $c > 0$, the function

$$p(t) = \sum_{n=0}^{\infty} u_n \pi_n(ct) \tag{49}$$

belongs to $\mathscr{PM}$, and so also to $\mathscr{P}$.

Not every $p \in \mathscr{PM}$ is expressible in the form of (49), since (49) implies that, for instance,

$$p'(0) = -c(1 - u_1) > -\infty.$$

If, therefore, $\mathscr{Q}$ denotes the class of functions (49) with $(u_n) \in \mathscr{R}$ and $c > 0$, we have the strict inclusions

$$\mathscr{Q} \subset \mathscr{PM} \subset \mathscr{P}. \tag{50}$$

For any $p \in \mathscr{P}$, and any integer k, the sequence $(p(nk^{-1});\ n = 0, 1, 2, \ldots)$ belongs to $\mathscr{R}$, and hence the function

$$p_k(t) = \sum_{n=0}^{\infty} p(nk^{-1}) \pi_n(kt) \tag{51}$$

belongs to $\mathscr{P}$. If $\zeta_1, \zeta_2, \ldots$ are independent Poisson variables with mean t, then

$$p_k(t) = \mathbf{E}\{p[k^{-1}(\zeta_1 + \zeta_2 + \cdots + \zeta_k)]\},$$

and the weak law of large numbers therefore implies that

$$\lim_{k \to \infty} p_k(t) = p(t) \tag{52}$$

for all $t \geqq 0$. Thus $\mathscr{Q}$ is dense in $\mathscr{P}$ (in the product topology), and so *a fortiori* is $\mathscr{PM}$.

It is the fact that $\mathscr{PM}$ is not closed in $\mathscr{P}$ that makes its identification such a delicate problem. It is of course possible that interesting topologies for $\mathscr{P}$ exist in which $\mathscr{PM}$ is closed, but such topologies would need to be strong enough to prevent p_k converging to p, for all $p \in \mathscr{P} - \mathscr{PM}$.

If p_1 and p_2 are two p-functions, construct as in Theorem 1 processes Z_1, Z_2 on distinct probability spaces Ω_1, Ω_2 to satisfy (19) with $p = p_1$ and $p = p_2$, respectively. On the product space $\Omega = \Omega_1 \times \Omega_2$ define a new process Z by

$$Z_t(\omega_1, \omega_2) = Z_{1t}(\omega_1) Z_{2t}(\omega_2). \tag{53}$$

Using the product measure on Ω, it is immediately apparent that Z satisfies (19), with

$$p(t) = p_1(t) p_2(t). \tag{54}$$

Thus the product of two p-functions is itself a p-function, and consequently the product of two members of $\mathscr{P}$ is in $\mathscr{P}$. It follows that, under the operation of pointwise multiplication, $\mathscr{P}$ is a commutative Hausdorff topological semigroup with identity. The arithmetical properties of this semigroup have been studied by Kendall [8] and Davidson [2], who have shown that it exhibits many of the properties classically associated with the convolution semigroup of probability distributions on the line. The latter is of course isomorphic to the semigroup of positive-definite functions (under pointwise multiplication) of which $\mathscr{P}$ is a subsemigroup.

The fact that $\mathscr{P}$ is closed under pointwise multiplication permits the construction of further examples of p-functions. A particularly important use of this device is due to Kendall [7]. Let λ, a be positive numbers, and construct a Poisson process Π of rate λ on $(0, \infty)$. Define Z_t to be equal to 1 if no points of Π lie in the interval $(t - a, t)$, and 0 otherwise. Then, for $0 = t_0 < t_1 < t_2 < \cdots < t_n$,

$$\begin{aligned}\mathbf{P}\{Z_{t_\alpha} = 1\ (\alpha = 1, 2, \ldots, n)\} &= \mathbf{P}\left\{\text{no points of } \Pi \text{ in } \bigcup_{\alpha=1}^{n} (t_\alpha - a, t_\alpha)\right\} \\ &= e^{-\lambda L},\end{aligned}$$

where L is the measure of the set

$$\bigcup_{\alpha=1}^{n} (t_\alpha - a, t_\alpha) \cap (0, \infty).$$

It is not difficult to check that

$$L = \sum_{\alpha=1}^{n} \min (t_\alpha - t_{\alpha-1}, a),$$

so that Z satisfies (19) with

$$p(t) = \exp\{-\lambda \min (t, a)\}. \tag{55}$$

Thus the function (55) belongs to $\mathscr{P}$ for all $\lambda, a > 0$ (though not to $\mathscr{PM}$ because it is not differentiable at $t = a$). Since $\mathscr{P}$ is closed under multiplication, it contains every function of the form

$$p(t) = \exp\left\{-\sum_{i=1}^{m} \lambda_i \min (t, a_i)\right\}$$

with $\lambda_i, a_i > 0$. The compactness of the space of p-functions then implies that, if λ is any measure on $(0, \infty]$ with

$$\int \min (1, x)\lambda(dx) < \infty,$$

then

$$p(t) = \exp\left\{-\int \min (t, x)\lambda(dx)\right\} \tag{56}$$

defines an element of $\mathscr{P}$. But the functions of the form

$$\phi(t) = \int \min (t, x)\lambda(dx)$$

are exactly the continuous, non-negative, concave functions on $[0, \infty)$ with $\phi(0) = 0$. We have therefore proved that, for all such ϕ, the function

$$p(t) = e^{-\phi(t)} \tag{57}$$

belongs to $\mathscr{P}$. Notice that the set of functions of this form is a convex subset of the nonconvex set $\mathscr{P}$.

The p-functions (57) play a central role in Kendall's theory, for they are the infinitely divisible elements of $\mathscr{P}$, those possessing nth roots for all n. That they have this property is obvious; to see that they are the only ones let f be any function with the property that $f^\alpha \in \mathscr{P}$ for arbitrarily small values of α. Then $f > 0$ and we may define $\phi = -\log f$. From (18) with $n = 3$,

$$\begin{aligned} 0 &\leqq F(t_1, t_2, t_3; f^\alpha) \\ &= f^\alpha(t_3) - f^\alpha(t_1)f^\alpha(t_3 - t_1) - f^\alpha(t_2)f^\alpha(t_3 - t_2) + f^\alpha(t_1)f^\alpha(t_2 - t_1)f^\alpha(t_3 - t_2) \\ &= \alpha\{-\phi(t_3) + \phi(t_3 - t_1) + \phi(t_2) - \phi(t_2 - t_1)\} + O(\alpha^2). \end{aligned}$$

Since this holds for arbitrarily small a, we must have

$$\phi(t_3) - \phi(t_3 - t_1) \leqq \phi(t_2) - \phi(t_2 - t_1),$$

which implies that ϕ is concave.

Among the functions of the form (57) are the completely monotonic functions, which can be written in the form

$$p(t) = \int e^{-ty}\nu(dy) \tag{58}$$

for probability measures ν on $[0, \infty)$. It is shown in [13, II] that these belong not merely to $\mathscr{P}$ but also to $\mathscr{PM}$; they are indeed exactly the functions p_{aa} which can arise from *reversible* Markov chains.

Thus far the only p-functions discussed have been standard, but there are interesting examples of p-functions which fail to satisfy (15). For example, if the variables Z_t are independent, with $\mathbf{P}(Z_t = 1) = b$, then (19) is satisfied with

$$p(t) = b \qquad (t > 0). \tag{59}$$

Hence (59) defines a p-function whenever $0 \leqq b \leqq 1$, and this is not standard unless $b = 1$. Since the product of p-functions is a p-function, the function

$$p(t) = b\bar{p}(t) \tag{60}$$

is a nonstandard p-function for $0 \leqq b < 1$ and $\bar{p} \in \mathscr{P}$.

These nonstandard p-functions are continuous except at the origin, but there are more irregular ones. Suppose for instance that G is any proper additive subgroup of the line, and define Z_t to be 1 with probability one if $t \in G$, and 0 otherwise. Then (19) is satisfied with p the indicator function of $G \cap [0, \infty)$. If G is measurable, it necessarily has measure zero, so that $p = 0$ almost everywhere.

In a sense these are the only two possibilities for measurable p-functions. More precisely, the following result is proved in [14].

THEOREM 3. *Let p be any p-function. Then exactly one of the following statements is true:*

(i) *p is standard,*

(ii) *there is a standard p-function* $\bar{p}$ *and a constant* $0 < b < 1$ *such that* $p(t) = b\bar{p}(t)$,

(iii) p *is almost everywhere zero,*

(iv) p *is not Lebesgue measurable.*

It is also worth remarking that, in case (ii), as, more obviously, in case (iv), the process Z in Theorem 1 cannot be chosen to be measurable.

It has already been noted that $\mathscr{P}$ is not closed as a subset of the compact space of p-functions, and it becomes relevant to identify the closure $\overline{\mathscr{P}}$ of $\mathscr{P}$. Since

$$b = \lim_{n \to \infty} p_n(t),$$

where

$$p_n(t) = \exp\{-n \min(t, -n^{-1} \log b)\}$$

is of the form (55), the constant p-function (59) belongs to $\overline{\mathscr{P}}$, as does the general p-function

$$b\bar{p}(t) = \lim_{n \to \infty} p_n(t)\bar{p}(t)$$

of type (ii). The important question is therefore which p-functions of type (iii) belong to $\overline{\mathscr{P}}$.

If in (42) we set $a = 1 - q^{-1}\alpha$ and let $q \to \infty$, we obtain the p-function

$$\begin{aligned} p(t) &= \pi_t(\alpha t) && (t \text{ integral}) \\ &= 0 && (\text{otherwise}), \end{aligned} \tag{61}$$

which is thus a p-function of type (iii) belonging to $\overline{\mathscr{P}}$.

On the other hand, not every p-function belongs to $\mathscr{P}$. To see this, it suffices to note the inequality, discovered independently by Freedman and Davidson, which states that for every $p \in \mathscr{P}$ and $s < t$,

$$p(s) \geqq \tfrac{1}{2} + \{p(t) - \tfrac{3}{4}\}^{1/2}, \tag{62}$$

so long as $p(t) > \frac{3}{4}$. This must therefore hold also for all $p \in \overline{\mathscr{P}}$. In particular, any $p \in \overline{\mathscr{P}}$ of type (iii) must satisfy

$$p(t) \leqq \tfrac{3}{4}, \tag{63}$$

for all $t > 0$. This is certainly not true of all p-functions of type (iii).

It seems very likely that the upper bound $\frac{3}{4}$ in (63) can be improved considerably. The maximum value of (61) is e^{-1} (when $\alpha = t = 1$), and all the evidence suggests that this is the sharp upper bound to $p(t)$ for $p \in \overline{\mathscr{P}}$ of type (iii). It will be seen, therefore, that the identification of $\overline{\mathscr{P}}$ is closely involved with establishment of inequalities between the values of p, true for all $p \in \mathscr{P}$. These problems are extremely difficult, mainly because of the lack of convexity of $\mathscr{P}$; the exact form of the set

(64) $$\{(p(1), p(2)); p \in \mathscr{P}\} \subseteq [0, 1] \times [0, 1]$$

is not even known.

The most important unsolved problem of the theory remains, however, the determination of the class $\mathscr{PM}$ of those standard p-functions which can arise from Markov chains. In the one-to-one correspondence (30), $\mathscr{PM}$ (as a subset of $\mathscr{P}$) corresponds to a subset $\mathscr{M}'$ of the class of measures satisfying (29). The elements of $\mathscr{M}'$ are measures on $(0, \infty]$, but it is easy to show that the value of the atom at infinity is irrelevant to the question of membership of $\mathscr{M}'$. Thus attention may be restricted to those μ with $\mu\{\infty\} = 0$ (that is, to recurrent states); the corresponding subset of $\mathscr{M}'$ is denoted by $\mathscr{M}$.

There are of course many ways of combining Markov chains to form new ones, and these may be used to establish structural properties of the class $\mathscr{M}$. Thus it is shown in [13, I] that $\mathscr{M}$ is a convex cone, closed under countable sums, where these are consistent with (29). There are of course measures in $\mathscr{M}$ having infinite total mass, but these cause no real difficulty, since it can be shown (see [13, IV]) that every measure in $\mathscr{M}$ is the sum of a countable number of totally finite measures in $\mathscr{M}$. The convolution of a finite (or even infinite, under suitable convergence conditions) number of finite members of $\mathscr{M}$ is again in $\mathscr{M}$. At the same time it must be stressed that $\mathscr{M}$ is not closed in any obvious topology, since $\mathscr{PM}$ is dense in $\mathscr{P}$.

Using these properties of $\mathscr{M}$, we can at once exhibit large classes of elements of $\mathscr{M}$. It is, for instance, immediately apparent that $\mathscr{M}$ contains any measure with density of the form

(65) $$\sum_{m=0}^{\infty} \sum_{n=1}^{\infty} a_{mn} t^m e^{-nt}$$

with $a_{mn} \geqq 0$, so long as (29) is satisfied.

On the other hand, we have already seen that measures in $\mathscr{M}$ must be absolutely continuous (with respect to Lebesgue measure). Moreover, it is a consequence of another result of Austin and Ornstein ([1], p. 149) that the density of any nonzero measure in $\mathscr{M}$ must be everywhere strictly positive, and this density may also be shown to be lower-semicontinuous. These facts go a long way toward determining the class $\mathscr{M}$, and thus $\mathscr{PM}$, but a complete solution remains elusive.

The class $\mathscr{PM}$ consists, of course, of all diagonal Markov transition functions p_{ii}, and it may be asked whether corresponding results are available for the nondiagonal functions p_{ij} $(i \neq j)$. To describe the theory which has been developed to answer this and related questions, which depends on the concept of a *quasi-Markov chain* (see [11], [13]), would take us too far afield, but the answer may be stated thus. The nondiagonal Markov transition functions p_{ij} are exactly those which may be expressed as convolutions

$$p_1 * d\lambda * p_2, \tag{66}$$

where p_1 and p_2 belong to $\mathscr{P}\mathscr{M}$,

$$a = \int_0^\infty p_1(t)\, dt < \infty,$$

and λ is a measure on $[0, \infty)$ with

$$a\lambda[0, \infty) \leqq 1,$$

which, apart from a possible atom at the origin, belongs to $\mathscr{M}$. The whole characterization problem for Markov transition probabilities therefore depends on the cone $\mathscr{M}$.

NOTE ADDED IN PROOF (May 1970): The problem of characterizing $\mathscr{P}\mathscr{M}$ is now solved, and simple necessary and sufficient conditions are known for a measure to belong to $\mathscr{M}$. An account will appear as number V in the series [13].

UNIVERSITY OF SUSSEX

REFERENCES

1. CHUNG, K. L., *Markov Chains with Stationary Transition Probabilities*. Berlin: Springer-Verlag, 1967.
2. DAVIDSON, R., "Arithmetic and other properties of certain Delphic semigroups," *Z. fur Wahrscheinlichkeitstheorie und verwandte Gebiete*, *10* (1968), 120–172.
3. ERDÖS, P., W. FELLER, and H. POLLARD, "A theorem on power series," *Bull. Amer. Math. Soc.*, *55* (1949), 201–204.
4. FELLER, W., "Fluctuation theory of recurrent events," *Trans. Amer. Math. Soc.*, *67* (1949), 98–119.
5. ———, *An Introduction to Probability Theory and its Applications*. New York: John Wiley & Sons, Inc., 1957.
6. KENDALL, D. G., "Unitary dilations of one-parameter semigroups of Markov transition operators, and the corresponding integral representations for Markov processes with a countable infinity of states," *Proc. London Math. Soc.*, *9* (1959), 417–431.
7. ———, "Renewal sequences and their arithmetic," *Symposium on Probability Methods in Analysis*. Berlin: Springer-Verlag, 1967, pp. 147–175.
8. ———, "Delphic semigroups, infinitely divisible regenerative phenomena, and the arithmetic of *p*-functions," *Z. fur Wahrscheinlichkeitstheorie und verwandte Gebiete*, *9* (1968), 163–195.
9. KINGMAN, J. F. C., "Ergodic properties of continuous-time Markov processes and their discrete skeletons," *Proc. London Math. Soc.*, *13* (1963), 593–604.
10. ———, "The stochastic theory of regenerative events," *Z. fur Wahrscheinlichkeitstheorie und verwandte Gebiete*, *2* (1964), 180–224.
11. ———, "Linked systems of regenerative events," *Proc. London Math. Soc.*, *15* (1965), 125–150.

12. KINGMAN, J. F. C. "Some further analytical results in the theory of regenerative events," *J. Math. Anal. Appl.*, *11* (1965), 422–433.
13. ———, "Markov transition probabilities," *Z. fur Wahrscheinlichkeitstheorie und verwandte Gebiete*, I: *7* (1967), 248–270; II: *9* (1967), 1–9; III: *10* (1968), 87–101; IV: *11* (1968), 9–17.
14. ———, "On measurable *p*-functions," *Z. fur Wahrscheinlichkeitstheorie und verwandte Gebiete*, *11* (1968), 1–8.
15. LOYNES, R. M., "On certain applications of the spectral representation of stationary processes," *Z. fur Wahrscheinlichkeitstheorie und verwandte Gebiete*, *5* (1966), 180–186.

Local Noncommutative Analysis

IRVING SEGAL

1. Introduction

Despite the inroads of linear functional analysis, much of analysis still deals ultimately not with vectors in a linear space, but with functions defined on a suitably structured point set, generally endowed with properties reminiscent of physical space. It seems as if some notion of physical space is quite possibly a primordial concept in the human mind and inevitably colors all our perceptions of nature and formulations of natural law. The mathematical emphasis on equations which are *local* in geometrical space (or transforms of such equations) and the complementary physical idea that the primary forces in nature are exerted through local interactions (i.e., by a kind of physical contact in the underlying space) may similarly reflect the apparent lack of firm roots in human perception of any other equally broadly applicable type of model. In any event, the fact is that mathematicians have always concerned themselves largely with local processes, structures, and equations, a notable instance being the theory of partial differential equations; and that physically, locality of the interaction is probably the most general nontrivial guiding principle which has as yet been found effective.

A classical situation in which ideas of locality have been quite successfully merged with ideas of causality, relativistic invariance, and other such foundational theoretical physical desiderata, is the theory of hyperbolic partial differential equations. In this connection, as well as quite generally in connection with partial differential equations, one deals with an entirely explicit and well developed notion of a local function of a given function. Specifically, a "local" function of a given class of functions f is a mapping from this class to another class, which has the form: $f \to g$, where $g(x) = \varphi(f(x))$ for all x, and φ is a given function defined on the range of values of f. This notion of locality may be characterized intrinsically in various ways, depending on the mathematical contexts of the functions involved. For example, within the context of abstract measure theory, g is a local function of f if and only if it is measurable with respect to every sigma-ring with respect to which f is measurable. Within a linear C^∞ framework, a sheaf-theoretical characterization has been given by Peetre.

It seems logically natural and is scientifically relevant to extend the treatment of local functions, and the differential equations involving them, to the case of functions whose values are not necessarily numerical but may lie in a noncommutative algebra. This extension must be distinguished from the familiar and important extensions to the cases of functions having values in either a given linear space or in a given vector bundle over the space. A natural type of algebra in the present connection, by virtue of its simplicity, technical effectiveness, and physical relevance in the quantum theory of radiation, is an algebra of operators in a Hilbert space. When the algebra is Abelian, one is more or less back in the familiar vector-valued function case. When the algebra is not commutative, and there is no a priori information concerning the commutators of the basic operators, one is in a very general situation. With relatively strong regularity assumptions, such as compactness of the underlying space, boundedness of the operators involved, and uniform continuity of the mapping from the space to the operators, one has a framework explored by Atiyah and shown by him to clarify greatly the Grothendieck theory of vector bundles, from an analytical point of view. Within this context, it would be straightforward to extend the usual treatment of local functions, indeed in such a way that the idea in the theory of hyperbolic equations that the intervention of local lower-order terms does not essentially enlarge the region of influence of the basic equation carries over. However, even considerably weaker regularity assumptions are too strong to appear compatible with either of two additional desiderata which are mathematically interesting and physically relevant: (i) invariance under an open simple Lie group, such as the Lorentz group; (ii) simple nontrivial assumptions on the form of commutators, such as the assumption that $[f(x), f(x')]$ should lie in the center of the algebra, and not be identically zero, for any two points x and x'. I shall treat here aspects of a theory which is compatible with, and indeed makes essential use of such desiderata, for two reasons: (a) mathematically interesting, indeed quite unexpected, results emerge; (b) three centuries of the development of the mathematical theory of light can be summarized in the statement (now relatively very well established) that light is described in a categorical way by Maxwell's equations for operator-valued functions satisfying these desiderata.

More specifically, I shall first treat the question of the meaning and existence of nonlinear local functions of the generalized and relatively singular functions which are the only ones satisfying the indicated desiderata. So-called "white noise" is an example of the type of singular function whose powers have no a priori meaning, but arise naturally in a nonlinear local theory compatible with (i) and (ii); it is more singular than heretofore treated functions, but its suitably *renormalized* powers will be shown to exist in a definite mathematical sense, and indeed to exist as a strict function at individual points of the space in question; and these powers remain local functions, despite the formally infinite renormalizations involved. Second, an initial step in the "solution" (again in a sense which is radically generalized, but basically conceptually simple) of differential equations involving such renormalized local products will be treated in a somewhat more regular case than that of white noise. This case is nevertheless interesting in that it involves singularities which are comparable to apparently crucial ones of "quantum electrodynamics," as suitably mathematically formulated.

There is little that is distinctive here about white noise, and logically the present considerations should be extended to generalized stochastic processes from the standpoint pioneered by Bochner [1] and Wiener, and developed by their students, among whom the work of Cameron and Martin has been especially relevant. Historically, the cited work was the chief mathematical source for analysis in function space, but the conception of quantum field theory initiated by Dirac, Heisenberg, Pauli, and others was scientifically stimulating if not mathematically entirely cogent. Local noncommutative analysis should help to round out and connect these rather distinct developments.

2. Nonlinear functions of white noise

A. The intuitive concept of "white noise" may be formulated in a mathematically general way as follows. Let M be an abstract Lebesgue measure space: $M = (R, \mathbf{R}, r)$, where R is a given set, $\mathbf{R}$ is a given sigma-ring of subsets of M, and r is a countably additive measure on $\mathbf{R}$. The *white noise* on M is the isonormal process over the space $\mathbf{H} = RL_2(M)$ of all real square-integrable functions on M, where this process is defined as follows. If $\mathbf{H}$ is a given real Hilbert space, a *concrete isonormal process* over $\mathbf{H}$ is a linear mapping, say ϕ, from $\mathbf{H}$ into the random variables (i.e., measurable functions modulo null functions) on a probability measure space (i.e., one of total measure one), having the property that if $x_1, \ldots, x_n$ is an arbitrary finite set of mutually orthogonal vectors in $\mathbf{H}$, then the random variables $\phi(x_1), \ldots, \phi(x_n)$ are mutually independent. This is equivalent to the property that for every finite set of vectors $y_1, \ldots, y_m$ in $\mathbf{H}$, the random variables $\phi(y_1). \ldots, \phi(y_m)$ have a joint normal distribution, and that $E(\phi(x)) = 0$, $E(\phi(x)\phi(y)) = c\langle x, y\rangle$ for arbitrary x and y in $\mathbf{H}$, where c is a constant. It will be no essential loss of generality for present purposes to assume that $c = 1$.

It may be suggestive, although entirely unnecessary mathematically, to employ the symbolism

$$\phi(x) \sim \int \varphi(a)x(a)\, da, \qquad x \in L_2(M),$$

where the purely symbolic function φ corresponds to the intuitive white noise, i.e., the $\varphi(a)$ are for different a, mutually independent, identically distributed, normal random variables, of mean 0, and of variance of such "infinity" that the L_2-norm of $\phi(x)$ in L_2 over the probability space is the same as the L_2-norm of x in $L_2(M)$. It should perhaps be emphasized that, a priori, neither $\varphi(\cdot)$ nor $\varphi(x)$ exists in a strict mathematical sense; $\varphi(\cdot)$ is a "generalized random function" or "stochastic distribution." Nevertheless, as is well known, formal linear operations on the $\varphi(\cdot)$, including differential and integral operators, can be given mathematically effective formulation and treatment by their transformation into suitable mathematical operations on the unexceptionable object $\phi(\cdot)$.

These methods of treatment of linear aspects of "weak" functions break down when it is attempted to apply them to nonlinear considerations. Indeed, one might be tempted to believe that no effective notion of nonlinear function of a white noise can be given, since the apparently simpler concept of a nonlinear function of a Schwartzian distribution has proved incapable of effective nontrivial development. The fact is, however, that there is an additional structure which is frequently present in the case of weak stochastic or operational functions, which is lacking in the familiar case of a numerical function, on which such a concept may be based, and indeed, the function and its powers may be represented by mathematical, i.e., nonsymbolic, strict (i.e., nongeneralized) functions on M, in these cases.

B. The linear theory of weak processes can to a large extent be pursued independently of the precise space M on which the process is a generalized function. Thus the white noise, as defined earlier, is primarily a pure Hilbert space construct, and its basic theory can proceed quite independently of whether this space is a function space, as in Wiener's theory of the "homogeneous chaos," and if so, of how it is represented. The nonlinear theory, however, depends in an essential way on the multiplicative structure in the algebra of numerical functions on M. One could deal with this in part by introducing suitable topological linear algebras, but it is simpler, particularly in connection with the determination of an actual function on M representing the processes in question, to give a concrete treatment for functions on the given space M, endowed with appropriate additional structure to be indicated later.

The problem is that of legitimizing in a mathematically effective and formally valid way expressions of the form $F(\varphi(a))$, where $\varphi(a)$ is the previously indicated symbolic white noise, and F is a given function of a real variable. It is too well known that this cannot be done in line with the usual mathematical ideas to require elaboration here; even the simplest nontrivial case $F(\lambda) = \lambda^2$ is totally elusive, for the square of the white noise on a measure space without atoms appears as an infinite constant on the space. I shall show that if one gives up for the moment the literal interpretation of the application of the nonlinear function F to a weak function as the limit of its applications to approximating strict functions, and insists only that the *resulting generalized function* $\psi(a) \sim F(\varphi(a))$ *should be a "local" function of* $\varphi(a)$, *in an intrinsically characterized sense, whose transformation properties under vector displacements in function space are in formal agreement with those of the intuitive function* $F(\varphi(a))$, then one obtains essentially unique results in the crucial cases of the powers, $F(\lambda) = \lambda^p$. These powers behave in many ways as do conventional powers, and are the limits of *renormalized* powers of strict functions, in a way which clarifies their nature and illustrates the natural occurrence of additive "divergences," or apparent infinities.

My method is a development from one used earlier in connection with quantum fields and Brownian motion. In terms of the conventional mathematical representation $x(t)$ of the latter, the situation may be described as follows. It was shown by Wiener that a suitable version of $x(t)$ has the property of possessing fractional derivatives $x^{(e)}(t)$ of all orders $e < \frac{1}{2}$; for orders $e \geqq \frac{1}{2}$, the fractional derivative exists only in a generalized sense, as a stochastic distribution for example; for $e = 1$, the derivative is simply white noise on R^1. Thus there is no problem in the definition of a power of $x^{(e)}(t)$ if $e < \frac{1}{2}$; when $e = \frac{1}{2}$, a literal interpretation leads to the result that $x^{(1/2)}(t)^2 = \infty$ with probability one; but with a suitable "renormalization," the powers of $x^{(1/2)}(t)$ are well-defined, finite, nontrivial stochastic distributions. For $\frac{1}{2} < e < 1$, the powers cannot exist in the same sense as in the case $e = \frac{1}{2}$ (see [4]). However, they exist essentially as generalized functions ("distributions") on the probability space. This conclusion does not appear valid in the still more singular case $e = 1$. It is shown here, by a method applicable to a variety of processes similar to white noise, that with a suitable increase in the regularity of the test functions employed, a similar conclusion holds. Due, however, to the failure of the Soboleff inequalities in an infinite-dimensional space, it is convenient to emphasize the use of Hilbert space methods, and to treat the generalized functions as sesquilinear forms on a dense domain of regular vectors in $L_2(\Omega)$, where Ω

denotes the probability space in question, rather than as linear functionals. The formal correspondence is this: if f is a function, the associated sesquilinear form F is given by the equation: $F(g, h) = \int fg\bar{h}$.

The case $e = \frac{1}{2}$ has essentially the same singularity with respect to the formation of renormalized powers as a so-called scalar relativistic quantum process in a two-dimensional space time, *as a process in space at a fixed time*. The cases $\frac{1}{2} < e < 1$ similarly involve singularities quite comparable to those of n-dimensional such processes for $n > 2$. White noise is thus more singular than any such quantum process, as a process in space at a fixed time.

C. A natural mathematical category in which to treat renormalized powers of generalized stochastic processes is that of the ergodically quasi-invariant (EQI) processes. A linear map ϕ from a real (topological) linear vector space **L** to the random variables on a probability measure space, say Ω, will here be called a *process* (it is also known as a "weak distribution," or as a "generalized random process," in the dual space to **L**); more precisely, it is a *version* of a process, the process itself being the class of all equivalent versions, relative to the usual equivalence relation (see [5]) of agreement of probabilities defined by finite sets of conditions. It is no essential loss of generality to suppose that the random variables $\phi(\mathbf{L})$ form a separating collection for the space Ω, i.e., the minimal sigma-ring with respect to which they are all measurable is, modulo the ideal of all null sets, the full sigma-ring of Ω; and this supposition will be made throughout.

Such a process is called *quasi-invariant* if any of the following equivalent conditions holds:

(i) For every $y \in \mathbf{L}^*$ (the dual space), there exists an invertible measurable transformation on Ω, carrying null sets into null sets, and such that $\phi(x) \to \phi(x) + \langle x, y \rangle$ (for suitably chosen Ω; otherwise the measurable transformation need not exist as a point transformation on Ω, but only as an automorphism of the Boolean ring of measurable sets modulo the ideal of null sets).

(ii) For every $y \in \mathbf{L}^*$, there exists a unitary operator U on the Hilbert space $\mathbf{K} = L_2(\Omega)$, such that $U\Psi(x)U^{-1} = \Psi(x) + \langle x, y \rangle I$, where $\Psi(x)$ denotes the operator in **K** consisting of multiplication by $\phi(x)$ (in general, an unbounded but always normal operator), I being the identity operator on **K**.

(iii) There exists an automorphism of the algebra of measurable functions on Ω, modulo the ideal of null functions, which carries (the equivalence class of) $\phi(x)$ into (that of) $\phi(x) + \langle x, y \rangle$.

These are various ways of asserting the "absolute continuity" of the transformations $z \to z + y$ in the dual space $\mathbf{L}^*$ relative to the weak probability measure in **L*** corresponding to ϕ; when the process arises from a conventional countably additive probability measure in **L*** in the natural fashion they are equivalent to absolute continuity in the conventional sense; it was in this form that absolute continuity was first shown in function space, in the work of Cameron and Martin [2]. More specifically, this work shows that the transformation $g \to g + f$ is absolutely continuous in $C[0, 1]$ with respect to Wiener measure, provided f satisfies a certain regularity condition. This result is closely related to (and follows from) the absolute continuity of arbitrary vector displacements in Hilbert space, relative to the isonormal process. When **L** is finite-dimensional, every process corresponds to a probability measure on **L***, which is quasi-invariant if and only if it is absolutely continuous with nonvanishing density function.

Ergodicity is similar generalization of the ergodic property of vector displacements in R^n, i.e., the nonexistence of invariant measurable sets other than null sets or their complements. More specifically, an EQI process is a quasi-invariant one such that only the constants are invariant under all the automorphisms of the algebra of random variables induced from vector displacements in **L***. Again, the ergodicity of Wiener space relative to vector translations was shown by Cameron and Martin; and this is closely related to (and deducible from) the ergodicity of the isonormal process in a Hilbert space relative to vector displacements in the space.

Suppose then that ϕ is an ergodic quasi-invariant process on a space **L** of functions $x(\cdot)$, so that symbolically $\phi(x) \sim \int \varphi(a)x(a)\,da$, and consider in an intuitive preliminary way the problem of treating the "square" $\phi^{(2)}$ such that

$$\phi^{(2)}(x) \sim \int (\varphi(a))^2 x(a)\,da.$$

Mathematically, the indicated integral is in general entirely nebulous, but in a *formal* way it is immediately visible that it has the following properties:

(i) Multiplication by $\phi^{(2)}(x)$ commutes with multiplication by $\phi(x')$, for all x and x' (this follows from the commutativity of the operations of multiplication by given numerical functions).

(ii) Under the automorphism of the algebra of random variables on Ω which carries $\phi(x)$ into $\phi(x) + F(x)$, where $F(x) = \int x(a)f(a)\,da$ (so that, symbolically, $\varphi(\cdot) \to \varphi(\cdot) + f(\cdot)$), $\phi^{(2)}(x)$ is carried into $\phi^{(2)}(x) + 2\phi(fx) + \int f^2 x$.

Now ergodic quasi-invariance implies formally that if any such function $\phi^{(2)}$ exists, it is unique within an additive constant; it is totally unique if it is integrable, and the normalization is made that its expectation value vanishes. It develops that such a notion of renormalized power can be applied to generalized as well as strict functions on Ω, and that in this way one is led to an effective definition of the renormalized powers of weak EQI processes through their intrinsic characterization by properties similar to (i) and (ii) and their normalization by vanishing expectation value.

Before this can be done, however, a suitable dense space of regular "test" functions on Ω must be specified. A natural space to consider in this connection, invariantly attached to the underlying Hilbert space $\mathbf{H} = L_2(M)$, and one studied intensively by Kristensen, Mejlbo, and Poulsen (see [3]), is a direct analogue to the space of all infinitely differentiable functions with infinitely differentiable Fourier transforms, all in L_2, over R^n (the so-called "Schwartz" space); however, the vectors of this space are insufficiently regular to serve as test functions, either in connection with a treatment of generalized functions as linear functionals or a treatment as sesquilinear forms. A suitable space of test functions may be derived from the imposition of suitable additional structure to the space M.

There are two natural generalizations of R^n from the standpoint of white noise theory: (i) the replacement of R^n by an arbitrary complete Riemannian manifold, and (ii) its replacement by an arbitrary locally compact Abelian group. In generalization (i) there is a natural distinguished operator, the Laplacian, which induces in a natural way a test function domain in $L_2(\Omega)$ which is effective at least in the case of R^n. It is probably effective quite generally, but the present methods, designed to treat generalization (ii), leave this question open. In place of the Laplacian, an arbitrary translation invariant operator B in $L_2(G)$, G being the given group, having the property that its spectral function $B(\cdot)$ on the dual group G^* has the property that $B(\cdot)^{-1} \in L_p(G^*)$ for all sufficiently large p, will be used. The corresponding test functions are not necessarily bounded on Ω, and for this and other reasons it is convenient to deal with sesquilinear forms, rather than linear functionals, on this domain of test functions.

A mathematical statement of the existence of renormalized powers of white noise is conveniently made with the aid of the concept of the isonormal symmetric (quantum) process over a given complex Hilbert space $\mathbf{H}$. This is the system $(\Psi, \mathbf{K}, v, \Gamma)$ composed of a complex Hilbert space $\mathbf{K}$, a mapping Ψ from $\mathbf{H}$ to the self-adjoint operators in $\mathbf{K}$ having the property that $e^{i\Psi(x)}e^{i\Psi(y)} = e^{(i/2)\operatorname{Im}\langle x,y\rangle}e^{i\Psi(x+y)}$ for arbitrary vectors x and y in $\mathbf{H}$, a unit vector v in $\mathbf{K}$, and a continuous unitary representation Γ of the full unitary group on $\mathbf{H}$ by unitary operators on $\mathbf{K}$ such that

$$\Gamma(U)v = v, \qquad \Gamma(U)\Psi(x)\Gamma(U)^{-1} = \Psi(Ux)$$

for all unitary operators U on $\mathbf{H}$ and vectors x in $\mathbf{H}$, such that v is a cyclic vector for the operators $e^{i\Psi(x)}$, $x \in \mathbf{H}$; and uniquely characterized among such systems, within unitary equivalence, by the property that if for any self-adjoint operator X in $\mathbf{H}$, $d\Gamma(X)$ denotes the self-adjoint generator of the one-parameter unitary group $\Gamma(e^{itX})$, $t \in R^1$, then $d\Gamma(X) \geqq 0$ whenever $X \geqq 0$.

The connection with white noise comes about from the circumstance that if $\mathbf{H}_r$ is any real subspace of $\mathbf{H}$ such that $\mathbf{H}$ is the direct sum, over the real field, of $\mathbf{H}_r$ and $i\mathbf{H}_r$, then the restriction map $\Psi|\mathbf{H}_r$, relative to the expectation values defined by the functional $E(A) = \langle Av, v\rangle$ for any operator A in the ring of operators generated by the $e^{i\Psi(x)}$, $x \in \mathbf{H}_r$, is, within a certain isomorphism, the isonormal (stochastic) process over $\mathbf{H}_r$, apart from a factor of $2^{1/2}$. For the nature of this isomorphism, see [6]; it will suffice here to point out that the $\Psi(x)$ for $x \in \mathbf{H}_r$ are mutually commutative and determine a maximal Abelian ring of operators; this then implies by general theory that they may be identified with the multiplication operators associated with certain real measurable functions acting on $L_2(\Omega)$ for a suitable measure space Ω; from the circumstance that $E(e^{i\Psi(x)}) = e^{-(1/4)\|x\|^2} = E(e^{i\phi(x)/2^{1/2}})$, Ω may be chosen as the probability space associated with white noise in such a way that respective expectation values, on the one hand for operators in $\mathbf{K}$ and on the other for random variables on Ω, are equal.

NOTATION. For any positive operator A in a Hilbert space $\mathbf{H}$, $\mathbf{D}_\infty(A)$ denotes the common parts of the domains of the e^{nA} $(n = 1, 2, \ldots)$; while $[\mathbf{D}_\infty(A)]$ denotes this set as a linear topological space in the topology in which a generic neighborhood of 0 consists of all x such that $\|A^k x\| < \epsilon$ for $k < k_0$, for some ϵ and k_0.

THEOREM 2.1. *Let G be a given locally compact Abelian group and B a given real, positive, translation-invariant operator in $L_2(G)$ such that $B(\cdot)^{-1} \in L_p(G^*)$ for all sufficiently large p, where G^* denotes the dual (character) group of G. Let $(\Psi, \mathbf{K}, v, \Gamma)$ denote the standard normal symmetric process over the Hilbert space $\mathbf{H} = L_2(G)$. Let H denote $d\Gamma(B)$. Then for arbitrary $a \in G$ and $j = 0, 1, 2, \ldots$ there exists a sesquilinear form $\varphi^{(j)}(a)$ on $[\mathbf{D}_\infty(H)]$ such that:*

(i) *The map $(a, u, u') \to \varphi^{(j)}(a)(u, u')$ is continuous from $G \times [\mathbf{D}_\infty(H)] \times [\mathbf{D}_\infty(H)]$ into C^1.*

(ii) $\varphi^{(0)}(a)(u, u') = \langle u, u'\rangle$, $\varphi^{(j)}(a)(v, v) = 0$ *(for all a and nonzero j).*

(iii) *If x is in $\mathbf{D}_\infty(B)$, then, denoting the sesquilinear form: $(u, u') \to \varphi^{(j)}(a)(e^{i\Psi(x)}u, e^{i\Psi(x)}u')$, as $F_j(a, x)$, the following relations hold when x is real and in $\mathbf{D}_\infty(B)$:*

$$F_j(a, x) = F_j(a, 0)$$

$$F_j(a, ix) = \sum_{k=0}^{j} \binom{j}{k} F_{j-k}(a, 0)x(a)^k.$$

Furthermore these conditions uniquely characterize the $\varphi^{(j)}(a)$.

REMARK. It may be useful to indicate once again the connection with the intuitive ideas concerning the present situation. The form $\varphi^{(j)}(a)$ may be thought of as the result of "renormalizing," by suitable subtractions of "infinite quantities," the formal but mathematically nonexistent power $\varphi(a)^j$. Thus $\varphi^{(1)}(a)$ is merely the white noise $\varphi(a)$ at the point a, and thus in the first instance a generalized function; this generalized function is, however, when regarded as a sesquilinear form on a suitable domain in $L_2(\Omega)$, a strict function. Condition (i) is to the effect that this strict function as well as the strict functions corresponding to higher renormalized powers than the first, is continuous. Condition (ii) asserts, first, the foregoing interpretation of $\varphi^{(j)}(a)$ in the case $j = 0$, and gives, second, the key normalization condition that the expectation values of all the renormalized powers vanish. Condition (iii) then asserts, first, that the $\varphi^{(j)}(a)$ are indeed in a generalized sense functions of the white noise in that multiplication by them, acting on $L_2(\Omega)$, commutes with multiplication by the white noise; and, second, that these functions are indeed renormalized powers in the sense that they have exactly the same transformation properties under vector displacements as do the formal powers (i.e., the conclusion of the binomial theorem is essentially applicable to the renormalized powers).

PROOF OF THEOREM 2.1. The proof is parallel to the proof of Theorem 4.1 in [7] and is based on the idea that the operator B which occurs in that theorem plays two separate roles, one of which is here taken over by the operator I, the covariance operator of the white noise process. As shown in [II], any element g of $\mathbf{D}_\infty(B)$ differs on a null set from a (unique) continuous function, and it is this continuous function which enters into condition (iii). The uniqueness part of the theorem is a corollary to results in [II] and is proved by the same method as the uniqueness for the cited theorem in [II].

The proof of the existence then proceeds by the establishment of the existence and properties of the limit of $\langle :\phi(x_a)^n: u, u' \rangle$, as x converges to the "delta-function" on G, where $x_a(b) = x(a^{-1}b)$ and u and u' are arbitrary elements of $\mathbf{D}_\infty(H)$. Lemmas 4.1 and 4.2 of [II] and the associated notation will be employed, except that the ϕ and φ of [II] have been replaced by Ψ and ψ, for the present Ψ, unlike the ϕ and Ψ of [II], has no special connection with the operator B. (Colloquially speaking, the Ψ is an isonormal process and has no special dynamics, while ϕ is the quantized scalar field corresponding to the differential equation $\phi'' + B^2\phi = 0$. In addition, at a fixed time $\Psi|H_r$ has covariance operator I, while ϕ has covariance operator B^{-1}.) The general idea of the proof may be considered to be to begin with an examination of the case of sufficiently smooth vectors in Wiener's polynomial chaos, as adapted by Anzai and Kakutani, as test functions in $L_2(\Omega)$. The main lemma is then the following:

LEMMA 2.1. *There exists a constant K and a positive integer t such that for arbitrary u and u′ in $\mathbf{D}_\infty(H)$ and $x \in \mathbf{D}_\infty(B)$,*

$$\langle :\Psi(x)^r : u, u' \rangle = \int \prod_{i=1}^{r} \hat{x}(k_i) M_{u,u'}(k_1, \ldots, k_r) \prod_i dk_i,$$

where $M_{u,u'}$ is integrable and

$$\|M_{u,u'}\|_1 \leqq K \|(I + H)^t u\| \, \|(I + H)^t u'\|.$$

PROOF OF LEMMA 2.1. Consider first the special case in which u and u' have the forms

$$u = \left(\int :\hat{\psi}(p_1) \cdots \hat{\psi}(p_n): F(p_1, \ldots, p_n) \prod_i dp_i \right) v$$

$$u' = \left(\int :\hat{\psi}(q_1) \cdots \hat{\psi}(q_{n'}): F'(q_1, \ldots, q_{n'}) \prod_i dq_i \right) v,$$

where the integrals are over G^{*n} and $G^{*n'}$, respectively, and F and F' are symmetric measurable functions on these spaces such that $B_n^j F$ and $B_{n'}^j F'$ are square-integrable for all positive integral j. According to the Lemma 4.1 cited from [II], with B taken as I,

$$\begin{aligned} \langle :\Psi(x)^r : u, u' \rangle = {} & [n!\, n'!\, r!/(n - s)!\, (n' - s)!\, s!] \\ & \times \int \prod_{i=1}^{r} \hat{x}(k_i) F(p_1, \ldots, p_s, k_1, \ldots, k_{n-s}) \\ & \times \bar{F}'(p_1, \ldots, p_s, k_{n-s+1}, \ldots, k_{n'}) \prod dk_i \, dp_j, \end{aligned}$$

where $2s = n + n' - r$, the integral in question vanishing if s is not integral. On the other hand,

$$\begin{aligned} \|(I + H)^t u\|^2 &= n! \int |F(p_1, \ldots, p_n)|^2 (1 + B(p_1) + \cdots + B(p_n))^{2t} \prod_i dp_i \\ &= n! \, \|F_1\|_2^2, \end{aligned}$$

where $F_1(p_1, \ldots, p_n) = F(p_1, \ldots, p_n)(1 + B(p_1) + \cdots + B(p_n))^t$; and similarly, $\|(I + H)^t u'\|^2 = n'! \, \|F_1'\|_2^2$. Thus, provided the following integrand is absolutely integrable,

$$\begin{aligned} \langle :\Psi(x)^r : u, u' \rangle = \int \prod_{i=1}^{r} \hat{x}(k_i) F_1(p, k') \bar{F}_1'(p, k'') (1 + B_s(p) + B_{n'-s}(k'))^{-t} \\ \cdot (1 + B_s(p) + B_{n'-s}(k''))^{-t}, \end{aligned}$$

where $k' = (k_1, \ldots, k_{n-s})$ and $k'' = (k_{n-s+1}, \ldots, k_r)$. This means that the expression in question has the form indicated in Lemma 2.1 with

$$M_{u,u'}(k) = \int F_1(p, k')\bar{F}_1'(p, k'')(1 + B_s(p) + B_{n-s}(k'))^{-t} \cdot (1 + B_s(p) + B_{n'-s}(k''))^{-t},$$

provided the indicated absolute integrability holds. This integral for $M_{u,u'}(k)$ is of the form to which Lemma 4.2 of [II] is applicable; it implies

$$\|M_{u,u'}\|_1 \leqq K\|F_1\|_2\|F_1'\|_2,$$

where

$$K = \text{const. } \sup_p \int (1 + B_s(p) + B_{n-s}(k'))^{-t} \cdot (1 + B_s(p) + B_{n'-s}(k''))^{-t}\, dk'\, dk''.$$

It then follows from Lemma 4.2 of [II] by the same argument as in [II] that K is finite and independent of n and n' for sufficiently large t. This shows that $M_{u,u'}$ is appropriately majorized in the special case in question; when u and u' are general in $D_\infty(H)$, the same argument as in [II] leads to the conclusion of Lemma 2.1.

Completion of Proof of Theorem 2.1. The Lebesgue dominated convergence theorem shows the existence of the putative limit defining $\varphi^{(j)}(a)$; the inequality of Lemma 2.1 shows that this sesquilinear form is continuous on $[D_\infty(H)]$. The proofs that the conditions (ii) and (iii) are satisfied are virtually identical to those given in [II] for parallel results in the case treated there.

As earlier noted, the foregoing argument is based on the idea that two conceptually distinct operators happen to be the same in the case of a dynamical quantum process treated in [II] and that white noise can be treated as the case in which one of the two operators is I and the other is unchanged. The role of the two different operators is clarified by the following result, which gives a basis for the combination of the present theorem with Theorem 4.1 of [II]. It is essentially to the effect that renormalized powers similar to those for white noise exist for any centered translation-invariant normal process on a locally compact Abelian group, provided the covariance operator of the process is bounded.

Corollary 2.1. *Let Δ denote any centered bounded normal group translation-invariant (stochastic) process defined on all of $L_2(G)$. Then, with the notation of Theorem* 2.1, *there exist sesquilinear forms $\delta^{(j)}(a)$ having the same properties as the $\psi^{(j)}(a)$, except that in place of the given equation for $F_j(a, ix)$ one replaces the $x(a)$ on the right side of the equation by $x'(a)$,* *where x' is the inverse Fourier transform of $C(\cdot)^{1/2}\hat{x}(\cdot)$, C being the covariance operator of the process Δ.*

PROOF. A normal process Δ defined on a Hilbert space **H** is bounded if $\|\Delta(x)\| \leqq \text{const}\,\|x\|$ for all $x \in \mathbf{H}$. This means that $\Delta(x) \sim \psi(C^{1/2}x)$, for all (real) $x \in L_2(G)$, C being the covariance operator of the process. The argument given is affected only through the intervention of products of terms $C(k_i)$, which are the same as certain terms given in [II]; as long as the $C(k)$ are bounded, the same estimates are obtained. The boundedness of $C(k)$ implies also that $C(\cdot)^{1/2}\hat{x}(\cdot)$ is in $L_1(G^*)$, so that x' is a continuous function on G.

REMARK. The locality of the $\varphi^{(j)}$, as functions of the given $\varphi^{(1)}$, is indicated by the following intuitive paraphrase of condition (iii): the transformation $\varphi(x) \to \varphi(x) + f(x)$, where f is a smooth function vanishing in the vicinity of a point x_0, does not affect $\varphi^{(j)}(x)$ in the vicinity of this point, but only outside this neighborhood, in accordance with the equation given under (iii).

3. Local noncommutative differential equations

There is as yet no treatment of nonlinear differential equations involving processes as singular as white noise. I shall treat here a first step towards the solution of less singular equations arising in the study of the concept of a relativistic quantum field.

The mathematical origin of the question to be considered may be indicated briefly as follows. The basic dynamical equations of relativistic quantum field theory may be reduced to a complex of equations, each of which has the following form (see [8]):

$$(*) \qquad u'(t) = iH_I(t)u(t), \qquad u(0) = u_0,$$

where $H_I(t)$ is for each t a generalized self-adjoint operator, of a type which need not be specified here, but which may be described as an analogue to a space average of a renormalized power of a white noise, except that the $H_I(t)$ are for different values of t noncommutative.

A solution of equation (*) is in the first instance a function $S(t, t')$ on $R^1 \times R^1$ to the unitary operator on the Hilbert space **K** which is in question, such that $S(t, t) = I$ for all t, and

$$(**) \qquad (\partial/\partial t)S(t, t') = iH_I(t)S(t, t').$$

Suitable technical qualifications intervene, for which see [8]. Equation (**) determines a materially more general type of dynamics than that deriving from the perturbation of a self-adjoint operator by another self-adjoint operator, say $H_0 \to H_0 + H_I$, in the case that all the operators H_0, H_I, and $H_0 + H_I$ have unique self-adjoint extensions. In this case, $S(t, t')$ exists and is given by the equation

$$S(t, t') = e^{-itH_0}e^{i(t-t')(H_0+H_I)}e^{it'H_0},$$

where the convention is employed that a function of an essentially self-adjoint operator is the same function of its closure.

The presently crucial properties of $S(\cdot,\cdot)$ are:

(i) The map $(t, t') \to S(t, t')$ is continuous from $R^1 \times R^1$ to the unitary operators on $\mathbf{K}$, in their strong topology.

(ii) For arbitrary a, b, and c in R^1,

$$S(c, b)S(b, a) = S(c, a).$$

(iii) $U_0(-a)S(b, c)U_0(a) = S(b + a, c + a)$, where $U_0(t) = e^{itH_0}$.

Conversely, given such a pair $[U_0(\cdot), S(\cdot,\cdot)]$, the equation

$$U(t) = U_0(t + t')S(t + t', t')U_0(-t')$$

defines a continuous one-parameter unitary group. In the analytically particularly tractable case indicated earlier the self-adjoint generator H of this group is the closure of $H_0 + H_I$. In general, however, the operators H and H_0 arising in this way from a given pair $[U_0(\cdot), S(\cdot,\cdot)]$ need not have any vectors other than 0 in their common domain. Moreover, even when $H - H_0$ has a unique self-adjoint extension, say H_I, it need not be true that H is the unique self-adjoint extension of $H_0 + H_I$.

Thus a given pair $[U_0(\cdot), S(\cdot,\cdot)]$ consisting of a continuous one-parameter unitary group $U_0(\cdot)$ on a Hilbert space $\mathbf{K}$, together with a unitary "propagator" $S(\cdot,\cdot)$ on this space, satisfying the (so-called "covariance") condition earlier indicated relating S and U_0, generalize the familiar situation, which the Schrödinger equation exemplifies, in which one is given a self-adjoint operator H_0, and a "perturbation" H_I such that $H_0 + H_I$ has a unique self-adjoint extension. This generalization permits the treatment of differential equations of the form $u' = i(\Delta + V)u$, where $u(t)$ has its values in $L_2(R^n)$, and V is in a certain class of generalized (nonstrict) functions.

There is a further generalization which is appropriate to quantum field dynamics as well as to differential operators in R^n of the indicated type, involving, however, more singular generalization functions V. This is the notion of "distribution propagator," in which $S(t, t')$ is replaced by $S(f)$, f being the characteristic function of the interval $[t', t]$; f is then replaced by an element of $C_0^\infty(R^1)$ (the notation $C_0^\infty(S)$ indicating the collection of all infinitely differentiable functions of compact support in the manifold S). The basic propagator identity, $S(c, b)S(b, a) = S(c, a)$, is then replaced by the equation

$$S(f)S(g) = S(f + g)$$

if the support of g is strictly less than that of f (intuitively, this means that the times involved in g are all earlier than those involved in f). Unlike the notion of "distribution semigroup" treated by Schwartz, Lions, Mosco, and others, which, when its values are unitary operators in Hilbert space, is not materially more general than the conventional one of a unitary group, this notion, of distribution propagator, is wider than that of the unitary propagator $S(t, t')$ indicated earlier. This mathematical circumstance may be regarded as in part correlative with the intuitive idea that there exist dynamical situations in which the system undergoing temporal evolution cannot be physically fixed in time, but can be fixed only in an arbitrarily small time interval; the characteristic function of the time interval $[t, t']$ must be replaced by a function in $C_0^\infty(R^1)$ which smooths out the ends of the interval. A major problem in the indicated applications is that of establishing in an appropriate sense the asymptotic constancy of $S(t, t')$ as $t' \to -\infty$ and $t \to +\infty$, the limiting operator being the so-called "scattering," "collision," or "dispersion" operator; and this question would not be affected by the replacement of a strict by a distribution propagator.

Distribution propagators will be treated elsewhere; they are described here in order to explain the origin of the operators involving local noncommutative functions that will be treated here and to indicate the direction of the research. Returning now to the problem of solving equation (*), when the sesquilinear form $\int H_I(t)f(t)\,dt$ is sufficiently regular (for $f \in C_0^\infty(R^1)$) as to: (i) correspond to a densely defined Hilbert-space operator; (ii) have the property that any two self-adjoint extensions of this operator coincide; and (iii) be such that the unique self-adjoint extension depends in a suitably regular manner on $f \in C_0^\infty(R^1)$, there is a natural approach to the construction of the operator $S(f)$. This is first to represent f in the form $f = f_1 + \cdots + f_n$, where each f_i is a non-negative element of $C_0^\infty(R^1)$, of support ultimately becoming arbitrarily small, and so ordered that if f_i is supported by the interval $[a_i, b_i]$, then the inequality $i < j$ implies that $a_i < a_j$, $b_i < b_j$. Then $S(f)$ should be approximated by $\exp(iH_I[f_n]) \cdots \exp(iH_I[f_1])$, where $H_I[f]$ denotes the closure of the densely defined Hilbert-space operator corresponding to $\int H_I(t)f(t)\,dt$. This is substantiated in the special case in which H_0, H_I, and $H_0 + H_I$ all have unique self-adjoint extensions by a well-known theorem due to Trotter.

I shall next treat question (i), of the denseness of the domains of the (strict, i.e., nongeneralized) Hilbert-space operators associated with the sesquilinear forms $\int H_I(t)f(t)\,dt$. The methods are basically general but will be expounded only in the case of the "scalar" processes treated in [7].

THEOREM 3.1. *With the notations and assumptions of Theorem* 1 *and the additional assumptions:* (i) $B(\cdot)^{-3} \in L_p(G^*)$ *for all* $p > 1$; (ii) $g \in L_1(G)$, $\hat{g} \in L_1(G^*)$; (iii) $f \in L_1(R^1)$, $|\hat{f}(l)| = 0(|l|^{-r})$, $|l| \to \infty$, *the Hilbert-space operator associated with the sesquilinear form*

$$\int \left(\int :\phi(x, t)^r : g(x)\, dx \right) f(t)\, dt,$$

where ϕ *indicates the associated scalar quantum process, contains* $D_\infty(H)$ *in its domain.*

PROOF. Recall that if $\Gamma(\cdot)$ denotes the one-parameter group of temporal displacements on the Hilbert space $\mathbf{K}$ representing the dynamics of the process symbolized as $\phi(x, t)$, i.e., symbolically $\phi(x, t) = \Gamma(t)^*\phi(x, 0)\Gamma(t)$, then

$$\int :\phi(x, t)^r : g(x)\, dx = \Gamma(t)^{-1} \int :\phi(x, 0)^r : g(x)\, dx\Gamma(t).$$

This means that if $L_t(w, w')$ is the value of the form $\int :\phi(x, t)^r : g(x)\, dx$ on the pair of vectors w and w' in $\mathbf{D}_\infty(H)$, where H is the self-adjoint generator of the one-parameter group $\Gamma(\cdot)$, then

$$L_t(w, w') = L_0(\Gamma(t)w, \Gamma(t)w').$$

Since L_0 is continuous on $[\mathbf{D}_\infty(H)]$, $L_t(w, w')$ is a continuous function of t. The form $M(\cdot,\cdot)$ given by the equation

$$M(w, w') = \int L_t(w, w') f(t)\, dt$$

is therefore well defined. In order to show that the associated Hilbert-space operator, to the form $M(\cdot,\cdot)$, contains $\mathbf{D}_\infty(H)$ in its domain, it suffices to establish the inequality

$$|M(w, w')| \leqq c(w)\|w'\|$$

for all $w \in \mathbf{D}_\infty(H)$ and $w' \in \mathbf{K}$. It suffices, by arguments given in [II] depending on linearity and the circumstance that $L_0(w, w') = 0$ if $w \in \mathbf{K}_i$, $w' \in \mathbf{K}_j$, and $|i - j| > r$, to assume that $w \in \mathbf{K}_n$ and $w' \in \mathbf{K}_{n'}$ for certain n and n'. In this case w may be represented in the form

$$w = \left(\int :\hat{\phi}(p_1) \cdots \hat{\phi}(p_n) : F(p_1, \ldots, p_n)\, dp \right) v,$$

and w' may be similarly represented in terms of a symmetric function F' on $G^{*n'}$. It is necessary for this reduction that $c(w)$ satisfy the inequality: $c(w) \leqq \text{const} \|(I + H)^a w\|$, for a suitable constant and exponent a, both of which must be independent of n.

Since the action of $\Gamma(t)$ on w is to replace the corresponding function F by its product with $\exp\,(it \sum_j B(p_j))$, the computation given as Lemma 4.1 in [II] shows that $M(w, w')$ is a sum of integrals of the following type, and that it suffices to bound this typical integral in the indicated fashion:

$$\int \left[\prod_{i=1}^{s} B(p_i)\right]^{-1} \left[\prod_{i=1}^{r} B(k_i)\right]^{-1} \exp\left[it\left(\sum_{i=1}^{n-s} B(k_i) - \sum_{i=n-s+1}^{r} B(k_i)\right)\right]$$
$$\times F(p_1, \ldots, p_s, -k_1, \ldots, -k_{n-s})\bar{F}(p_1, \ldots, p_s, k_{n-s+1}, \ldots, k_r)$$
$$\times \bar{\hat{g}}\left(\sum_{i=1}^{r} k_i\right) f(t)\, dp_1 \cdots dp_s\, dk_1 \cdots dk_r\, dt.$$

On introducing the square-integrable functions F_1 and F_1' as in [II] (differing from F and F' by appropriate power-products of the $B(p_i)$ and $B(k_i)$), this integral takes the form

$$\int \left[\prod_{i=1}^{r} B(k_i)\right]^{-1/2} \hat{f}\left(\sum_{i=1}^{n-s} B(k_i) - \sum_{i=n-s+1}^{r} B(k_i)\right) \bar{\hat{g}}(k_1 + \cdots + k_r)$$
$$\times F_1(p_1, \ldots, p_s, k_1, \ldots, k_{n-s}) F_1'(p_1, \ldots, p_s, k_{n-s+1}, \ldots, k_r)\, dp\, dk.$$

Since $w \in \mathbf{D}_\infty(H)$, $K^a F_1 \in L_2$, where $K(q_1, \ldots, q_n) = 1 + B(q_1) + \cdots + B(q_n)$, for arbitrary real a. Now replace F_1 by $K^a F_1$ and introduce a compensating factor of K^{-a} at the beginning of the integrand; the integral resulting has a form to which Lemma 4.2 of [II] is applicable, with $x = (k_1, \ldots, k_{n-s})$, $y = (k_{n-s+1}, \ldots, k_r)$, and $z = (p_1, \ldots, p_s)$. It follows that the integral in question exists if the following expression is finite and is then bounded by this expression:

$$\|K^a F_1\|_2 \|F_1'\|_2$$
$$\times \sup_{p_1,\ldots,p_s} \left[\int (1 + B(p_1) + \cdots + B(p_s) + B(k_1) + \cdots + B(k_{n-s}))\right]^{2a}$$
$$\times \left[\prod_{i=1}^{r} B(k_i)\right]^{-1} \left|\hat{f}\left(\sum_{i=1}^{n-s} B(k_i) - \sum_{i=n-s+1}^{r} B(k_i)\right)\right|^2 |\hat{g}(k_1 + \cdots + k_r)|^2\, dk.$$

The quantities $\|K^a F_1\|_2$ and $\|F_1'\|_2$ are identical with $\|(1 + H)^a w\|$ and $\|w'\|$, within factors dependent only on n and n', which scale as in [II] in such a way that it remains only to show that the indicated supremum over $p_1, \ldots, p_s$ is finite and bounded independently of n and n'. Since $B(p) \geqq 0$

for all p, this supremum is bounded by the corresponding integral with all $B(p)$ replaced by 0. The resulting integral

$$\int (1 + B(k_1) + \cdots + B(k_{n-s}))^{-2a}\left[\prod_{i=1}^{r} B(k_i)\right]^{-1} \times \left|\hat{f}\left(\sum_{i=1}^{n-s} B(k_i) - \sum_{i=n-s+1}^{r} B(k_i)\right)\right|^2 |\hat{g}(k_1 + \cdots + k_r)|^2\, dk$$

is no longer dependent on n or n', so that it suffices to show that it is finite.

To treat this integral, set

$$d(k_1, \ldots, k_r) = \sum_{i=1}^{n-s} B(k_i), \qquad e(k_1, \ldots, k_r) = \sum_{i=n-s+1}^{r} B(k_i),$$

and consider separately the integrals over the two regions, I: $|e - d| \geqq e/2$, and II: $|e - d| < e/2$. Since $\hat{f}$ is bounded and $|\hat{f}(l)| = 0(|l|^{-r})$ for $|l| \to \infty$, $|\hat{f}(l)| \leqq C(1 + |l|)^{-r}$ for a suitable constant C. In particular,

$$\begin{aligned} |\hat{f}(d - e)| \leqq C(1 + |d - e|)^{-r} &\leqq 2Ce^{-r} \quad \text{in region I} \\ &\leqq C \qquad\quad \text{in region II.} \end{aligned}$$

The integral over region I is then bounded by

$$2C \int (1 + B(k_1) + \cdots + B(k_{n-s}))^{-2a}\left[\prod_{i=1}^{r} B(k_i)\right]^{-1} \times \left(\sum_{i=n-s+1}^{r} B(k_i)\right)^{-2r} |\hat{g}(k_1 + \cdots + k_r)|^2\, dk.$$

Since the arithmetic mean dominates the geometric mean,

$$\left(\sum_{i=n-s+1}^{r} B(k_i)\right)^{-2r} \leqq C' \prod_{i=n-s+1}^{r} B(k_i)^{-2},$$

and $(1 + B(k_1) + \cdots + B(k_{n-s}))^{-2a}$ may be similarly dominated. Thus the integral in question is dominated by

$$C'' \int \left(\prod_{i=1}^{n-s} B(k_i)\right)^{-[2(a/r)+1]} \left(\prod_{i=n-s+1}^{r} B(k_i)\right)^{-3} |\hat{g}(k_1 + \cdots + k_r)|^2\, dk.$$

If $a \geqq r$, this integral is finite; for, taking, as suffices, the case $a = r$, it has the form

$$(*) \qquad C'' \int \prod_{i=1}^{r} N(k_i) |\hat{g}(k_1 + \cdots + k_r)|^2\, dk; \qquad N(k) = B(k)^{-3}.$$

By Fubini's theorem, this is the same as $C'' \int (N * N * \cdots * N)|g|^2$, the inner product of the r-fold convolution of N with itself, with $|g|^2$, in the sense that if this convolution is finite a.e. and the indicated inner product with $|g|^2$ is finite, then the two integrals are the same. By hypothesis, $N \in L_p(G^*)$ for all $p > 1$, so that the indicated convolution exists and is again in this space, by the Hausdorff–Young Theorem. On the other hand, $|\hat{g}|^2$ is also in all $L_p(G^*)$, under the indicated assumption on g, so that the inner product is indeed finite.

It remains to consider the integral over region II. In this region, $e < 2d$; the integration over $k_{n-s+1}, \ldots, k_r$ therefore contributes at most const. $d^{\text{const.}}$, if $\hat{f}$, $|\hat{g}|$, and the $B(k_j)^{-1}$ for $j \geqq n - s + 1$ are replaced by constants which bound them. The resulting integral over the $k_1, \ldots, k_{n-s}$ is then bounded by one of the form: const. $\int d^{-2(a/r)+\text{const.}}\, dk_1 \cdots dk_{n-s}$, which is finite if a is chosen sufficiently large. Taking the greater of the two a's involved, the required inequality has been established.

REMARK 3.1. The exponent r in the required decay for $\hat{f}$, while possibly best possible in the generality of Theorem 2, is not best possible in the relativistic case. Indeed, an appropriate specialization of the foregoing argument shows that in the case $r = 2$, the conclusion of Theorem 3.1 is valid if only $|\hat{f}(l)| = 0(|l|^{-1})$. More specifically, the argument is the same except that because of the slower decay for $\hat{f}$, $N(k)$ must be taken as $B(k)^{-2}$. In this relativistic case, this is in L_p for $p > \frac{3}{2}$; the two-fold convolution of such a function with itself is in L_q for $q > 3$ and hence has finite inner product with $|g|^2$. This shows that $\int :\phi(x, t)^2 : g(x)\, dx$ is, as a function of t, and as a Hilbert-space operator, a distribution of first order; and it is known that it is not a distribution of zero order, i.e., a strict function.

Similar considerations are applicable to higher powers in the relativistic case. If, for example, $r = 3$ and $|\hat{f}(l)| = 0(|l|^{-2})$, the method given shows that the conclusion of Theorem 3.1 is implied by the convergence of the integral (*) with $N(k)$ taken as $B(k)^{-7/3}$. In the relativistic case, this function is in L_p for $p > \frac{9}{7}$ (in four space-time dimensions); the three-fold convolution with itself of a function in this case is in L_q for $q > 3$; it therefore has a finite inner product with $|g|^2$.

COROLLARY 3.1. *If, in addition, f is a real even function (or a translate of an even function), then the operator indicated in Theorem* 3.1 *has a self-adjoint extension.*

PROOF. The case of a translate of an even function reduces to that of an even function via transformation with $\Gamma(t')$, t' being the translation in question. It is known (and easily seen) that there exists a unique conjugation J on $\mathbf{K}$ such that $Jv = v$ and $J\phi(x, t)J^{-1} = \phi(x, -t)$; when f is even and g is real, the cited operator is real relative to J, and, being densely defined, has a self-adjoint extension.

REMARK 3.2. In colloquial language, Corollary 3.1 and Remark 3.1 imply that if ϕ and ψ are independent scalar quantum fields (in four-dimensional space-time), with periodic boundary conditions in space, and if $f(t) = 1 - |t/\varepsilon|$ for $|t| < \varepsilon$ and is 0 otherwise, then $\int :\phi(x, t)\psi(x, t)^2: f(t)\, dx\, dt$ has a self-adjoint extension as a conventional operator on the domain $\mathbf{D}_\infty(H)$. The same is perhaps true for "quantum electrodynamics," which has a somewhat similar "interaction Hamiltonian," but this has not yet been verified. The cited functions $f(t)$ are particularly convenient for generalized Riemann product integration of the type indicated earlier, for a simple sum of translates can represent a function identically one in an arbitrarily large interval; similar functions are available for higher values of r.

MASSACHUSETTS INSTITUTE OF TECHNOLOGY

REFERENCES

1. BOCHNER, S., *Harmonic Analysis and the Theory of Probability*. Berkeley, Cal.: University of California Press, 1955.
2. CAMERON, R. H., and W. T. MARTIN, "Transformations of Wiener integrals under translations," *Annals of Mathematics* (2), *45* (1944), 386–396.
3. KRISTENSEN, P., *Tempered Distributions in Functional Space*, Proceedings of the Conference on Analysis in Function Space, W. T. Martin and I. Segal, eds. Cambridge, Mass.: M.I.T. Press, 1964, Chap. 5, pp. 69–86.
4. SEGAL, I., "Transformations in Wiener space and squares of quantum fields," *Advances in Mathematics 4* (1970), 91–108.
5. ———, "Tensor algebras over Hilbert spaces, II," *Annals of Mathematics*, *63* (1956), 160–175.
6. ———, "Tensor algebras, I," *Trans. Amer. Math. Soc.*, *81* (1956), 106–134.
7. ———, "Nonlinear functions of weak processes, I and II," *Journal of Functional Analysis 4* (1969), 404–456, and in press. (These articles are referred to as "I" and "II".)
8. ———, *Local Nonlinear Functions of Quantum Fields*, Proceedings of a Conference in Honor of M. H. Stone. Berlin: Springer (in press).

Linearization of the Product of Orthogonal Polynomials

RICHARD ASKEY[1]

In [3] Bochner showed that there is a convolution structure associated with ultraspherical polynomials which generalizes the classical L^1 convolution algebra of even functions on the circle. Bochner uses the addition formula to obtain the essential positivity result. In [18] Weinberger shows that this positivity property follows from a maximum principle for a class of hyperbolic equations. Hirschman [8] has dualized this convolution structure and has proven the required positivity result by means of a formula of Dougall which linearizes the product of two ultraspherical polynomials. We will prove a theorem which gives most of Hirschman's results as well as other positivity results which were not previously known. In particular we obtain a positivity result for most Jacobi polynomials. In view of the known duality for the classical polynomials as functions of n and x, this suggests strongly that the Bochner–Weinberger result can be extended to Jacobi polynomials. This is the only missing step in proving the positivity of some Cesaro mean for most Jacobi series, and this result can then be used to solve some L^p convergence problems for Lagrange interpolation.

The theorem we prove is concerned with the problem of when an orthogonal polynomial sequence $p_n(x)$ satisfies

$$(1) \qquad p_n(x)p_m(x) = \sum_{k=|n-m|}^{n+m} \alpha_k p_k(x), \qquad \alpha_k \geqq 0.$$

Any sequence of orthogonal polynomials satisfies a recurrence formula

$$(2) \qquad xp_n(x) = \alpha_n p_{n+1}(x) + \beta_n p_n(x) + \gamma_n p_{n-1}(x),$$

$p_{-1}(x) = 0$, β_n is real, $\alpha_n\gamma_{n+1} > 0$. We normalize our orthogonal polynomials $p_n(x)$ by $p_n(x) = x^n + \cdots$. Then (2) takes the form

$$(3) \qquad p_1(x)p_n(x) = p_{n+1}(x) + a_n p_n(x) + b_n p_{n-1}(x).$$

[1] Supported in part by NSF grant GP-6764.

In order to have (1) we must have $a_n \geqq 0$. $b_n > 0$ is a general property of orthogonal polynomials. Our main theorem is

THEOREM 1. *If* (3) *holds for* $n = 1, 2, \ldots, a_n \geqq 0, b_n > 0$ *and* $a_{n+1} \geqq a_n$, $b_{n+1} \geqq b_n$, *then* (1) *holds for* $n, m = 0, 1, \ldots$.

By symmetry we may assume $m \leqq n$ and we will prove the theorem by induction on m. Since it holds for $m = 1, n = 1, 2, \ldots$ we may assume it holds for $m = 1, 2, \ldots, l$ and prove it for $m = l + 1, l < n$. We have

$$\begin{aligned} p_{l+1}(x)p_n(x) &= p_1(x)p_l(x)p_n(x) - a_l p_l(x)p_n(x) - b_l p_{l-1}(x)p_n(x) \\ &= p_l(x)p_{n+1}(x) + a_n p_l(x)p_n(x) + b_n p_l(x)p_{n-1}(x) \\ &\quad - a_l p_l(x)p_n(x) - b_l p_{l-1}(x)p_n(x) \\ &= p_l(x)p_{n+1}(x) + (a_n - a_l)p_l(x)p_n(x) \\ &\quad + (b_n - b_l)p_l(x)p_{n-1}(x) + b_l[p_l(x)p_{n-1}(x) - p_{l-1}(x)p_n(x)]. \end{aligned}$$

Since $a_n - a_l \geqq 0$ and $b_n - b_l \geqq 0$ and $b_l > 0$ by the induction assumption we are finished if we can take care of the last term. Using (3) again, we see that

$$\begin{aligned} &p_l(x)p_{n-1}(x) - p_{l-1}(x)p_n(x) \\ &\quad = p_{n-1}(x)[p_1(x)p_{l-1}(x) - a_{l-1}p_{l-1}(x) - b_{l-1}p_{l-2}(x)] \\ &\qquad\qquad - p_{l-1}(x)[p_1(x)p_{n-1}(x) - a_{n-1}p_{n-1}(x) - b_{n-1}p_{n-2}(x)] \\ &\quad = (a_{n-1} - a_{l-1})p_{n-1}(x)p_{l-1}(x) + b_{l-1} \\ &\qquad \times [p_{l-1}(x)p_{n-2}(x) - p_{n-1}(x)p_{l-2}(x)] \\ &\qquad\qquad + (b_{n-1} - b_{l-1})p_{l-1}(x)p_{n-2}(x). \end{aligned}$$

Continuing in this fashion we have terms that have positive coefficients except possibly for the last one $p_1(x)p_{n-l}(x) - p_{n-l+1}(x)$. But we use (3) again to see that this term is $a_{n-l}p_{n-l}(x) + b_{n-l}p_{n-l-1}(x)$ and $a_{n-l} \geqq 0$, $b_{n-l} > 0$. This completes the proof of Theorem 1.

The Charlier polynomials (see [6]) normalized in this way satisfy

$$(4) \quad d_1(x; a)d_n(x; a) = d_{n+1}(x; a) + nd_n(x; a) + and_{n-1}(x; a), \qquad a > 0,$$

and Theorem 1 is immediately applicable. Similarly, Theorem 1 is applicable to Hermite polynomials since

$$(5) \qquad H_1(x)H_n(x) = H_{n+1}(x) + 2nH_{n-1}(x).$$

Here $H_n(x) = 2^n x^n + \cdots$; thus the normalized polynomials $K_n(x)$ satisfy

$$(6) \qquad K_1(x)K_n(x) = K_{n+1}(x) + \frac{n}{2} K_{n-1}(x).$$

For Laguerre polynomials $L_n^\alpha(x)$ we let $Q_n^\alpha(x) = (-1)^n L_n^\alpha(x)/n!$ Then the recurrence formula

$$(7)\quad (n + 1)L_{n+1}^\alpha(x) - (2n + \alpha + 1 - x)L_n^\alpha(x) + (n + \alpha)L_{n-1}^\alpha(x) = 0$$

becomes

$$(8)\qquad Q_1^\alpha(x)Q_n^\alpha(x) = Q_{n+1}^\alpha(x) + 2nQ_n^\alpha(x) + n(n + \alpha)Q_{n-1}^\alpha(x),$$

and Theorem 1 applies if $\alpha > -1$.

The Meixner polynomials $\varphi_n(x) = F(-n, -x, \beta; 1 - 1/\gamma)$, $\beta > 0$, $0 < \gamma < 1$ satisfy

$$(9)\quad \frac{-x(1 - \gamma)}{\gamma}\varphi_n(x) = (n + \beta)\varphi_{n+1}(x) - \left(n + \frac{n}{\gamma} + \beta\right)\varphi_n(x) + \frac{n}{\gamma}\varphi_{n-1}(x).$$

See [12]. If we normalize these polynomials accordingly

$$K_n(x) = \frac{(\beta)_n}{[1 - (1/\gamma)]^n}\varphi_n(x) = x^n + \cdots,$$

then (9) becomes

$$(10)\quad K_1(x)K_n(x) = K_{n+1}(x) + \frac{n(1 + \gamma)}{(1 - \gamma)}K_n(x) + \frac{n\gamma(n + \beta - 1)}{(1 - \gamma)^2}K_{n-1}(x)$$

and the assumptions of Theorem 1 are satisfied for $0 < \gamma < 1$, $\beta > 0$.

We now consider the most important special case, that of the Jacobi polynomials $P_n^{(\alpha,\beta)}(x)$. They satisfy

$$(11)\qquad P_n^{(\alpha,\beta)}(x) = \frac{\Gamma(2n + \alpha + \beta + 1)}{2^n\Gamma(n + \alpha + \beta + 1)n!}x^n + \cdots$$

and

$$(12)\quad \begin{aligned} &2(n + 1)(n + \alpha + \beta + 1)(2n + \alpha + \beta)P_{n+1}^{(\alpha,\beta)}(x) \\ &= (2n + \alpha + \beta + 1)[(2n + \alpha + \beta)(2n + \alpha + \beta + 2)x + \alpha^2 - \beta^2] \\ &\quad \times P_n^{(\alpha,\beta)}(x) - 2(n + \alpha)(n + \beta)(2n + \alpha + \beta + 2)P_{n-1}^{(\alpha,\beta)}(x). \end{aligned}$$

If we define

$$R_n^{(\alpha,\beta)}(x) = \frac{2^n\Gamma(n + \alpha + \beta + 1)n!}{\Gamma(2n + \alpha + \beta + 1)}P_n^{(\alpha,\beta)}(x) = x^n + \cdots$$

we have

$$(13)\quad \begin{aligned} &R_1^{(\alpha,\beta)}(x)R_n^{(\alpha,\beta)}(x) \\ &= R_{n+1}^{(\alpha,\beta)}(x) + \frac{\alpha - \beta}{\alpha + \beta + 2}\left[1 - \frac{(\alpha + \beta + 2)(\alpha + \beta)}{(2n + \alpha + \beta + 2)(2n + \alpha + \beta)}\right]R_n^{(\alpha,\beta)}(x) \\ &\quad + \frac{4(n + \alpha + \beta)(n + \alpha)(n + \beta)n}{(2n + \alpha + \beta + 1)(2n + \alpha + \beta)^2(2n + \alpha + \beta - 1)}R_{n-1}^{(\alpha,\beta)}(x) \\ &= R_{n+1}^{(\alpha,\beta)}(x) + \alpha_n R_n^{(\alpha,\beta)}(x) + \beta_n R_{n-1}^{(\alpha,\beta)}(x). \end{aligned}$$

It is clear that $\alpha_n \geqq 0$, $\beta_n > 0$ if $\alpha \geqq \beta$ and that $\alpha_{n+1} \geqq \alpha_n$ if either $\alpha + \beta \geqq 0$ or $\alpha = \beta$. We take first the case $\alpha = \beta$ which is much easier. Then

$$\beta_n = \frac{(n + 2\alpha)n}{(2n + 2\alpha + 1)(2n + 2\alpha - 1)} = \frac{1}{4}\left[\frac{4n^2 + 8\alpha n}{4n^2 + 8\alpha n + 4\alpha^2 - 1}\right]$$
$$= \frac{1}{4}\left[1 - \frac{(4\alpha^2 + 1)}{4n^2 + 8\alpha n + 4\alpha^2 + 1}\right]$$

and clearly $\beta_{n+1} \geqq \beta_n$ if $\alpha \geqq \frac{1}{2}$. (1) fails for $-1 < \alpha = \beta < -\frac{1}{2}$ since the coefficient of $p_3(x)$ in $p_2(x)p_3(x) = \sum_{k=0}^{5} \alpha_k p_k(x)$ is negative. This still leaves us with an open problem because the theorem is known to hold for $\alpha = \beta \geqq -\frac{1}{2}$ (see [9]), and it would be desirable to have this result follow from a general theorem. It is not surprising that the proof fails for these values of α, since the author has long suspected that the theorem should be easier to prove for large α. This is because $P_n^{(\alpha,\beta)}(1)$ is much larger than $P_n^{(\alpha,\beta)}(x)$, $-1 < x < 1$, x fixed, for α large and large n.

There are two other special cases which we can handle easily, $\alpha + \beta = 0$ and $\alpha + \beta = 1$. If $\alpha + \beta = 0$ then

$$\beta_n = \frac{4n^2(n^2 - \alpha^2)}{(4n^2 - 1)4n^2} = \frac{n^2 - \alpha^2}{4n^2 - 1} = \frac{1}{4}\left[1 - \frac{\alpha^2 - \frac{1}{4}}{n^2 - \frac{1}{4}}\right]$$

and $\beta_{n+1} \geqq \beta_n$ if $\alpha^2 - \frac{1}{4} \geqq 0$ or $\alpha \geqq \frac{1}{2}$. If $\alpha + \beta = 1$ then

$$\beta_n = \frac{4(n + 1)n(n + \alpha)(n + 1 - \alpha)}{(2n + 2)(2n + 1)^2(2n)} = \frac{(n + \alpha)(n + 1 - \alpha)}{4(n + \frac{1}{2})^2}$$
$$= \frac{(n + \frac{1}{2})^2 - (\frac{1}{2} - \alpha)^2}{4(n + \frac{1}{2})^2}$$

and $\beta_{n+1} \geqq \beta_n$ if $\alpha \geqq \frac{1}{2}$, $\alpha = 1 - \beta$.

A natural conjecture is that $\beta_{n+1} \geqq \beta_n$ if $\alpha \geqq \frac{1}{2}$. However, to have $\beta_{n+1} \geqq \beta_n$ and $\beta_n \to \frac{1}{4}$ we must have $\beta_n < \frac{1}{4}$ and it is easy to check that $\beta_n > \frac{1}{4}$ for large n if $\alpha^2 + \beta^2 < \frac{1}{2}$. In spite of the seeming complexity of the problem of finding out when $\beta_{n+1} \geqq \beta_n$, it is relatively easy. We let $k = (\alpha + \beta)/2$. Then

$$4\beta_n = \frac{(n^2 + 2kn)(n^2 + 2kn + \alpha\beta)}{(n + k)^2[(n + k)^2 + \frac{1}{4}]} = \left[1 - \frac{k^2}{(n + k)^2}\right]\left[1 - \frac{(k^2 - \frac{1}{4} - \alpha\beta)}{(n + k)^2 - \frac{1}{4}}\right].$$

If $k^2 \geqq \frac{1}{4} + \alpha\beta$, or $\alpha - \beta \geqq 1$, we are finished since each of the factors on the right-hand side is an increasing function of n. A long routine calculation allows us to conclude that $\beta_{n+1} \geqq \beta_n$ if $\alpha - \beta < 1$, $\alpha + \beta \geqq 1$ or if $\alpha - \beta < 1$, $\alpha + \beta < 1$ and

$$(2\alpha^2 + 2\beta^2 - 1)[(\alpha + \beta + 2)^2(\alpha + \beta + 4)^2] \\ \geqq [(\alpha + \beta + 4)^2 + (\alpha + \beta + 2)^2 - 1](\alpha - \beta)^2(\alpha + \beta)^2.$$

However, the correct conclusion for the non-negativity of the coefficients in (1) for Jacobi polynomials is probably $\alpha \geqq \beta$ and $\alpha + \beta \geqq -1$; therefore we shall spare the reader the calculations which led to these results. The last result adds very little to $|\alpha \pm \beta| \geqq 1$, $\alpha \geqq |\beta|$ in any case.

As was pointed out in [2] there is a symmetry between (α, β) and (β, α). We have

$$P_n^{(\alpha,\beta)}(-x) = (-1)^n P_n^{(\beta,\alpha)}(x);$$

thus the results we obtained for $\alpha \geqq \beta$ by normalizing so that $P_n^{(\alpha,\beta)}(1) > 0$ could be obtained for $\alpha \leqq \beta$ if we normalize so that $P_n^{(\alpha,\beta)}(-1) > 0$.

The methods used to prove Theorem 1 can be used to prove other theorems as well. Theorem 2 is equivalent to Theorem 1, but it is worth stating because it has the same form as Weinberger's theorem.

THEOREM 2. *Let* $\Delta_n k(n) = k(n+1) + \alpha_n k(n) + \beta_n k(n-1)$, $n = 0, 1, 2, \ldots$, $k(-1) = 0$. *If* $a(n)$, $n = 0, 1, \ldots$, *is given we consider the difference equation*

$$\Delta_n a(n, m) = \Delta_m a(n, m), \qquad n, m = 0, 1, \ldots, \tag{14}$$

where $a(n, 0) = a(0, n) = a(n)$, $a(n, -1) = a(-1, n) = 0$. *Then if* $a(n) \geqq 0$, $n = 0, 1, \ldots$, *and* $\alpha_n \geqq 0$, $\beta_{n+1} \geqq 0$, $\alpha_{n+1} \geqq \alpha_n$, $\beta_{n+2} \geqq \beta_{n+1}$, $n = 0, 1, \ldots$, *we have* $a(n, m) \geqq 0$, $n, m = 0, 1, \ldots$.

The proof is exactly the same as the proof of Theorem 1. (14) can be considered as a strange type of hyperbolic difference equation and this raises the question of which other hyperbolic difference equations have a maximum principle.

We give one other application of this type of argument. About twenty years ago P. Turan established a very interesting inequality for Legendre polynomials

$$P_{n-1}(x)P_{n+1}(x) - [P_n(x)]^2 < 0, \qquad -1 < x < 1. \tag{15}$$

Inequalities of this sort were exhaustively studied by Karlin and Szegö for most of the classical polynomials in [13]. It has been thought that these inequalities were in some sense restricted to the classical polynomials because they are the only polynomials which satisfy second-order equations in both n and x. We shall obtain a Turan-type inequality for a wide class of orthogonal polynomials which will contain the known Turan inequality for Hermite polynomials.

THEOREM 3. *Let $p_n(x)$ satisfy $p_n(x) = x^n + \cdots$ and*

$$p_1(x)p_n(x) = p_{n+1}(x) + b_n p_{n-1}(x) \tag{16}$$

where $p_1(x) = x + a$ and $b_n > 0$, $b_{n+1} \geqq b_n$, $n = 1, 2, \ldots$. Then

$$p_{n-1}(x)p_{n+1}(x) - p_n^2(x) < 0, \qquad -\infty < x < \infty, \qquad n = 1, 2, \ldots. \tag{17}$$

For $n = 1$, (17) is obvious since $p_0(x)p_2(x) - p_1^2(x) = -b_1 < 0$. We complete the proof by induction.

$$\begin{aligned} p_{n+1}(x)p_{n-1}(x) - p_n^2(x) &= p_1 p_n p_{n-1} - b_n p_{n-1}^2 - p_n^2 \\ &= p_n[p_n + b_{n-1}p_{n-2}] - b_n p_{n-1}^2 - p_n^2 \\ &= (b_{n-1} - b_n)p_{n-1}^2 + b_{n-1}[p_n p_{n-2} - p_{n-1}^2]. \end{aligned} \tag{18}$$

It is interesting to observe that Theorem 3 is in one sense a best possible theorem. In [13, p. 131] it was pointed out that if $p_n(x)$ satisfies (17) and $r_n(x) = k_n p_n(x)$, then $r_n(x)$ satisfies (17) if $k_n k_{n+2} > 0$, $k_n k_{n+2} - k_{n+1}^2 \leqq 0$, $n = 0, 1, \ldots$. In [1] it was observed that these conditions are necessary if $p_n(x) = a^n x^n + \cdots$. Thus the polynomials $p_n(x)$ with the normalization $p_n(x) = x^n + \cdots$ are a natural class to consider. For this class, $b_n > 0$ and $b_{n+1} \geqq b_n$ are necessary conditions for (17) to hold. In (18) if we take x to be a zero of $p_{n-1}(x)$ we see that $b_{n-1} > 0$ is necessary. Also in (18), if we look at the highest power of x which appears on the right-hand side it is $(b_{n-1} - b_n)x^{2n-2}$; thus $b_{n-1} \leqq b_n$ is necessary.

Whenever non-negative numbers occur in a problem it is natural to see if any probability is lurking in the background. We have no probabilistic meaning for the conclusion, but the hypothesis does have a probabilistic interpretation. Karlin and McGregor [11] have shown that to each birth and death process on the non-negative integers with zero a reflecting barrier there corresponds a sequence of orthogonal polynomials whose spectral measure is concentrated on $[0, \infty)$. These polynomials $Q_n(x)$ satisfy

$$\begin{aligned} Q_0(x) &= 1, \\ -xQ_n(x) &= \mu_n Q_{n-1}(x) - (\lambda_n + \mu_n)Q_n(x) + \lambda_n Q_{n+1}(x), \\ &\qquad \mu_0 = 0,\ \lambda_n > 0,\ \mu_{n+1} > 0,\ n = 0, 1, \ldots. \end{aligned} \tag{19}$$

From (19) it is easy to see that

$$Q_n(x) = \frac{(-1)^n}{\lambda_0\lambda_1 \cdots \lambda_{n-1}} x^n + \cdots;$$

thus

$$R_n(x) = \lambda_0\lambda_1 \cdots \lambda_{n-1}(-1)^n Q^n(x)$$

is normalized in the same way we normalized our polynomials. (19) then becomes

$$(20) \qquad R_1(x)R_n(x) = R_{n+1}(x) + (\lambda_n + \mu_n + \lambda_0)R_n(x) + \mu_n\lambda_{n-1}R_{n-1}(x).$$

A sufficient condition that the coefficients in (20) satisfy the hypothesis of Theorem 1 is that $\lambda_{n+1} \geqq \lambda_n > 0$, $\mu_{n+1} \geqq \mu_n > 0$. In terms of birth and death processes this condition says that the rate of absorption from state n to state $n + 1$ which is given by λ_n does not decrease with n nor does the rate of absorption from state n to state $n - 1$ which is given by μ_n.

We conclude with some references to earlier work on Theorem 1. For Legendre polynomials (1) was stated by Ferrers [7] and proofs were given shortly thereafter by a number of people. Dougall [4] stated (1) for ultraspherical polynomials and a proof was first given by Hsü [9]. For Hermite polynomials (1) was given by Nielson [15] and for Laguerre polynomials the α_k were first computed by Watson [17] and given in a different form so that it was obvious that (1) holds by Erdelyi [5]. Hylleraas [10] proved (1) for Jacobi polynomials for $\alpha = \beta + 1$ and in unpublished work Gangolli has remarked that (1) holds for $\alpha = k$, $\beta = 0$; $\alpha = 2k + 1$, $\beta = 1$; $\alpha = 7$, $\beta = 3$, $k = 1, 2, \ldots$. The case $\beta = -\frac{1}{2}$, $\alpha \geqq \beta$ follows from the case $\alpha = \beta$ in a standard way. Thus we have new proofs of (1) in all the cases that have previously been established except $\alpha = \beta + 1$, $-1 < \beta < -\frac{1}{2}$ and $-\frac{1}{2} < \alpha = \beta < \frac{1}{2}$. However, in all the cases except the ones considered by Gangolli, α_k was explicitly found, and it is sometimes necessary to have it exactly to use (1). α_k can be computed explicitly in the case of Jacobi polynomials, but it seems impossible to use it in the form that it has been found [14, (3.7)]. In a preprint just received, G. Gasper has proved (1) for Jacobi polynomials, $P_n^{(\alpha,\beta)}(x)$, $\alpha + \beta + 1 \geqq 0$, $\alpha \geqq \beta$.

University of Wisconsin
Madison

REFERENCES

1. Askey, R., "On some problems posed by Karlin and Szegö concerning orthogonal polynomials," *Boll. U.M.I.*, (3) *XX* (1965), 125–127.
2. Askey, R., and S. Wainger, "A dual convolution structure for Jacobi polynomials," in *Orthogonal Expansions and Their Continuous Analogues*. Carbondale, Ill.: Southern Illinois Press, 1968, pp. 25–26.
3. Bochner, S., "Positive zonal functions on spheres," *Proc. Nat. Acad. Sci.*, *40* (1954), 1141–1147.
4. Dougall, J. "A theorem of Sonine in Bessel functions, with two extensions to spherical harmonics," *Proc. Edinburgh Math. Soc.*, *37* (1919), 33–47.
5. Erdelyi, A., "On some expansions in Laguerre polynomials," *J. London Math. Soc.*, *13* (1938), 154–156.

6. ERDELYI, A., *Higher Transcendental Functions*, Vol. 2. New York: McGraw-Hill, Inc., 1953.
7. FERRERS, N. M., *An Elementary Treatise on Spherical Harmonics and Subjects Connected With Them*. London, 1877.
8. HIRSCHMAN, I. I., Jr., "Harmonic Analysis and Ultraspherical Polynomials," *Symposium on Harmonic Analysis and Related Integral Transforms*. Cornell, 1956.
9. HSÜ, H. Y., "Certain integrals and infinite series involving ultraspherical polynomials and Bessel functions," *Duke Math. J.*, *4* (1938), 374–383.
10. HYLLERAAS, EGIL A., "Linearization of products of Jacobi polynomials," *Math. Scand.*, *10* (1962), 189–200.
11. KARLIN, S., and J. L. MCGREGOR, "The differential equations of birth and death processes and the Stieltjes moment problem," *Trans. Amer. Math. Soc.*, *86* (1957), 489–546.
12. ———, "Linear growth, birth and death processes," *J. Math. Mech.*, *7* (1958), 643–662.
13. KARLIN, S., and G. SZEGÖ, "On certain determinants whose elements are orthogonal polynomials," *J. d'Anal. Math.*, *8* (1961), 1–157.
14. MILLER, W., "Special functions and the complex Euclidean group in 3-space, II," *J. Math. Phys.*, *9* (1968), 1175–1187.
15. NIELSEN, N. "Recherches sur les polynomes d'Hermite," *Det. Kgl. Danske Viden. Selskab. Math. fys. Medd.* I, 6 (1918).
16. SZEGÖ,G. *Orthogonal Polynomials. Amer. Math. Soc. Colloq. Pub.* 23. Providence, R.I.: American Mathematical Society, 1959.
17. WATSON, G. N., "A note on the polynomials of Hermite and Laguerre," *J. London Math. Soc.*, *13* (1938), 29–32.
18. WEINBERGER, H., "A maximum property of Cauchy's problem," *Ann. of Math.*, (2) *64* (1956), 505–513.

Eisenstein Series on Tube Domains

WALTER L. BAILY, JR.[1]

Introduction

We wish to prove here that under certain conditions the Eisenstein series for an arithmetic group acting on a tube domain [8], [9] in $\mathbf{C}^m$ generate the field of automorphic functions for that group. The conditions are that the domain be equivalent to a symmetric bounded domain having a 0-dimensional rational boundary component (with respect to the arithmetic group) [3, Section 3] and that the arithmetic group be maximal discrete in the (possibly not connected) Lie group of all holomorphic automorphisms of the domain. More precisely, under the same conditions, we prove that certain types of polynomials in certain linear combinations of the Eisenstein series generate a graded ring $\mathfrak{R}$ such that the graded ring of all automorphic forms of weights which are multiples of a certain integer with respect to the given arithmetic group, satisfying a certain "growth condition at infinity," is the integral closure of $\mathfrak{R}$ in its quotient field. The objective in proving this result is that by using it one can prove that the field generated by the Fourier coefficients of certain linear combinations of Eisenstein series is also a field of definition for the Satake compactification of the quotient of the domain by the arithmetic group [3]; when that field is an algebraic number field, and especially when it is the rational number field, one may hope for arithmetic results to follow. Our methods here have been adapted largely from the proof of this result in a special case (see [12]).

We now describe our results in more detail.

Let G be a connected, semisimple, linear algebraic group defined over the rational number field $\mathbf{Q}$. Let $\mathbf{R}$ be the real number field. We assume at the outset that G is $\mathbf{Q}$-simple, that $G^0_{\mathbf{R}}$ (the identity component of $G_{\mathbf{R}}$) is centerless (i.e., has center reduced to $\{e\}$) and has no compact simple factors, and that if K is a maximal compact subgroup of $G^0_{\mathbf{R}}$, then $\mathfrak{X} = K\backslash G^0_{\mathbf{R}}$ is a Hermitian symmetric space. (Later it will be easy to relax

[1] The author wishes to acknowledge support for research on the subject matter of this paper from NSF grant GP 6654.

these assumptions somewhat.) Then G is centerless (see [3, Section 11.5]), the absolutely simple factors of G (all centerless) are defined over the algebraic closure of $\mathbf{Q}$ in $\mathbf{R}$, and we may write (see [3, Section 3.7]) $G = \mathscr{R}_{\mathbf{k}/\mathbf{Q}}G'$, where G' is absolutely simple and $\mathbf{k}$ is a totally real number field. Moreover, $\mathfrak{X}$ is isomorphic to a symmetric bounded domain D in $\mathbf{C}^m$. We make the further assumption that $\mathfrak{X}$ is isomorphic to a tube domain

$$\mathfrak{T} = \{X + iY \in \mathbf{C}^m \mid Y \in \mathfrak{R}\}, \tag{1}$$

where $\mathfrak{R}$ is a homogeneous, self-adjoint cone in $\mathbf{R}^m$; according to [8, Sections 4.5, 4.9, and 6.8 (Remark 1)], this can also be expressed by saying that the relative $\mathbf{R}$-root system ${}_{\mathbf{R}}\Sigma$ of G does not contain the double of any element of ${}_{\mathbf{R}}\Sigma$, i.e., that ${}_{\mathbf{R}}\Sigma$ is a sum of simple root systems of type C.

In [3, Section 3], we have defined the concept of "rational boundary component of D" and have proved that a boundary component F of D is rational if and only if the complexification P of

$$N(F) = \{g \in G^0_{\mathbf{R}} \mid Fg = F\}$$

is defined over $\mathbf{Q}$. We now add the assumption that there exists a rational boundary component F_0 of D such that $\dim F_0 = 0$. Then $\mathfrak{X}$ may be identified with the tube domain $\mathfrak{T}$ in such a way that every element of $N(F_0)$ acts by a linear affine transformation of the ambient vector space $\mathbf{C}^m$ of $\mathfrak{T}$ and such that every element of the unipotent radical $U = U(F_0)$ of $N(F_0)$ acts by real translations. If F is any boundary component of D, let $N_h(F)$ be the normalizer of $N(F)$ in the group G_h of all holomorphic automorphisms of $\mathfrak{X}$; in particular, let $N_h(F_0) = N_h$.

Let Γ be an arithmetic subgroup of G_h; i.e., $\Gamma \cap G_{\mathbf{Z}}$ is of finite index in Γ and in $G_{\mathbf{Z}}$. Put $\Gamma' = G^0_{\mathbf{R}} \cap \Gamma$ and let $\Gamma_0 = \Gamma \cap N_h$. Clearly Γ' is a normal subgroup of Γ. If $g \in G_h$ and $Z \in \mathfrak{T}$, let $j(Z, g)$ denote the functional determinant of g at Z. We shall see that for $\gamma \in \Gamma_0, j(z, \gamma)$ is a root of unity. Let $G'_{\mathbf{Q}} = G_{\mathbf{Q}} \cap G^0_{\mathbf{R}}$ and if $a \in G'_{\mathbf{Q}}$, let $\Gamma_{0,a} = \Gamma \cap aN_ha^{-1}$ and let l_a be the least common multiple of the orders of all the roots of unity which occur in the form $j(*, \gamma)$ for $\gamma \in a^{-1}\Gamma_{0,a}a$. Then for any sufficiently large positive integer l, divisible by l_a, the series

$$E_{l,a}(Z) = \sum_{\nu \in \Gamma/\Gamma_{0,a}} j(Z, \gamma a)^l \tag{2}$$

converges absolutely and uniformly on compact subsets of $\mathfrak{T}$ and represents there an automorphic form with respect to Γ.

Let $G^*_{\mathbf{Q}}$ be the normalizer in G_h of $G'_{\mathbf{Q}}$. We shall see that $\Gamma \subset G^*_{\mathbf{Q}}$. Let $N_{\mathbf{Q}} = N_h \cap G^*_{\mathbf{Q}}$. Then we shall also see that $G^*_{\mathbf{Q}}$ is the disjoint union of a finite number of double cosets $\Gamma a N_{\mathbf{Q}}$, where a runs over a finite subset A

of $G'_{\mathbf{Q}}$. For each $a \in A$, let c_a be a complex number and denote the assignment $a \mapsto c_a$ by c. We put

(3) $$E_{l,c} = \sum_{a \in A} c_a^l \cdot E_{l,a}.$$

Our main interest will be for the case when $c_a \neq 0$ for all $a \in A$. Let $\mathfrak{V} = \mathfrak{T}/\Gamma$ and let $\mathfrak{V}^*$ be the Satake compactification (see [3]) of $\mathfrak{V}$. The main result of this paper is the following theorem:

THEOREM. *Let Γ be a maximal discrete arithmetic subgroup of G_h, let other notation be as above, and fix the mapping c such that $c_a \neq 0$ for all $a \in A$. Then every meromorphic function on $\mathfrak{V}^*$ can be expressed as the quotient of two isobaric polynomials in the Eisenstein series $\{E_{l,c}\}_{l_0 | l}$, $l > 0$, where l_0 is the l.c.m. of all l_a for $a \in A$. If* **B** *is a basis for the module of isobaric polynomials of sufficiently high weight in these, then the elements of* **B** *may be used as the coordinates of a well defined mapping Ψ of $\mathfrak{V}^*$ into complex projective space. The variety $\mathfrak{W} = \Psi(\mathfrak{V}^*)$ is birationally equivalent to $\mathfrak{V}^*$, $\mathfrak{V}^*$ is the normal model of $\mathfrak{W}$, and if* **k** *is a field of definition for $\mathfrak{W}$ (e.g., the field generated by all Fourier coefficients of all elements of* **B**), *then there exists a projective variety defined over* **k** *which is biregularly equivalent to $\mathfrak{V}^*$.*

1. Notational conventions

In what follows, if H is a group, g, an element of H, and X, a subset of H, let ${}^gX = gXg^{-1}$. If P is a subgroup of H, the phrase "P is self-normalizing" will be used to mean that P is its own normalizer in H. If H is topological, H^0 will denote the component of H containing the identity. Use **C** and **Z** to denote the complex numbers and rational integers, respectively.

2. Tube domains

In this section only, G is a centerless, connected, *simple* linear algebraic group defined over **R**. With notation as in the introduction, the space $\mathfrak{X} = K\backslash G^0_{\mathbf{R}}$ is isomorphic to a tube domain given by (1). The cone $\mathfrak{K}$ may be described as the interior of the set of squares of a real Jordan algebra J with $\mathbf{R}^m$ as underlying vector space. We denote by N the Jordan algebra norm in J. We have $[G_h : G^0_{\mathbf{R}}] = 1$ or 2, and the domains included under each case are described in [3, Section 11.4]. In every case, $G^0_{\mathbf{R}}$ contains an element σ which acts on $\mathfrak{T}$ by $\sigma(Z) = -Z^{-1}$ (Jordan algebra inverse), and $j(Z, \sigma) = \pm \mathrm{N}(Z)^{-\nu}$ for a certain positive integer ν. In each case where $[G_h : G^0_{\mathbf{R}}] = 2$, G_h contains an element τ of order two, not in $G^0_{\mathbf{R}}$, such that

τ operates on $\mathfrak{T}$ by a linear transformation of $\mathbf{C}^m$. Hence, $j(Z, \tau) = \pm 1$ (a constant function of Z).

Let ${}_{\mathbf{R}}T$ be a maximal $\mathbf{R}$-trivial torus of G with relative $\mathbf{R}$-root system ${}_{\mathbf{R}}\Sigma$ and simple $\mathbf{R}$-root system ${}_{\mathbf{R}}\Delta$. Then ${}_{\mathbf{R}}\Sigma$ is of type C and we may speak of the compact and noncompact roots in ${}_{\mathbf{R}}\Sigma$ (see [3, Section 1]). Precisely one simple root α is noncompact. Let S be the 1-dimensional subtorus of ${}_{\mathbf{R}}T$ on which all simple $\mathbf{R}$-roots vanish except α. The centralizer $\mathscr{Z}(S)$ of S and the positive $\mathbf{R}$-root groups in G generate a maximal $\mathbf{R}$-parabolic subgroup P of G. Let $P_1 = P \cap G^0_{\mathbf{R}}$. Then $P_1 = N(F_0)$ for some 0-dimensional boundary component F_0 of (the bounded realization of) $\mathfrak{X}$, and $\mathfrak{X}$ may be identified with $\mathfrak{T}$ in such a way that every element of P_1 acts by a linear affine transformation on the ambient affine space $\mathbf{C}^m$ of $\mathfrak{T}$ and the unipotent radical U of P_1, by real translations. Using the Bruhat decomposition relative to $\mathbf{R}$, one may prove that if $g \in G^0_{\mathbf{R}}$, then $g \in P_1$ is also necessary in order for g to act by a linear transformation of $\mathbf{C}^m$. Direct calculation shows that τ normalizes P_1 in each case where $[G_h : G^0_{\mathbf{R}}] = 2$, and since P_1 is self-normalizing in $G^0_{\mathbf{R}}$ (see [3, Section 1.5(1)]), it follows that τ and P_1 generate the normalizer N_h of P_1 in G_h and that $N_h = P_1 \cup \tau P_1$. Hence, $j(Z, g)$ is a constant function of Z if and only if $g \in N_h$; and $N(F_0) = N_h \cap G^0_{\mathbf{R}}$. Moreover, $\sigma \in \mathscr{N}({}_{\mathbf{R}}T)$, and viewed as an element of the relative $\mathbf{R}$-Weyl group of G, sends every $\mathbf{R}$-root into its negative. Then if $g \in \mathscr{Z}({}_{\mathbf{R}}T)\sigma$, we have $j(Z, g) = c \cdot \mathrm{N}(Z)^{-\nu}$, where c is a constant, since $\mathscr{Z}({}_{\mathbf{R}}T) \subset \mathscr{Z}(S)$ acts linearly on $\mathfrak{T}$ and in particular each element of it multiplies the norm function N by a constant.

3. Relative root systems

We now lift the assumption that G be absolutely simple and return to the general assumptions of the introduction. Then $G = \mathscr{R}_{\mathbf{k}/\mathbf{Q}}G'$, where $\mathbf{k}$ is a totally real algebraic number field and G' is simple and defined over $\mathbf{k}$. According to [3, Section 2.5], we may choose a maximal $\mathbf{Q}$-trivial torus ${}_{\mathbf{Q}}T$, a maximal $\mathbf{R}$-trivial torus ${}_{\mathbf{R}}T$, and a maximal torus T in G such that T is defined over $\mathbf{Q}$ and ${}_{\mathbf{Q}}T \subset {}_{\mathbf{R}}T \subset T$. Then for suitable maximal $\mathbf{k}$-trivial torus ${}_{\mathbf{k}}T'$, $\mathbf{R}$-trivial torus ${}_{\mathbf{R}}T'$, and maximal $\mathbf{k}$-torus T' in G' we have ${}_{\mathbf{k}}T' \subset {}_{\mathbf{R}}T' \subset T'$, ${}_{\mathbf{Q}}T \subset \mathscr{R}_{\mathbf{k}/\mathbf{Q}}({}_{\mathbf{k}}T')$, $T = \mathscr{R}_{\mathbf{k}/\mathbf{Q}}T'$. We take compatible orderings (see [3, Section 2.4]) on all the root systems. If P is a $\mathbf{Q}$-parabolic subgroup of G and U is the unipotent radical of P, then $P = \mathscr{R}_{\mathbf{k}/\mathbf{Q}}P'$ and $U = \mathscr{R}_{\mathbf{k}/\mathbf{Q}}U'$, where P' is a $\mathbf{k}$-parabolic subgroup of G' and U' is its unipotent radical; if P is maximal $\mathbf{Q}$-parabolic, then P' is maximal $\mathbf{k}$-parabolic, and even maximal $\mathbf{R}$-parabolic in G'. If $\boldsymbol{\alpha} \in {}_{\mathbf{R}}\Sigma$, then the restriction of $\boldsymbol{\alpha}$ to ${}_{\mathbf{Q}}T$ is either zero or a simple $\mathbf{Q}$-root; in the latter case, $\boldsymbol{\alpha}$ is called *critical*; each simple $\mathbf{Q}$-root is the restriction of precisely one critical **R**-root for each irreducible factor (see [3, Sections 2.9, 2.10]). If F is a rational boundary component, then $N(F)_{\mathbf{C}} = P$ is a maximal **Q**-parabolic subgroup of G (see [3, Section 3.7]). Until further notice, let F be the 0-dimensional rational boundary component F_0 of the introduction, let $P = N(F)_{\mathbf{C}}$, and let U be the unipotent radical of P. We may assume the relationship between P, ${}_{\mathbf{Q}}T$, and ${}_{\mathbf{Q}}\Delta$ to be that P contains the minimal **Q**-parabolic subgroup of G generated by $\mathscr{Z}({}_{\mathbf{Q}}T)$ and the unipotent subgroup of G generated by the positive **Q**-root spaces. By [5, Section 4.3], this determines P within its $G_{\mathbf{Q}}$-conjugacy class. Since $\dim F = 0$, it follows that the noncompact simple root $\boldsymbol{\alpha}_i$ in each irreducible factor G_i of G is critical, that ${}_{\mathbf{Q}}\Sigma$ is of type C, and, using the latter fact to define the compact and noncompact **Q**-roots, that each $\boldsymbol{\alpha}_i$ restricts onto the noncompact, simple **Q**-root $\boldsymbol{\beta}$. It follows from this that the noncompact positive (respectively noncompact negative, respectively compact) **R**-roots are those in whose restriction to ${}_{\mathbf{Q}}T$, expressed as a sum of simple **Q**-roots, $\boldsymbol{\beta}$ appears with coefficient $+1$ (respectively -1, respectively 0).

If S denotes the 1-dimensional subtorus of ${}_{\mathbf{Q}}T$ annihilated by all simple **Q**-roots except $\boldsymbol{\beta}$, then P is generated by $\mathscr{Z}(S)$ and U. We have ${}_{\mathbf{R}}T \subset \mathscr{Z}({}_{\mathbf{Q}}T) \subset \mathscr{Z}(S)$, and all elements of $\mathscr{Z}(S) \cap G^0_{\mathbf{R}}$ act by linear transformations on $\mathfrak{T}$.

Let $g \in \mathscr{N}({}_{\mathbf{Q}}T)_{\mathbf{R}}$ represent the element of the relative Weyl group ${}_{\mathbf{Q}}\mathbf{W}$ which sends every **Q**-root into its negative. Then g normalizes $\mathscr{Z}({}_{\mathbf{Q}}T)$ and sends ${}_{\mathbf{R}}T$ into another maximal **R**-trivial torus of $\mathscr{Z}({}_{\mathbf{Q}}T)$. Hence, we may find $h \in \mathscr{Z}({}_{\mathbf{Q}}T)_{\mathbf{R}}$ such that $gh \in \mathscr{N}({}_{\mathbf{R}}T)$, and we may assume $gh \in G^0_{\mathbf{R}}$ because ${}_{\mathbf{R}}T_{\mathbf{R}}$ meets every component of $G_{\mathbf{R}}$ (see [5, Section 14.4]). Replace g by gh. By our preceding considerations, it is clear that $\operatorname{Ad} g$ transforms every positive noncompact **R**-root into a negative one.

On the other hand, for each irreducible factor G_i of G, let σ_i be the element of $G^0_{i\mathbf{R}}$ sending Z_i into $-Z_i^{-1}$, and let $\sigma = \prod_i \sigma_i$. Then $\operatorname{Ad} \sigma$ sends every **R**-root into its negative. Hence, σg normalizes and so belongs to P_1. It follows that $j(Z, g) = \text{const} \cdot \prod_i \mathrm{N}_i(Z_i)^{-\nu_i}$, where N_i is the Jordan algebra norm for the ith irreducible factor, Z_i is the component of Z in the ith irreducible factor of $\mathfrak{T}$, and ν_i are certain strictly positive integers.

4. The Bruhat decomposition

Retaining the notation of Section 3, let $\mathbf{N}^+$ denote the unipotent subgroup of G generated by the subgroups of G associated to the positive **Q**-roots. Let $\mathbf{N}^-$ be the similarly defined group associated to the set of negative **Q**-roots. Of course, $\mathbf{N}^+_{\mathbf{R}}$ is connected and therefore $\mathbf{N}^+_{\mathbf{Q}} \subset G^0_{\mathbf{R}}$. Let $G^0_{\mathbf{Q}}$

denote the subgroup of $G_{\mathbf{Q}}$ generated by all the groups ${}^{g}\mathbf{N}^{+}_{\mathbf{Q}}$ for $g \in G_{\mathbf{Q}}$ (see [13]). Obviously $G^{0}_{\mathbf{Q}} \subset G^{0}_{\mathbf{R}}$.

The main facts we need are the following (see [13, Section 3.2 (18) et al.]): (i) The group $G_{\mathbf{Q}}$ can be written as the disjoint union of the sets

$$\text{(4)} \qquad \mathbf{N}^{+}_{\mathbf{Q}} w \mathscr{Z}({}_{\mathbf{Q}}T)_{\mathbf{Q}} \mathbf{N}^{+}_{\mathbf{Q}} \qquad \text{(Bruhat decomposition)},$$

where w runs over a complete set of representatives of the cosets of $\mathscr{Z}({}_{\mathbf{Q}}T)_{\mathbf{Q}}$ in $\mathscr{N}({}_{\mathbf{Q}}T)_{\mathbf{Q}}$, and (ii) $G_{\mathbf{Q}} = \mathscr{Z}({}_{\mathbf{Q}}T)_{\mathbf{Q}} \cdot G^{0}_{\mathbf{Q}}$. It follows from (ii) that each of the elements w appearing in (i) can be taken in $G^{0}_{\mathbf{Q}} \subset G^{0}_{\mathbf{R}}$. Since $P_0 = \mathscr{Z}({}_{\mathbf{Q}}T) \cdot \mathbf{N}^{+}$ is a minimal $\mathbf{Q}$-parabolic subgroup of G, it follows from the remark just made and from [5, Section 4.13] that if two $\mathbf{Q}$-parabolic subgroups of G are conjugate, then they are conjugate by an element of $G^{0}_{\mathbf{Q}} \subset G'_{\mathbf{Q}}$.

Let Γ be an arithmetic subgroup of $G^{0}_{\mathbf{R}}$. Since $G^{0}_{\mathbf{R}}$ has no compact simple factors, Γ is Zariski-dense in G. Let $\mathscr{A}(\Gamma)$ be the algebra of all finite linear combinations with rational coefficients of elements of Γ (viewed as matrices from the matrix representation of G).

In the Bruhat decomposition (4), if $w \in \mathscr{N}({}_{\mathbf{Q}}T)$ is such that $\operatorname{Ad} w(\mathbf{N}^{+}) \cap \mathbf{N}^{-}$ has dimension strictly smaller than that of $\mathbf{N}^{+}$, then the Zariski closure of $\mathbf{N}^{+}_{\mathbf{Q}} w \mathscr{Z}({}_{\mathbf{Q}}T)_{\mathbf{Q}} \mathbf{N}^{+}_{\mathbf{Q}}$ in G is a proper algebraic subset of G. Since ${}_{\mathbf{Q}}\Sigma$ is of type C, there is one element w_0 of ${}_{\mathbf{Q}}\mathbf{W}_{\mathbf{Q}}$ which sends every positive $\mathbf{Q}$-root into its negative; thus, $\operatorname{Ad} w_0(\mathbf{N}^{+}) \cap \mathbf{N}^{-}$ has the same dimension as $\mathbf{N}^{+}$; and w_0 is the only element of ${}_{\mathbf{Q}}\mathbf{W}_{\mathbf{Q}}$ for which this is true. Therefore, since Γ' is Zariski-dense in G (see [4, Theorem 1]), we have that

$$\Gamma' \cap \mathbf{N}^{+}_{\mathbf{Q}} w_0 \mathscr{Z}({}_{\mathbf{Q}}T)_{\mathbf{Q}} \mathbf{N}^{+}_{\mathbf{Q}}$$

is nonempty. Every element of $\mathbf{N}^{+}_{\mathbf{Q}}$ can be written in the form $u \cdot n'$, where $u \in U_{\mathbf{Q}}$ and $n' \in \mathbf{N}^{+}_{\mathbf{Q}} \cap \mathscr{Z}(S)$. Since w_0 normalizes $\mathscr{Z}(S)$ (by taking every compact $\mathbf{Q}$-root into another), it follows that if $\gamma_0 \in \Gamma' \cap \mathbf{N}^{+}_{\mathbf{Q}} w_0 \mathscr{Z}({}_{\mathbf{Q}}T)_{\mathbf{Q}} \mathbf{N}^{+}_{\mathbf{Q}}$, then we may write $\gamma_0 = u w_0 p$, where $p \in P_{\mathbf{Q}}$, and we may assume $w_0, p \in G'_{\mathbf{Q}}$, adjusting by an element of $\mathscr{Z}({}_{\mathbf{Q}}T)_{\mathbf{Q}}$ if necessary.

In the notation of the introduction, we see that $\Gamma' \subset G_{\mathbf{Q}}$, since G is centerless (see [4, Theorem 2]). Also Γ normalizes Γ' and hence normalizes the group algebra $\mathscr{A}(\Gamma')$ (over $\mathbf{Q}$) of Γ'. That group algebra contains $G_{\mathbf{Q}}$. In fact, consider the left regular representation of $G_{\mathbf{C}}$ on its group algebra over $\mathbf{C}$, let V be the vector space over $\mathbf{C}$ spanned by all the matrices in Γ', and let W be the vector space over $\mathbf{C}$ spanned by all the matrices in $G_{\mathbf{Q}}$; clearly, $V_{\mathbf{Q}} = \mathscr{A}(\Gamma')$ and $V \subset W$; since $\Gamma' \subset G_{\mathbf{Q}}$, $\Gamma' \cdot V_{\mathbf{Q}} = V_{\mathbf{Q}}$, and since Γ' is Zariski-dense in G, it follows that $G \cdot V = V$; since the identity matrix $e \in V$, we have $G_{\mathbf{C}} \subset V_{\mathbf{C}}$, hence $W = V$ and $G_{\mathbf{Q}} \subset W_{\mathbf{Q}} = V_{\mathbf{Q}}$. Moreover, since the adjoint representation of G on V is faithful (G is centerless) and defined over $\mathbf{Q}$, it is an isomorphism of algebraic groups, defined over $\mathbf{Q}$, between G and its image. Hence, $G_{\mathbf{Q}}$ may be identified with the set of $\mathbf{Q}$-rational points in the image of G. Therefore, $G_{\mathbf{Q}}$ is its own normalizer in $G_{\mathbf{C}}$. From the foregoing, we also see easily that Γ normalizes $G'_{\mathbf{Q}}$, i.e., $\Gamma \subset G^*_{\mathbf{Q}}$, and that $G^*_{\mathbf{Q}} \cap G^0_{\mathbf{R}} = G'_{\mathbf{Q}}$. Hence, $G'_{\mathbf{Q}}$ is of finite index in $G^*_{\mathbf{Q}}$.

If P_* is any $\mathbf{Q}$-parabolic subgroup of G and if H is any arithmetic subgroup of G, then $G_{\mathbf{Q}}$ is the union of finitely many disjoint double cosets $HaP_{*\mathbf{Q}}$. Thus in particular we may write

(5) $$G_{\mathbf{Q}} = \bigcup_{a \in A_1} \Gamma' a P_{\mathbf{Q}},$$

from which we have $G'_{\mathbf{Q}} = \bigcup_{a \in A_1} \Gamma' a N(F_0)_{\mathbf{Q}}$. Since P is self-normalizing, we see from property (ii) of $G^0_{\mathbf{Q}}$ that $G^*_{\mathbf{Q}} = G^0_{\mathbf{Q}} \cdot N_{\mathbf{Q}} = G'_{\mathbf{Q}} \cdot N_{\mathbf{Q}}$. Then it is obvious that we have

(6) $$G^*_{\mathbf{Q}} = \bigcup_{a \in A} \Gamma a N_{\mathbf{Q}} \qquad \text{(disjoint union)}$$

for some finite subset A of $G^0_{\mathbf{Q}}$.

5. Eisenstein series

We know that $G_h/G^0_{\mathbf{R}}$ is a finite Abelian group of type $(2, 2, \ldots, 2)$, if not trivial. With notation as before and $a \in G^*_{\mathbf{Q}}$, define $\Gamma_{0,a} = \Gamma \cap {}^aN_h$, $\Gamma'_{0,a} = \Gamma' \cap {}^aN_h$, $\Gamma_0 = \Gamma_{0,e}$, $\Gamma'_0 = \Gamma'_{0,e}$. We know that ${}^a\Gamma' \subset G'_{\mathbf{Q}}$ and we also know (see [3, Section 3.14]) that there exists a positive integer d_a such that if $\gamma \in a^{-1}\Gamma' a \cap N_h$, then $j(Z, \gamma)^{d_a} = 1$, and by Section 2, if $p \in N_h$, then $j(Z, p)$ is constant as a function of Z. Hence, since $\Gamma_{0,a}/\Gamma'_{0,a}$ is Abelian of type $(2, 2, \ldots, 2)$, if not trivial, we have $j(Z, \gamma)^{2d_a} = 1$ if $\gamma \in {}^{a^{-1}}\Gamma_{0,a}$. Let the set A be as in (6) and let d be the l.c.m. of all d_a, $a \in A$. Let l be a positive integer divisible by d. For each a that we have chosen, select a complex number c_a, and then form the series $E_{l,c}$ given in (3). By our choice of l, it is clear that the values of the individual terms $j(Z, \gamma a)^l$ do not depend on the choice of representatives γ of the cosets of $\Gamma_{0,a}$ in Γ. If this series converges, it clearly defines a holomorphic automorphic form on $\mathfrak{T}$ with respect to Γ.

If $b \in \Gamma$, write $b = gp$, with $g = g_b \in G^0_{\mathbf{Q}}$, $p = p_b \in N_{\mathbf{Q}}$. Write Γ as a disjoint union of double cosets $\Gamma' b \Gamma_{0,a}$, $b \in B$. Since Γ' is normal in Γ, we have $\Gamma' b \Gamma_{0,a} = b\Gamma'\Gamma_{0,a}$ and

(7) $$E_{l,a}(Z) = \sum_{b \in B} \left(\sum_{\gamma' \in \Gamma'/\Gamma'_{0,a}} j(Z, b\gamma' a)^l \right).$$

Using the cocycle identity (see [3, Section 1.8 (1)]) for j, the convergence of $E_{l,a}$ for sufficiently large l then follows from known convergence results on Eisenstein series [3, Section 7.2]. Writing $b = gp$ as above, we have

$$j(Z, b\gamma a) = j(Z, g(p\gamma p^{-1})(pap^{-1})p) = c \cdot j(Z, g\gamma_1 a_1),$$

where c is the constant value of $j(Z, p)$, $a_1, g \in G'_{\mathbf{Q}}$ (since $N_{\mathbf{Q}}$ normalizes the latter), and $\gamma_1 \in \Gamma_b = p\Gamma' p^{-1} \subset G'_{\mathbf{Q}}$. Let $\Gamma_{b,a_1} = {}^{a_1}N_h \cap \Gamma_b$. Then the inner sum in (7) becomes

$$\sum_{\gamma_1 \in \Gamma_b/\Gamma_{b,a_1}} j(Z, g\gamma_1 a_1)^l = E_{a_1,b,l}(Z), \tag{8}$$

which is an automorphic form of weight l with respect to the arithmetic group ${}^g\Gamma_b$. Therefore, $\Phi_F E_{l,a}$ may be calculated for any rational boundary component F by applying information already available (see [3, Sections 3.12, 7.7]) on the limits of Poincaré–Eisenstein series for arithmetic groups contained in $G'_{\mathbf{Q}}$. If g is a holomorphic automorphism of a domain $D \subset \mathbf{C}^m$, then $j(Z, g)$ is a nowhere vanishing holomorphic function on D. Using this fact and the information just referred to, one sees easily that for any rational boundary component F, $\Phi_F E_{l,a}$, if not $\equiv 0$, is the sum of a normally convergent (in general, infinite) series of lth powers of holomorphic functions of which not all are identically zero, such that each term not identically zero is a nowhere vanishing holomorphic function on F.

If $F_a = F_0 \cdot a^{-1}$, $a \in A$, we have for $a' \in A$

$$\begin{aligned}
\Phi_{F_a}(E_{l,a'}) &= \lim_{Z \to F_0} j(Z, a^{-1})^l E_{l,a'}(Za^{-1}) \\
&= \lim_{Z \to F_0} \sum_{\gamma \in \Gamma/\Gamma_{0,a'}} (j(Z, a^{-1}) j(Za^{-1}, \gamma a'))^l \\
&= \lim_{Z \to F_0} \sum_{\gamma \in \Gamma/\Gamma_{0,a'}} j(Z, a^{-1}\gamma a')^l,
\end{aligned} \tag{9}$$

and all terms on the right go to zero except those for which $a^{-1}\gamma a' \in N_h$; the existence of such terms implies $a = a'$ and $\gamma \in \Gamma_{0,a}$, which leaves, in case $a = a'$, only one term with nonzero limit; it follows that we have

$$\Phi_{F_a}(E_{l,a'}) = \delta_{a'a} \qquad \text{(Kronecker delta symbol).} \tag{10}$$

Therefore,

$$\Phi_{F_a}(E_{l,c}) = c_a^l. \tag{10'}$$

Since each orbit of zero-dimensional rational boundary components under Γ contains F_a for precisely one $a \in A$, we see that if the mapping $c: A \to \mathbf{C}$ is nonzero on all of A, then $\Phi_{F_1} E_{l,c} \neq 0$ for every 0-dimensional, rational boundary component F_1. By the transitivity of the Φ-operator (see [3, Section 8]), it follows that for any rational boundary component F, $\Phi_F E_{l,c}$ is not identically zero. Consequently, by the previous remarks, $\Phi_F E_{l,c}$ is the sum of a series of lth powers of nowhere-vanishing holomorphic functions on F.

The following is easily proved (see [1, Lemmas V, VI]):

LEMMA 1. *Let $\sum a_i$ be an absolutely convergent series of complex numbers, not all zero. Then there exists a positive integer m such that $\sum a_i^m \neq 0$.*

If f is an integral automorphic form (see [3, Section 8.5]) with respect to Γ, then f may be viewed as a cross-section of an analytic coherent sheaf on $\mathfrak{B} = \mathfrak{T}/\Gamma$, such that that sheaf has a unique prolongation to an analytic coherent sheaf on the Satake compactification $\mathfrak{B}^*$ of $\mathfrak{B}$, and such that f has a unique extension, which we denote by f^*, to a cross-section of the extended sheaf (see [3, Section 10.14]). We know from [3, Section 8.6] that each of the series $E_{l,c}$ is an integral automorphic form with respect to Γ. From this, from Lemma 1, and from the discussion preceding the lemma, we obtain at once:

PROPOSITION 1. *There exists a finite set $\mathscr{L}$ of positive integers l such that the members of the family $\{E_{l,a}^*\}_{l \in \mathscr{L}, a \in A}$ have no common zero's on $\mathfrak{B}^*$. If $c: a \to c_a$ is given such that $c_a \neq 0$ for all $a \in A$, then, by increasing the size of $\mathscr{L}$ if necessary, we may assert that the members of the family $\{E_{l,c}^*\}_{l \in \mathscr{L}}$ (fixed c) have no common zero's on $\mathfrak{B}^*$.*

The proof is obvious.

Referring to the first paragraph of this section, let l_0 be a positive integer divisible by d such that for all positive integers l divisible by l_0 the series $E_{l,c}$ converge. Let n be a positive integer divisible by l_0 and let Λ_n be the linear space over $\mathbf{C}$ spanned by all expressions of the form

$$E_{l_1,a_1} \cdots E_{l_\mu,a_\mu}, \tag{11}$$

where $a_i \in A$, such that $l_i > 0$, $l_0 | l_i$, $i = 1, \ldots, \mu$, and $l_1 + \cdots + l_\mu = n$. An element of Λ_n will be called an isobaric polynomial (in the Eisenstein series) of degree n. Denote by Λ_n^* the linear space of the extensions to $\mathfrak{B}^*$ of the elements of Λ_n. We make the same set of definitions with the functions $E_{l,a}$ replaced by $E_{l,c}$ for a fixed, nowhere zero c and denote the spaces corresponding in this way to Λ_n and to Λ_n^* by $\Lambda_{n,c}$ and $\Lambda_{n,c}^*$, respectively. Clearly $\Lambda_{n,c}^* \subset \Lambda_n^*$. Then we have from the proposition immediately:

COROLLARY. *If $l_0 | n$ and if n is sufficiently large, then the elements of Λ_n^* (respectively of $\Lambda_{n,c}^*$) have no common zero's on $\mathfrak{B}^*$.*

Clearly, if $p, q \in \Lambda_n^*$, $q \neq 0$, then p/q is a meromorphic function on $\mathfrak{B}^*$; or it may be identified with a meromorphic function on $\mathfrak{T}$ invariant under Γ.

7. Projective imbeddings

Let Λ^* be any subspace of Λ_n^* such that the elements of Λ^* have no common zero's on $\mathfrak{B}^*$ and let $\lambda_0, \ldots, \lambda_M$ be a basis of Λ^*. Then the assignment

$$Z \mapsto [\lambda_0(Z): \cdots : \lambda_M(Z)]$$

determines well defined holomorphic mappings of $\mathfrak{T}$ and of $\mathfrak{B}^*$ into CP^M, which we denote by $\tilde{\psi}_{(\lambda)}$ and $\psi_{(\lambda)}$, respectively. Let $\mathfrak{W} = \psi_{(\lambda)}(\mathfrak{B}^*)$; then $\mathfrak{W}$ is a (complete) projective variety. The functions $\tilde{\lambda}_0, \ldots, \tilde{\lambda}_M$ induced on $\mathfrak{T}$ by $\lambda_0, \ldots, \lambda_M$ are automorphic forms of weight n. If k is a suitably large positive multiple of n, then a basis of automorphic forms of weight k affords an injective holomorphic imbedding of $\mathfrak{B}^*$ in some projective space such that $\mathfrak{B}^*$ is mapped biregularly onto its image. It follows (see [2, pp. 353–354]) that since the λ_i are nonconstant and without common zero's, the mapping $\psi_{(\lambda)}$ has the property that for every $x \in \mathfrak{W}$, $\psi_{(\lambda)}^{-1}(x)$ is a finite set. Hence, we may speak of the degree of the mapping $\psi_{(\lambda)}: \mathfrak{B}^* \to \mathfrak{W}$, which is equal to the degree of the algebraic extension of the field of rational functions on $\mathfrak{W}$ by the field of those on $\mathfrak{B}^*$. Moreover, if $l_0 | l$, $l > 0$, then the image of $E_{l,a}^*$ or of $E_{l,c}^*$ under restriction to the image of any rational boundary component F in $\mathfrak{B}^*$ is a nontrivial cross-section (if not identically zero) of an ample sheaf, and is hence nonconstant.

Furthermore, we see at the same time that $\tilde{\psi}_{(\lambda)}$ can be extended to a mapping of the union of all rational boundary components which is continuous in the Satake topology (see [3, Section 4.8]).

We now let $\Lambda^* = \Lambda_{n,c}^*$ for fixed, nowhere-vanishing c. Let ψ_n be the mapping associated to a basis of Λ^*, let $\mathfrak{W}_n = \psi_n(\mathfrak{B}^*)$, and let d_n be the degree of ψ_n. It is clear that $d_n \geqq d_{n+l_0}$. Therefore, there exists an n' such that $d_n = d_{n'}$ for all $n \geqq n'$. Also it is easily seen (by the Jacobian criterion) that if p is a regular point of $\mathfrak{B} \subset \mathfrak{B}^*$ and if ψ_{n_1} induces a biregular mapping of a neighborhood of p onto a neighborhood of a regular point of $\mathfrak{W}_{n_1}$, then the same is true for all $n \geqq n_1$. We wish to prove the following proposition:

PROPOSITION 2. *The degree $d_{n'}$ is equal to one.*

We proceed to the proof of this by stages.

If $g \in G_h$, for any real number r, let

$$L_{g,r} = \{Z \in \mathfrak{T} \mid |j(Z, g)| = r\};$$

this is a real analytic subset of $\mathfrak{T}$ and is different from $\mathfrak{T}$, and hence is of measure zero if $j(Z, g)$ is not constant as a function of Z, i.e., if $g \notin N_h$. Since the series (2) converges uniformly on compact sets, it follows that the number of the surfaces $L_{\gamma a,r}$ passing through any compact subset of $\mathfrak{T}$ for *fixed* r is finite. If $|j(Z, \gamma a)| \equiv d \cdot |j(Z, \gamma' a')|$, where d is a positive real constant, $\gamma, \gamma' \in \Gamma$, $a, a' \in A$, then we see easily from the cocycle relation that $a = a'$ and $\gamma = \gamma' \cdot \gamma_a$, $\gamma_a \in \Gamma_{0,a}$.

From the termwise majoration of the series (2) in a truncated Siegel domain, afforded by [3, Section 7.7(i)], we conclude that on an F_0-adapted truncated Siegel domain $\mathfrak{S}$ in $\mathfrak{T}$, there exists a series of positive constants which majorizes the series

$$\sum_{\gamma \in \Gamma/\Gamma_{0,a}} j(Z, b^{-1}\gamma a)^l \qquad (l_0|l, l > 0)$$

on $\mathfrak{S}$, for any $a, b \in A$ (to reduce this to the form considered in [3, Section 7.7(i)], write $b^{-1}\gamma a = b^{-1}a(a^{-1}\gamma a)$). Moreover, from this we conclude that, if $\mathfrak{S}$ is suitably chosen, then we have $|j(Z, b^{-1}\gamma a)| < 1$ if $b \neq a$ or if $b = a$ and $\gamma \notin \Gamma_{0,a}$. Define the subset $\mathscr{S}_a$ of $\mathfrak{T}$ by

$$\mathscr{S}_a \cdot a = \{Z \in \mathfrak{T} |\, |j(Z, b^{-1}\gamma a)| < 1 \text{ if } b \in A - \{a\} \text{ or if } b = a, \gamma \notin \Gamma_{0,a}\}.$$

Then from this discussion and from properties of the Satake topology (see [3, Section 4]) we have at once the following lemma:

LEMMA 2. *There exists a dense subset $\mathscr{E}$ of $\mathfrak{T}$ such that $\mathfrak{T} - \mathscr{E}$ is of measure zero and such that for $Z \in \mathscr{E}$, the terms in the series for $E_{l,c}$ are all distinct. For each $a \in A$, the set $\mathscr{S}_a$ is open, and nonempty, and contains an F_0-adapted Siegel domain. The closure of $\mathscr{S}_a$ in the Satake topology contains a neighborhood of F_0 in that topology.*

Consider only $n > 0$ divisible by l_0. Let Δ_n be the inverse image under $\psi_n \times \psi_n$ of the diagonal of $CP^{\mu_n} \times CP^{\mu_n}$, where $\mu_n + 1 = \dim \Lambda^*_{n,c}$. Clearly Δ_n contains the diagonal of $\mathfrak{B}^* \times \mathfrak{B}^*$, and $\Delta_n \supset \Delta_{n+l_0}$. Without loss of generality, we may assume $n' > 0$ chosen such that $\Delta_n = \Delta_{n'}$ for all $n \geqq n'$ (because $\{\Delta_n\}$ is a decreasing sequence of algebraic sets). Consider now only $n \geqq n'$. We have seen that $\psi_n^{-1}(x)$ is finite for all $x \in \mathfrak{W}$. We now prove the following lemma:

LEMMA 3. *If $x_0 = \tilde{\psi}_n(F_0)$, then $x_0 = \tilde{\psi}_n(F_1)$ for every 0-dimensional, rational boundary component F_1 and $\tilde{\psi}_n^{-1}(x_0)$ is precisely the set of all 0-dimensional, rational boundary components.*

PROOF. It is sufficient to consider the case $F_1 = F_a$ for some $a \in A$. Since the result is evidently independent of the choice of basis for $\Lambda^*_{n,c}$, we may assume the basis to consist of monomials in the series $E_{l,c}$. Then by (10′), the image under $\tilde{\psi}_n$ of F_a is the point in projective space represented by $[1:\cdots:1]$; thus $\tilde{\psi}_n(F_0) = \tilde{\psi}_n(F_a)$ for all a. The converse requires further considerations.

Let F be any rational boundary component of $\mathfrak{T}$ (we may have $F = \mathfrak{T}$) with $\dim F > 0$, and let $Z \in F$. The function $\Phi_F E_{l,c}$ on F is an automorphic form of positive weight on F with respect to the homomorphic image Γ_F of $N_h(F) \cap \Gamma$ in the full group of holomorphic automorphisms of F, and Γ_F is an arithmetically defined discontinuous group acting on F. Without loss of generality, we may assume that F_0 is a rational boundary component of F. We may view F (see [9, Section 4.11]) as a tube domain on which $N_{hF} = N_h(F) \cap N_h$ acts by linear transformations; with a certain abuse of notation, let $\Gamma_{F,a} = \Gamma_F \cap {}^aN_{hF}$. Then by [3, Section 7] it is easily seen that $\Phi_F E_{l,c}$ is the Eisenstein series

$$\sum_{a\in A_F} c_a^l \cdot \sum_{\gamma\in\Gamma_F/\Gamma_{F,a}} j_F(Z, \gamma a)^{q_F l}, \qquad Z \in F, \tag{12}$$

where q_F is a positive rational number (not a multi-index!), $j_F(Z, g)$ is a nonzero constant ε_g times the functional determinant of g as a transformation of F, and a runs over the set A_F of elements of A such that F_a is Γ-equivalent to a rational boundary component of F. By assuming *ab initio* suitable divisibility properties for l_0, we may assume that $q_F l$ has the necessary divisibility properties for the tube domain F. This having been said, we now prove that $\tilde{\psi}_n^{-1}(x_0) \cap \mathfrak{T}$ is empty and note that the same argument will prove that $\tilde{\psi}_n^{-1}(x_0) \cap F$ is empty by merely supplying the subscript F where needed.

If $Z \in \mathfrak{T}$ and $\tilde{\psi}_n(Z) = x_0$, then we have $E_{l,c}(Z) \neq 0$ and $f_k(Z) = E_{kl,c}(Z)/E_{l,c}(Z)^k$ will have the same value for *all* k (because $x_0 = [1 : \cdots : 1]$). Following [12, p. 126], we introduce the function

$$M_Z(\lambda) = \sum_{a\in A} \sum_{\gamma\in\Gamma/\Gamma_{0,a}} (\lambda - E_{l,c}(Z)(c_a \cdot j(Z, \gamma a))^{-l})^{-1}.$$

This series converges if λ is not one of the discrete set of points $E_{l,c}(Z) \cdot (c_a \cdot j(Z, \gamma a))^{-l}$ of $\mathbf{C}$, because of the convergence of the series for $E_{l,c}$ itself, and uniformly so on any bounded subset of $\mathbf{C}$ not meeting that discrete set. Hence, M_Z is a meromorphic function on $\mathbf{C}$ with infinitely many distinct poles. On the other hand, M_Z is holomorphic at the origin and its power series expansion is $-\sum_{k=1}^{\infty} f_k(Z)\lambda^{k-1}$. By hypothesis, all $f_k(Z)$ are equal, hence $M_Z(\lambda) = -f_k(Z)/(1 - \lambda)$ has at most one simple pole—which is absurd. This completes the proof of the assertion that $\tilde{\psi}_n^{-1}(x_0)$ is simply the set of 0-dimensional rational boundary components.

It follows from Lemmas 2 and 3 that there exists a neighborhood $\mathfrak{N}$ of x_0 such that $\mathfrak{T} \cap \tilde{\psi}_n^{-1}(\mathfrak{N}) \subset \left(\bigcup_a \mathcal{S}_a\right) \cdot \Gamma$. Let $\mathcal{S}_* = \left(\bigcup_a \mathcal{S}_a\right)$.

Suppose now that the degree $d_{n'}$ is greater than 1. We assume l_0 chosen such that $E_{l_0,c}$ is not identically zero. Henceforth, fix c and let $E_{l,c} = E_l$. Since $d_{n'} > 1$, one sees easily that there exist points $Z_1, Z_2 \in \mathcal{S}_*$, say, $Z_i \in \mathcal{S}_{a_i}$, $i = 1, 2$, in distinct orbits of Γ such that

(i) $E_{l_0}(Z_i) \neq 0$, $i = 1, 2$;
(ii) the canonical images of Z_1 and Z_2 are regular points of $\mathfrak{B} \subset \mathfrak{B}^*$;
(iii) $\tilde{\psi}_n(Z_1) = \tilde{\psi}_n(Z_2)$ and their common value w_n is a regular point of

$\mathfrak{W}_n$, and $\tilde{\psi}_n$ is a biregular mapping of a suitable neighborhood of Z_i onto a neighborhood of w_n, $i = 1, 2$;

(iv) $w_n \notin \psi_n(\mathfrak{B}^* - \mathfrak{B})$.

Defining $\mathbf{S}_n = \psi_n(\mathfrak{B}^* - \mathfrak{B})$, we may find a small neighborhood $\mathfrak{N}_i$ of Z_i contained in $\mathscr{S}_{a_i}$, $i = 1, 2$, such that $\mathfrak{N}_1 \cap \mathfrak{N}_2 \cdot \Gamma$ is empty, and such that $\tilde{\psi}_n$ is a biregular analytic mapping of $\mathfrak{N}_i$ onto a neighborhood $\mathfrak{N}$ of w_n in $\mathfrak{W}_n - \mathbf{S}_n$, $i = 1, 2$. Let φ denote the composed mapping $(\tilde{\psi}_n|\mathfrak{N}_2)^{-1} \circ (\tilde{\psi}_n|\mathfrak{N}_1)$ of $\mathfrak{N}_1$ onto $\mathfrak{N}_2$; because of the properties of $\tilde{\psi}_n$ and ψ_n which are stable for large n (q.v. supra), φ is independent of the choice of n for sufficiently large n, say $n \geqq n'$. Then φ is biregular and we may assume $E_{l_0} \neq 0$ on $\mathfrak{N}_1 \cup \mathfrak{N}_2$. Let $l = l_0$. With $f_k(Z) = E_l(Z)^{-k} \cdot E_{lk}(Z)$ as before, our conditions imply that $f_k(Z) = f_k(\varphi(Z))$, $k = 1, 2, \ldots, Z \in \mathfrak{N}_1$.

8. Extension of the mapping φ

We have the following lemma:

LEMMA 4. *The mapping φ may be extended to a holomorphic automorphism of $\mathfrak{T}$.*

PROOF. The main ideas here are those of [12, pp. 126–130]. Define $M_Z(\lambda)$ as before, when $E_l(Z) \neq 0$. By hypothesis, $E_l \neq 0$ in $\mathfrak{N}_1 \cup \mathfrak{N}_2$. If $Z \in \mathfrak{N}_1 \cup \mathfrak{N}_2$, then the set of points $\{E_l(Z)(c_a \cdot j(Z, \gamma a))^{-l}\}$ is a discrete set. Moreover, in any region of $\mathfrak{T} \times \mathbf{C}$ avoiding poles of $M_Z(\lambda)$, $M_Z(\lambda)$ is an analytic function, and its poles for any fixed $Z \in \mathfrak{T}$ such that $E_l(Z) \neq 0$ are the points of that discrete set, each of which is a simple pole with residue equal to the number of times a term of given value occurs in the Eisenstein series. As before, $-M_Z(\lambda) = \sum_{k=1}^{\infty} f_k(Z)\lambda^{k-1}$. Then the equations $f_k(Z) = f_k(\varphi(Z))$, $k = 1, 2, \ldots$, imply that the sets of poles of M_Z and of $M_{\varphi(Z)}$ are the same. If $Z \in \mathscr{S}_{a_i}$, then by the cocycle identity

$$j(Z, \gamma b) = j(Z_i, a_i^{-1})^{-1} \cdot j(Z_i, a_i^{-1}\gamma b), \qquad Z_i = Z \cdot a_i,$$

and $|j(Z_i, a_i^{-1}\gamma b)| < 1$, unless $b = a_i$ and $\gamma \in \Gamma_{0,a_i}$, in which case $|j(Z_i, a_i^{-1}\gamma a_i)| = 1$; therefore, $(c_{a_i} \cdot j(Z, a_i))^{-l}E_l(Z)$ is the pole of $M_Z(\lambda)$ having the smallest absolute value, $i = 1, 2$. Hence, we have

$$(13) \qquad (c_{a_1} j(Z, a_1))^{-l} E_l(Z) = (c_{a_2} j(\varphi(Z), a_2))^{-l} E_l(\varphi(Z)).$$

The equality of the other poles of M_Z and $M_{\varphi(Z)}$ gives the system of equations

$$(14) \qquad (c_{a_2} j(\varphi(Z), \gamma a_2))^{-l} E_l(\varphi(Z)) = (c_{b_Z} j(Z, \gamma_Z b_Z))^{-l} E_l(Z).$$

Dividing (13) by (14) and using the cocycle identity gives

$$j(\varphi(Z)\cdot a_2, a_2^{-1}\gamma a_2)^l = (c_{b_Z}j(Z, \gamma_Z b_Z)(c_{a_1}j(Z, a_1))^{-1})^l, \tag{15}$$

where $b_Z, \gamma_Z \in G_{\mathbf{Q}}^*$ depend, at first, on Z. However, the number of possibilities for each Z is at most countable, and using the fact that an analytic function not identically zero vanishes on at most a set of measure zero, we conclude that there exists a function f from the set of cosets $\Gamma/\Gamma_{0,a}$, for each a, into $G_{\mathbf{Q}}^*$ such that for suitable constants d_γ we have

$$j(\varphi(Z)a_2, a_2^{-1}\gamma a_2)^l \equiv d_\gamma(j(Z, f(\gamma))j(Z, a_1)^{-1})^l. \tag{16}$$

Since the set of lth roots of unity is finite and the function $j(Z, a)$, for fixed a, is a nonvanishing holomorphic function on all of $\mathfrak{T}$ and in particular on the neighborhood $\mathfrak{N}_1$, we may conclude that

$$j(\varphi(Z)a_2, a_2^{-1}\gamma a_2)^{-1} = h_\gamma(Z), \tag{17}$$

where h_γ is a nonvanishing holomorphic function on all of $\mathfrak{T}$. We now proceed to solve the system (17).

The group $a_2^{-1}\Gamma a_2$ is also arithmetic. We let χ be the one-to-one analytic mapping of $\mathfrak{N}_1$ into $\mathfrak{T}$ obtained by first applying φ and then translating by a_2. Our main problem becomes that of solving the equations

$$\cdot \quad j(\mathbf{a}(Z), \gamma)^{-1} \equiv h_\gamma(Z), \tag{18}$$

for an analytic mapping $\mathbf{a}$ of $\mathfrak{T}$ into itself such that $\mathbf{a} = \chi$ in a neighborhood of Z_1, where γ runs over an arithmetic group Γ_1 and h_γ are nonvanishing holomorphic functions on $\mathfrak{T}$. We now choose $\gamma_0 = uw_0p \in \Gamma_1$ as in Section 4 and let γ run over the elements $\lambda\gamma_0$, where λ runs over the lattice $\Lambda = U \cap \Gamma_1$ in $U_{\mathbf{R}}$. Identifying $U_{\mathbf{R}}$ with the group of real translations in $\mathfrak{T}$, the system (18) for these γ becomes

$$j(\mathbf{a}(Z) + u + \lambda, w_0p)^{-1} = g_\lambda^0(Z), \tag{19}$$

where g_λ^0 is holomorphic in $\mathfrak{T}$. By Section 3, this system becomes

$$\prod_i \mathfrak{N}_i((\mathbf{a}(Z) + u + \lambda)_i)^{\nu_i} = g_\lambda(Z), \tag{20}$$

where $g_\lambda = \text{const.}\ g_\lambda^0$ is holomorphic in $\mathfrak{T}$. Since we know that a solution $\chi = \mathbf{a}$ of the system (18) exists in a neighborhood of Z_1, it will follow that if a subsystem of the system (18) has a *unique* analytic solution for all $Z \in \mathfrak{T}$, then this solution must be an analytic continuation of χ, and hence must be a solution of the full system (18) for all $Z \in \mathfrak{T}$.

Let $n_1, \ldots, n_k$ be positive integers and $m = n_1 + \cdots + n_k$. If $z_i \in \mathbf{C}^{n_i}$, write $z_i = (z_{ij})_{j=1,\ldots,n_i}$, and if $z \in \mathbf{C}^m = \prod_i \mathbf{C}^{n_i}$, write $z = (z_i)_{i=1,\ldots,k}$. If also $z' \in \mathbf{C}^m$, define $z + z'$ by the usual component-by-component addition.

In what follows, let $\mathbf{p}_i$ be a homogeneous polynomial of degree $d_i > 0$ in n_i indeterminates with complex coefficients, viewed as a function on $\mathbf{C}^{n_i}$ in an obvious way. We assume that the first partial derivatives $\frac{\partial \mathbf{p}_i}{\partial x_{ij}}, j = 1, \ldots, n_i$ are linearly independent polynomials for each i. Let $\nu_1, \ldots, \nu_k$ be positive integers and put $\mathbf{p} = \mathbf{p}_1^{\nu_1} \cdots \mathbf{p}_k^{\nu_k}$. Let Λ be a Zariski-dense subset of $\mathbf{C}^m$.

LEMMA 5. *There exists a finite number of points $r^\alpha \in \Lambda \subset \mathbf{C}^m$, where α runs over a finite indexing set B', such that if D is a connected open subset of $\mathbf{C}^m$ and if $\{f_\alpha\}_{\alpha \in B'}$ are analytic functions on D for which the system*

$$\mathbf{p}(\mathbf{a}(z) + r^\alpha) = f_\alpha(z) \tag{21}$$

has an analytic solution **a** *mapping some open subset of D into $\mathbf{C}^m$, then the system* (21) *has a unique solution* **a** *defined on all of D, $\mathbf{a}\colon D \to \mathbf{C}^m$.*

PROOF. View the polynomials $\mathbf{p}_1, \ldots, \mathbf{p}_k$ as polynomials in independent sets of indeterminates $\{x_{ij}\}_{j=1,\ldots,n_i,\ i=1,\ldots,k}$. It is an elementary exercise to prove that if Π denotes the linear span, in the vector space of polynomials with complex coefficients, of the set of higher (i.e., of order higher than 1) partial derivatives of $\mathbf{p}$ and if $\pi_1, \ldots, \pi_N$ is a basis of Π, then $\pi_1, \ldots, \pi_N$, $\frac{\partial \mathbf{p}}{\partial x_{ij}}, j = 1, \ldots, n_i, i = 1, \ldots, k$ are linearly independent polynomials. It follows from Taylor's expansion that if $y \in \mathbf{C}^m$, then

$$\mathbf{p}(z + y) = \mathbf{p}(y) + \sum_{i,j} \frac{\partial \mathbf{p}}{\partial x_{ij}}(y) z_{ij} + \sum_{\beta \in B} P_\beta(y) Q_\beta(z), \tag{22}$$

where B is some finite indexing set, P_β and Q_β are homogeneous polynomials such that $\deg P_\beta + \deg Q_\beta = \deg \mathbf{p}$ and $P_\beta, \beta \in B$, are linearly independent elements of Π. Therefore we may find $r^\alpha \in \Lambda, \alpha \in B'$, card $B' = m + \text{card } B$, such that the determinant of the matrix

$$\left(\frac{\partial \mathbf{p}}{\partial x_{11}}(r^\alpha), \ldots, \frac{\partial \mathbf{p}}{\partial x_{kn_k}}(r^\alpha), \{P_\beta(r^\alpha)\}_{\beta \in B} \right)_{\alpha \in B'}$$

is nonzero (because the product of Zariski-dense sets is Zariski-dense in the product variety). Then it is possible to solve uniquely the system of equations

$$\sum_{i,j} \frac{\partial \mathbf{p}}{\partial x_{ij}}(r^\alpha) \mathbf{a}_{ij}(z) + \sum_{\beta \in B} P_\beta(r^\alpha) \mathbf{a}_\beta(z) = f_\alpha(z) - \mathbf{p}(r^\alpha) \tag{23}$$

for the "unknowns" $\mathbf{a}_{ij}(z)$ and $\mathbf{a}_\beta(z)$, and it is clear that $\mathbf{a}_{ij}(z)$ is analytic in all of D. Since a solution does exist in an open set and since D is connected, it follows that $\mathbf{a} = (\mathbf{a}_{ij})$ is analytic in all of D and is the unique solution there to the system (21). This proves the lemma.

To demonstrate the applicability of Lemma 5 to our situation, one notes the systems of equations (18) and (21) and observes (e.g., by an easy case-by-case verification) that the Jordan algebra norm $\mathfrak{N}_i$ always satisfies the requirements (i.e., that $\mathbf{p}_i$ be homogeneous and that the polynomials $\frac{\partial \mathbf{p}_i}{\partial x_{ij}}$ be linearly independent) imposed upon each of the polynomials $\mathbf{p}_i$ in Lemma 5. This computation is left to the reader. Finally we note that any translate $\Lambda + u$ of the lattice Λ of $U_{\mathbf{R}}$ is Zariski-dense in $\mathbf{C}^m$ (the complexification of $U_{\mathbf{R}}$).

Thus we have obtained an analytic mapping $\mathbf{a}$ of $\mathfrak{T}$ into the ambient space $\mathbf{C}^m$ of $\mathfrak{T}$ such that $\mathbf{a}$ coincides with χ in a small neighborhood of Z_1. Now view the noncompact Hermitian symmetric space $\mathfrak{T}$ as imbedded in its compact dual $\mathfrak{T}^c$ (see [14, Section 8.7.9]). Then $a_2 \in G^0_{\mathbf{R}}$ extends to a holomorphic automorphism of the compact, complex manifold $\mathfrak{T}^c$, and then $a_2^{-1} \circ \mathbf{a} = \mathbf{a}'$ is an analytic mapping of $\mathfrak{T}$ into $\mathfrak{T}^c$ which coincides with φ in a small neighborhood $\mathfrak{N}_1$ of Z_1. Our next step is to show that $\mathbf{a}'$ maps $\mathfrak{T}$ into the closure $\mathfrak{T}^*$ of $\mathfrak{T}$ in $\mathfrak{T}^c$. It is easy to see that for $n \geqq n'$ there exists a proper Zariski-closed subset $\mathcal{N}''_n$ of $\mathfrak{W}_n$ such that $\mathbf{S}_n \subset \mathcal{N}''_n$ and such that if $\mathcal{N}'_n = \psi_n^{-1}(\mathcal{N}''_n)$ and $\mathcal{N}_n = \tilde{\psi}_n^{-1}(\mathcal{N}''_n)$, then $\mathfrak{T} - \mathcal{N}_n$ is a covering manifold of $\mathfrak{V}^* - \mathcal{N}'_n$ and the latter is a covering manifold of $\mathfrak{W}_n - \mathcal{N}''_n$. (It is sufficient that $\mathcal{N}''_n$ include the singular points of $\mathfrak{W}_n$, the set $\mathbf{S}_n$, the images of the singular points of $\mathfrak{V}^*$, and the image of the set of regular points x of $\mathfrak{V}^*$ at which ψ_n is not locally biregular—the intersection of the last set with $\mathfrak{V}$ is, modulo singular points of $\mathfrak{V}$, a proper Zariski-closed subset of $\mathfrak{V}$ by the Jacobian criterion.) Let $\mathcal{N} = \mathcal{N}_n$.

The set $\mathfrak{T} - \mathcal{N}$ is connected and open and contains Z_1. Let Z_0 be any point of $\mathfrak{T} - \mathcal{N}$ and join Z_1 to Z_0 by a path ϕ. Let $\phi_{\mathbf{a}'}$ be the image of ϕ under $\mathbf{a}'$. We have $\tilde{\psi}_n \cdot \mathbf{a}' = \tilde{\psi}_n$ in a neighborhood of Z_1. It is clear that $\tilde{\psi}_n(\phi)$ is a path in $\mathfrak{W}_n - \mathcal{N}''_n$ and by the homotopy covering theorem (see [14, Sections 1.8.3–4]) can be raised to a unique path in $\mathfrak{T} - \mathcal{N}$ beginning at Z_2, since $\tilde{\psi}_n(Z_1) = \tilde{\psi}_n(Z_2)$. By analytic continuation of the relation $\tilde{\psi}_n \circ \mathbf{a}' = \tilde{\psi}_n$ it is clear that this covering path beginning at Z_2 can be no other than $\phi_{\mathbf{a}'}$. Hence, $\mathbf{a}'(Z_0)$ must be a point of $\mathfrak{T}$. Therefore, $\mathbf{a}'(\mathfrak{T} - \mathcal{N}) \subset \mathfrak{T} - \mathcal{N} \subset \mathfrak{T}$. Viewing $\mathfrak{T}$ as the bounded domain D, we have $\mathbf{a}'(D) \subset \bar{D}$, since $\mathbf{a}'$ is continuous in D and $\mathfrak{T} - \mathcal{N}$ is dense in $\mathfrak{T}$. By a known property (see [9, Section 4.8]) of the boundary components of D, since $\mathbf{a}'(D) \cap D$ is not empty, $\mathbf{a}'(D) \cap (\bar{D} - D)$ must be empty; thus $\mathbf{a}'(D) \subset D$. Let $\mathbf{a}''$ be the similarly defined mapping associated to φ^{-1}. Since, also, $\mathbf{a}''(D) \subset D$

and $\mathbf{a}'\mathbf{a}'' = \mathbf{a}''\mathbf{a}' =$ identity (because this is true in a small open set, $\mathfrak{N}_1$ or $\mathfrak{N}_2$, to begin with), we see that both $\mathbf{a}'$ and $\mathbf{a}''$ are bijective maps of D onto itself. This completes the proof of Lemma 4.

9. Conclusions

Thus we see that there exists a biholomorphic map $\mathbf{a}: D \to D$, $\mathbf{a} \in G_h - \Gamma$ such that $\tilde{\psi}_n \circ \mathbf{a} = \tilde{\psi}_n$. Let Γ^* be the subgroup of G_h generated by $\mathbf{a}$ and Γ. Clearly $\Gamma \subsetneqq \Gamma^*$, so that Γ^* cannot be a discrete subgroup of G_h. By construction, $\tilde{\psi}_n$ is constant on any orbit of Γ and also on any orbit of the group generated by $\mathbf{a}$ and hence is constant on any orbit of Γ^*. Since Γ^* is not discrete, it is not property discontinuous on $\mathfrak{T}$, and thus there is an orbit ω of Γ^* in $\mathfrak{T}$ with a limit point Z_0. Hence, in every neighborhood of Z_0, there exist infinitely many points mapped by $\tilde{\psi}_n$ onto the same point of $\mathfrak{W}_n$. But this contradicts what we already know about ψ_n. The contradiction having come from the assumption that $d_{n'} > 1$, we conclude that $d_n = 1$ for $n \geqq n'$. This proves Proposition 2. Therefore, ψ_n is a proper birational mapping of the normal variety $\mathfrak{V}^*$ onto $\mathfrak{W}_n$, $n \geqq n'$, and for $w \in \mathfrak{W}_n$, $\psi_n^{-1}(w)$ is finite. Since $d_n = 1$, we conclude from Zariski's main theorem (see [10, p. 124]) that $\mathfrak{V}^*$ is simply the normal model of $\mathfrak{W}_n$, and hence there exists a projective variety biregularly equivalent to $\mathfrak{V}^*$ and defined over any field of definition for $\mathfrak{W}_n$. Since by Chow's theorem (see [6]), any meromorphic function on $\mathfrak{W}_n$ is a rational function expressible as the quotient of two isobaric polynomials of like weight (see [3, Section 10.5]), *the proof of the theorem stated in the introduction is complete.*

COROLLARY 1. *Let G be a connected, semisimple algebraic group defined over $\mathbf{Q}$ such that $G_{\mathbf{R}}$ has no compact simple factors. Let K be a maximal compact subgroup of $G^0_{\mathbf{R}}$ and let Γ be a maximal discrete arithmetic subgroup of $G'^0_{\mathbf{R}}$. Assume that $\mathfrak{X} = K\backslash G^0_{\mathbf{R}}$ is Hermitian symmetric, and assume that G_h is connected. Then the same conclusions as those in the theorem hold.*

PROOF. Let Z be the center of G (which is not assumed to be centerless), and let $\pi: G \to G' = G/Z$. Then $\pi(\Gamma)$ is maximal, discrete, arithmetic in $G^0_{\mathbf{R}} = G_h$, and we may apply the theorem.

The cases where G_h is not connected are described in [3, Section 11.4].

COROLLARY 2. *Let G, $\mathfrak{X}$, and Γ be either as in the statement of the theorem or as in the statement of Corollary 1. Then the conclusions of the theorem, or of Corollary 1, respectively, are true if the module of isobaric polynomials in $\{E_{l,c}\}$ of sufficiently high weight is replaced by the module of isobaric polynomials in $\{E_{l,a}\}_{l_0 | l,\ l>0,\ a \in \dot{A}}$ of sufficiently high weight.*

PROOF. The proof is trivial because the first module mentioned is a submodule of the second.

The group Γ is called unicuspidal if all minimal $\mathbf{Q}$-parabolic subgroups of G are Γ-conjugate. Then, if P is as before and Γ is unicuspidal, we have $G_{\mathbf{Q}} = \Gamma \cdot P_{\mathbf{Q}} = P_{\mathbf{Q}} \cdot \Gamma$, and thus the set A may be chosen to consist of e alone. Let $E_l = E_{l,1}$ (i.e., $c_e = 1$). We then have the following corollary:

COROLLARY 3. *If Γ is unicuspidal, the cross-sections E_l^* have no common zero's on $\mathfrak{V}^*$, and the isobaric polynomials of sufficiently high weight in these induce a mapping ψ of $\mathfrak{V}^*$ onto a complete projective variety of which $\mathfrak{V}^*$ is the projective normal model.*

THE UNIVERSITY OF CHICAGO

REFERENCES

1. BAILY, W. L., Jr., "On Satake's compactification of V_n," *Amer. J. Math.*, *80* (1958), 348–364.
2. ———, "On the theory of θ-functions, the moduli of Abelian varieties, and the moduli of curves," *Ann. of Math.*, *75* (1962), 342–381.
3. BAILY, W. L., Jr., and A. BOREL, "Compactification of arithmetic quotients of bounded symmetric domains," *Ann. of Math.*, *84* (1966), 442–528.
4. BOREL, A., "Density and maximality of arithmetic groups," *J. f. reine u. ang. Mathematik*, *224* (1966), 78–89.
5. BOREL, A., and J. TITS, "Groupes réductifs," *Publ. Math. I.H.E.S.*, *27* (1965), 55–150.
6. CHOW, W. L., "On compact complex analytic varieties," *Am. J. Math.*, *71* (1949), 893–914.
7. HELGASON, S., *Differential Geometry and Symmetric Spaces*. New York: Academic Press, Inc., 1962.
8. KORÁNYI, A., and J. WOLF, "Realization of Hermitian symmetric spaces as generalized half-planes," *Ann. of Math.*, *81* (1965), 265–288.
9. ———, "Generalized Cayley transformations of bounded symmetric domains," *Am. J. Math.*, *87* (1965), 899–939.
10. LANG, S., *Introduction to Algebraic Geometry*. New York: Interscience Publishers, Inc., 1958.
11. RAMANATHAN, K. G., "Discontinuous groups, II," *Nachr. Akad. Wiss. Göttingen Math.-Phys. Klasse*, *22* (1964), 154–164.
12. SIEGEL, C. L., "Einführung in die Theorie der Modulfunktionen n-ten Grades," *Mathematische Annalen*, *116* (1938–1939), 617–657.
13. TITS, J., "Algebraic and abstract simple groups," *Ann. of Math.*, *80* (1964), 313–329.
14. WOLF, J., *Spaces of Constant Curvature*. New York: McGraw-Hill, Inc., 1967.

Laplace–Fourier Transformation, the Foundation for Quantum Information Theory and Linear Physics

JOHN L. BARNES

1. Two-dimensional choice information in a discrete function

For clarity this paper begins with the very simple examples shown in Figs. 1a, 1b, and 2. These illustrate the idea that a simple function can contain choice information (see [2]) in two ways. In these figures the elements chosen are shaded. To construct messages they are placed in positions located in intervals on the scale of abscissas called *cells* and on the scale of ordinates called *states*. It is assumed that they are chosen independently and that the binary locations are chosen with equal statistical frequency. In Fig. 1a the information is contained in the height, i.e., ordinate of the function. In Fig. 1b the information is contained in the position, i.e., the first or second place, in the cell. For the transmission of information the function illustrated in Fig. 1a would be said to be *height-modulated.* Sometimes "amplitude" or envelope modulation are the terms if the carrier is a sinusoidal wave. Fig. 1b would be said to be *position-* or *time-modulated.* Instead of position, time rate of change of position is often used. The terms are then pulse position or pulse rate modulation. Pulse radar, sonar, and neural axon spike transmission, as well as phase or frequency modulation in radio are in this class.

For height modulation over a 2-space-dimensional field one can select "black and white" still photographs. Here shades of gray give the height and are levels of energy.

Both height and position modulation may be used jointly as illustrated in Fig. 2. Here the message function has independent choice information carried in the height and horizontal position of the selected elements. If a coherent source such as a laser beam is used to form a 2-D space interference pattern on a photographic film, then called a *hologram*, by

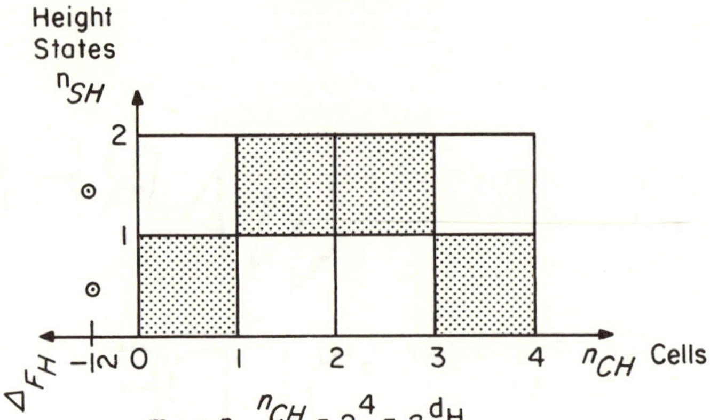

$$m_H = n_{SH}^{\;n_{CH}} = 2^4 = 2^{d_H}$$

m_H = 16 messages

d_H = 4 bits of information

Fig. 1a Height modulation

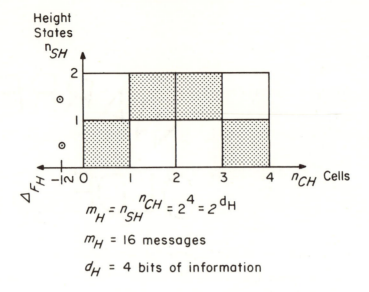

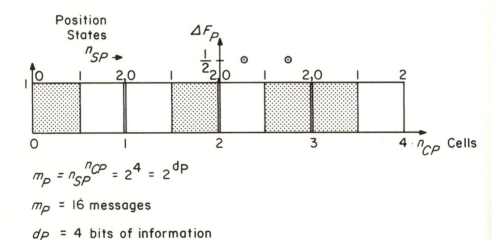

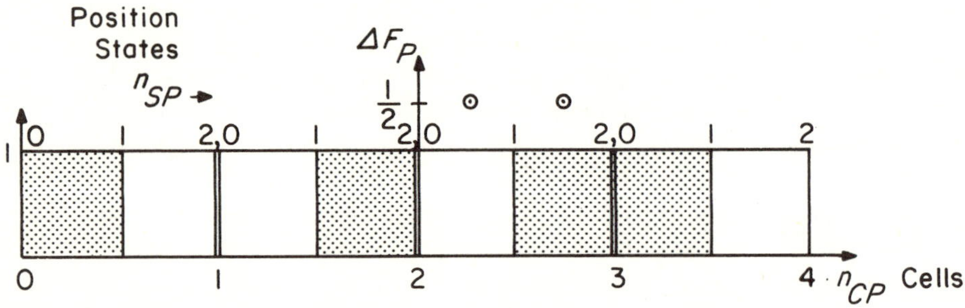

$$m_P = n_{SP}^{\;n_{CP}} = 2^4 = 2^{d_P}$$

m_P = 16 messages

d_P = 4 bits of information

Fig. 1b Position modulation

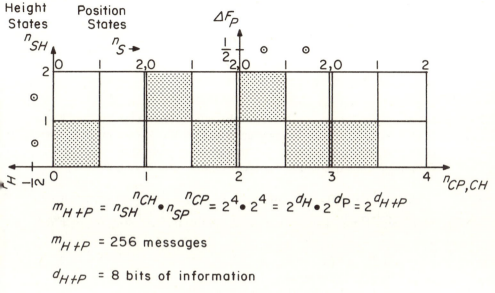

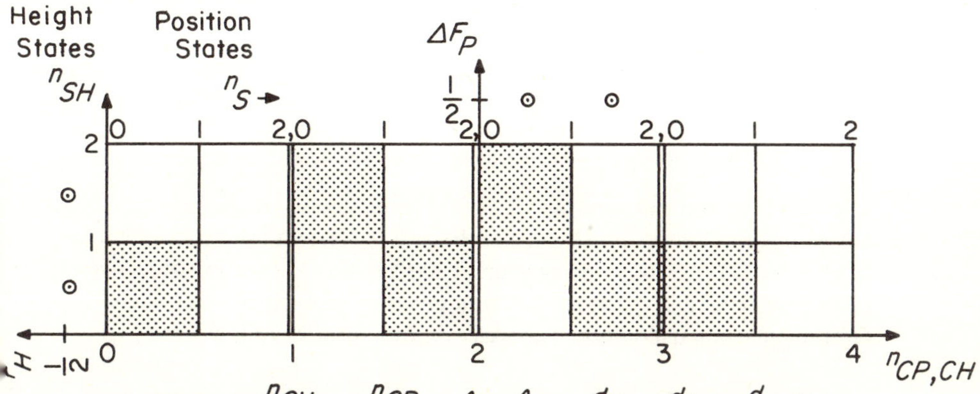

$$m_{H+P} = n_{SH}^{\;n_{CH}} \bullet n_{SP}^{\;n_{CP}} = 2^4 \bullet 2^4 = 2^{d_H} \bullet 2^{d_P} = 2^{d_{H+P}}$$

$$m_{H+P} = 256 \text{ messages}$$

$$d_{H+P} = 8 \text{ bits of information}$$

Fig. 2 Joint height and position modulation

summing exposure from a direct beam and from one reflected from a 3-D space object; and if then the direct beam is subtracted out by passing it through the hologram, there will result a presentation which retrieves the 3-D space picture from the 2-D space hologram. In this holographic process the retention of position, as well as height, modulation provides the depth information missing from an ordinary photograph. In line with this terminology, ordinary photography could be called "halfography" since it uses only half (approximately) the information available in the 3-D space physical world. It will be seen later that ordinary probability and statistics correspond to "halfography" since they often present approximately half the information available. By the use of (usually unique-derivative) complex analysis (see [9]) in place of real analysis, probability and statistics can be made whole.

2. Laplace–Fourier transformation as a path to the analytic complex domain

While a single set of symbols would suffice to show the mathematical relations to be discussed in what follows, particular symbols carry associated physical meaning to engineers and physicists beyond the form of the mathematical relations (see [10]). Hence the same mathematics will frequently be stated in several sets of symbols.

Professor Bochner's deep understanding of functional transform theory first influenced my thinking through his 1932 book [7] and his supervision of my graduate research at Princeton University.

3. Uncertainty theorems from the $\mathscr{L}$–$\mathscr{F}$ transformation

The fundamental uncertainty principle was known intuitively by the German composer Johann Sebastian Bach (see [2]) and used by him about 1730. Through the intervening years it has become known in successively more precise forms. A modern version given by Norbert Wiener at a 1924 Göttingen seminar could be expressed in our form as follows (see [3]):

WIENER'S UNCERTAINTY THEOREM. If $f(t), \dfrac{df(t)}{dt}, \dfrac{d^2f(t)}{dt^2}$ exist and $\in L_2$ on $\mathbf{R}$, and if the bilateral Fourier-integral transform $\mathscr{F}_{\text{II}}[f(t)] \triangleq F(i\omega)$, then

(a) $F(i\omega) \in L_2$ on $\mathbf{I}$,

(U_1) (b) $1 \leqq \Delta t \cdot \Delta|i\omega|$, and

(E_1) (c) $1 = \Delta t \cdot \Delta|i\omega|$ for $f(t)$ a Gaussian pulse,

in which $\Delta t \triangleq 2\sigma_t$, $\Delta\omega \triangleq 2\sigma_\omega \equiv \sigma_{|i\omega|}$, where energy

$$e(t) \triangleq \int_{t_1=-\infty}^{t} f^2(t_1)\, dt_1,$$

$$E(i\omega) \triangleq \int_{i\omega_1=-i\infty}^{i\omega} |F(i\omega_1)|^2 \frac{di\omega_1}{i2\pi},$$

$$e(\infty) < \infty,\ e_r(t) \triangleq \frac{e(t)}{e(\infty)}, \qquad E_r(i\omega) \triangleq \frac{E(i\omega)}{E(i\infty)},$$

$$\sigma_t \triangleq \left[\int_{t=-\infty}^{\infty} t^2\, de_r(t)\right]^{1/2},$$

$$\sigma_{i\omega} \triangleq \left[\int_{i\omega=-i\infty}^{i\infty} |i\omega|^2 \frac{dE_r(i\omega)}{i2\pi}\right]^{1/2}.$$

Note that by the Rayleigh–Parseval Theorem $|E(i\infty)| = e(\infty)$.

Two other physical representations of this theorem arise. See [2] for the first. Here time t_k is replaced by distance-space coordinate x_k, $k = 1, 2, 3$; time-angular frequency, $i\omega_k$, is replaced by distance-space angular frequency, $i\beta_k$; and energy $e_k(t_k)$ is replaced by linear momentum $p_k(t_k)$. Then the conclusion is:

(a′) $F_k(i\beta_k) \in L_2$ on $\mathbf{I}$,

(U_1') (b′) $1 \leqq \Delta x_k \cdot \Delta|i\beta_k|$, $k = 1, 2, 3$,

(E_1') (c′) $1 = \Delta x_k \cdot \Delta|i\beta_k|$ if $f_k(x_k)$ is a Gaussian pulse.

INTRODUCTION OF SYMBOLS THROUGH SEVERAL VERSIONS OF THE LAPLACE–FOURIER TRANSFORMATION

Original Domain	Transform Domain

Ordinary unilateral

$$\int_{t_k=0^+}^{\infty} f(t_k)e^{-s_k t_k}\,dt_k \triangleq \mathscr{L}_I[f(t_k)] \triangleq F(s_k),\ t_k \text{ is time},\ s_k \triangleq \sigma_k + i\omega_k,\ \omega_k \triangleq 2\pi f_k, \qquad k = 1, 2, 3,$$

$$\int_{x_k=0^+}^{\infty} f(x_k)e^{-\gamma_k x_k}\,d\chi_k \triangleq \mathscr{L}_I[f(x_k)] \triangleq F(\gamma_k),\ x_k \text{ is distance-space coordinate},\ \gamma_k \triangleq \alpha_k + i\beta_k,$$

$$\int_{\varphi_k=0^+}^{\infty} f(\varphi_k)e^{-\zeta_k \varphi_k}\,d\varphi_k \triangleq \mathscr{L}_I[f(\varphi_k)] \triangleq F(\zeta_k),\ \varphi_k \text{ is magnetic-flux coordinate},\ \zeta_k \triangleq \chi_k + i\psi_k.$$

Stieltjes (Type 1) and $\left.\begin{matrix} s_k \\ \gamma_k \\ \zeta_k \end{matrix}\right\}$*-multiplied*

$$\int_{t_k=0^-}^{\infty} f(t_k)\,d(1 - e^{-s_k t_k}) \triangleq \mathscr{L}_I[f(t_k)] \triangleq {}_{\mathscr{S}}F(s_k),$$

$$\int_{x_k-0^-}^{\infty} f(\chi_k)\,d(1 - e^{-\gamma_k x_k}) \triangleq \mathscr{L}_I[f(\chi_k)] \triangleq {}_{\mathscr{S}}F(\gamma_k), \qquad \text{Stieltjes (Type 1)}$$

$$\int_{\varphi_k=0^-}^{\infty} f(\varphi_k)\,d(1 - e^{-\zeta_k \varphi_k}) \triangleq \mathscr{L}_I[f(\varphi_k)] \triangleq {}_{\mathscr{S}}F(\zeta_k), \qquad k = 1, 2, 3,$$

$$\equiv s_k \int_{t_k=0^-}^{\infty} f(t_k)e^{-s_k t_k}\,dt_k \triangleq {}_{s_k}\mathscr{L}_I[f(t_k)] \triangleq {}_{s_k}F(s_k),$$

$$\equiv \gamma_k \int_{x_k=0^-}^{\infty} f(x_k)e^{-\gamma_k x_k}\,dx_k \triangleq {}_{\gamma_k}\mathscr{L}_I[f(x_k)] \triangleq {}_{\gamma_k}F(\gamma_k), \qquad \left.\begin{matrix} s_k \\ \gamma_k \\ \zeta_k \end{matrix}\right\}\text{-multiplied}$$

$$\equiv \zeta_k \int_{\phi_k=0^-}^{\infty} f(\varphi_k)e^{-\zeta_k \varphi_k}\,d\varphi_k \triangleq {}_{\zeta_k}\mathscr{L}_I[f(\varphi_k)] \triangleq {}_{\zeta_k}F(\zeta_k). \qquad k = 1, 2, 3.$$

* "$\triangleq$" means "equals by definition."

LAPLACE–FOURIER TRANSFORMATION AS THE BASIS OF DERIVATIVE-TRANSFORM THEOREMS

Original Domain

In rectangular coordinates let

$$\left.\begin{matrix} a_t(t_k) \\ a_\chi(x_k) \\ a_\varphi(\varphi_k) \end{matrix}\right\} \text{be action}$$

$$\text{in the } \left.\begin{matrix} t_k \\ x_k \\ \varphi_k \end{matrix}\right\} \text{direction}, \qquad k = 1, 2, 3.$$

Then by definition:

$$\frac{da_t}{dt_k} \triangleq e_k(t_k) \text{ energy},$$

$$\frac{da_\chi}{d\chi_k} \triangleq p_k(x_k)$$

linear momentum,

$$\frac{da_\varphi}{d\varphi_k} \triangleq q_k(\varphi_k)$$

electric charge,

$$\text{in direction} \left\{\begin{matrix} t_k, \\ x_k, \\ \varphi_k, \end{matrix}\right. \qquad k = 1, 2, 3.$$

Transform Domain

$$\text{Let} \quad \left\{\begin{matrix} {}_{s_k}\mathscr{L}_I[a_t(t_k)] \triangleq {}_{s_k}A(s_k), & {}_{s_k}\mathscr{L}_I[e_k(t_k)] \triangleq E_k(s_k), & s_k \triangleq \sigma_k + i\omega_k, \\ {}_{\gamma_k}\mathscr{L}_I[a_\chi(x_k)] \triangleq {}_{\gamma_k}A(\gamma_k), & {}_{\gamma_k}\mathscr{L}_I[p_k(x_k)] \triangleq P_k(\gamma_k), & \gamma_k \triangleq \alpha_k + i\beta_k, \\ {}_{\zeta_k}\mathscr{L}_I[a_\varphi(\varphi_k)] \triangleq {}_{\zeta_k}A(\zeta_k), & {}_{\zeta_k}\mathscr{L}_I[q_k(\varphi_k)] \triangleq Q_k(\zeta_k), & \zeta_k \triangleq \chi_k + i\psi_k, \end{matrix}\right. \qquad k = 1, 2, 3.$$

The Derivative Theorem [13] here becomes:

$$\text{If} \quad \left\{\begin{matrix} {}_{s_k}A(s_k) \\ {}_{\gamma_k}A(s_k) \end{matrix}\right. \text{exist, then} \left\{\begin{matrix} E_k(s_k) = s_k\, {}_{s_k}A(s_k) - s_k a_t(0), \\ P_k(\gamma_k) = \gamma_k\, {}_{\gamma_k}A(\gamma_k) - \gamma_k a\chi(0), \\ Q_k(\zeta_k) = \zeta_k\, {}_{\zeta_k}A(\zeta_k) - \zeta_k a_\varphi(0). \end{matrix}\right. \qquad \text{(Th)}$$

Next restrict consideration to the imaginary line in the complex domain:

$$\sigma_k = 0, \quad \alpha_k = 0, \quad x_k = 0, \qquad k = 1, 2, 3. \qquad \text{(I)}$$

Introduce the asymptotic physical assumptions for $n = 0, 1, 2, \ldots$:

$$\begin{matrix} \lim\limits_{i\omega_k \to i\infty} A(i\omega_k) \triangleq A(i\infty) = n\hbar, & \text{Planck (1900)} \\ \lim\limits_{i\beta_k \to i\infty} A(i\beta_k) \triangleq A(i\infty) = n\hbar, & \text{de Broglie (1923)} \\ \lim\limits_{i\psi_k \to i\infty} A(i\psi_k) \triangleq A(i\infty) = n\hbar, & \text{Dirac (1931)} \end{matrix} \qquad \text{(A)}$$

in which Planck's quantum of action $h \triangleq 2\pi\hbar$.

Choose the origins of $\left\{\begin{matrix} t_k \\ \chi_k \\ \varphi_k \end{matrix}\right.$ so far to the left that $\left.\begin{matrix} a_t(0) \\ a_x(0) \\ a_\varphi(0) \end{matrix}\right\} = 0$, since transient response is not of interest here. Then by Theorem (Th_1) under restrictions (**I**) and assumptions (A), and in terms of increments, for $n = 0, 1, 2, \ldots$:

$$\begin{matrix} \Delta E_k(i\omega_k) = n\hbar\,\Delta(i\omega_k), & \text{Planck's Law} \\ \Delta P_k(i\beta_k) = n\hbar\,\Delta(i\beta_k), & \text{de Broglie's Law} \\ \Delta Q_k(i\psi_k) = n\hbar\,\Delta(i\psi_k), & \text{Dirac's Law} \quad k = 1, 2, 3. \end{matrix} \qquad \text{(L)}$$

COMPLEX METRICS

Example: $d\tilde{p} \triangleq |d\tilde{p}|e^{i}\angle d\tilde{p}, \quad d\tilde{p}^{\dagger} \triangleq |d\tilde{p}|\cdot e^{-i}\angle d\tilde{p}, \quad (d\tilde{p})^2 \equiv d\tilde{p}\,d\tilde{p}^{\dagger}\cdot\frac{d\tilde{p}}{dP^{\dagger}} \equiv |d\tilde{p}|^2(e^{i}\angle d\tilde{p})^2 \equiv |d\tilde{p}|^2 e^{i2}\angle\tilde{p}, \qquad k, l = 1, 2, 3.$ (I)

	Original Domain		Transform Domain	
	Primal, Covariant	Dual, Contravariant	Primal, Covariant	Dual, Contravariant
Analogs	$\overrightarrow{dz}\cdot\overrightarrow{dz} = g_{kl}\,dz^k\,dz^l$ $\overrightarrow{d\tilde{\varphi}}\cdot\overrightarrow{d\tilde{\varphi}} = g_{kl}\,d\tilde{\varphi}_k\,d\tilde{\varphi}^l$	$\overrightarrow{d\tilde{p}}\cdot\overrightarrow{d\tilde{p}} = g^{kl}\,d\tilde{p}_k\,d\tilde{p}_l$ $\overrightarrow{d\tilde{q}}\cdot\overrightarrow{d\tilde{q}} = g^{kl}\,dq_k\,d\tilde{q}_l$	$\overrightarrow{d\gamma}\cdot\overrightarrow{d\gamma} = g_{kl}\,d\gamma^k\,d\gamma^l$ $\overrightarrow{d\tilde{\zeta}}\cdot\overrightarrow{d\tilde{\zeta}} = g_{kl}\,d\tilde{\zeta}^k\,d\tilde{\zeta}^l$	$\overrightarrow{d\tilde{P}}\cdot\overrightarrow{d\tilde{P}} = g^{kl}\,d\tilde{P}_k\,d\tilde{P}_l$ $\overrightarrow{d\tilde{Q}}\cdot\overrightarrow{d\tilde{Q}} = g^{kl}\,d\tilde{Q}_k\,d\tilde{Q}_l$
Analogs	(total distance magnitude)^2 $= g_{kl}\,dz^k\,dz^{\dagger l}$ (total flux magnitude)^2 $= g_{kl}\,d\tilde{\varphi}^k\,d\varphi^{\dagger l}$	(total momentum magnitude)^2 $= g^{kl}\,d\tilde{p}_k\,d\tilde{p}_l^{\dagger}$ (total charge magnitude)^2 $= g^{kl}\,d\tilde{q}_k\,d\tilde{q}_l^{\dagger}$	$\vert$total distance spectrum$\vert^2$ $= g_{kl}\,d\gamma^k\,d\gamma^{\dagger l}$ $\vert$total flux spectrum$\vert^2$ $= g_{kl}\,d\tilde{\zeta}^k\,d\zeta^{\dagger l}$	$\vert$total momentum spectrum$\vert^2$ $= g^{kl}\,d\tilde{P}_k\,d\tilde{P}_l^{\dagger}$ $\vert$total charge spectrum$\vert^2$ $= g^{kl}\,d\tilde{Q}_k\,d\tilde{Q}_l^{\dagger}$
		Rayleigh–Parseval Theorem Cross-Domain Invariance		
	$\vert$total distance$\vert$	=	$\vert$total distance spectrum$\vert$	
		$\vert$total momentum$\vert$ =		$\vert$total momentum spectrum$\vert$
	$\vert$total flux$\vert$	=	$\vert$total flux spectrum$\vert$	
		$\vert$total charge$\vert$ =		$\vert$total charge spectrum$\vert$

FIG. 3 ACTION FUNCTIONS FROM ENERGY, LINEAR MOMENTUM, LAPLACE–FOURIER

Original Domain

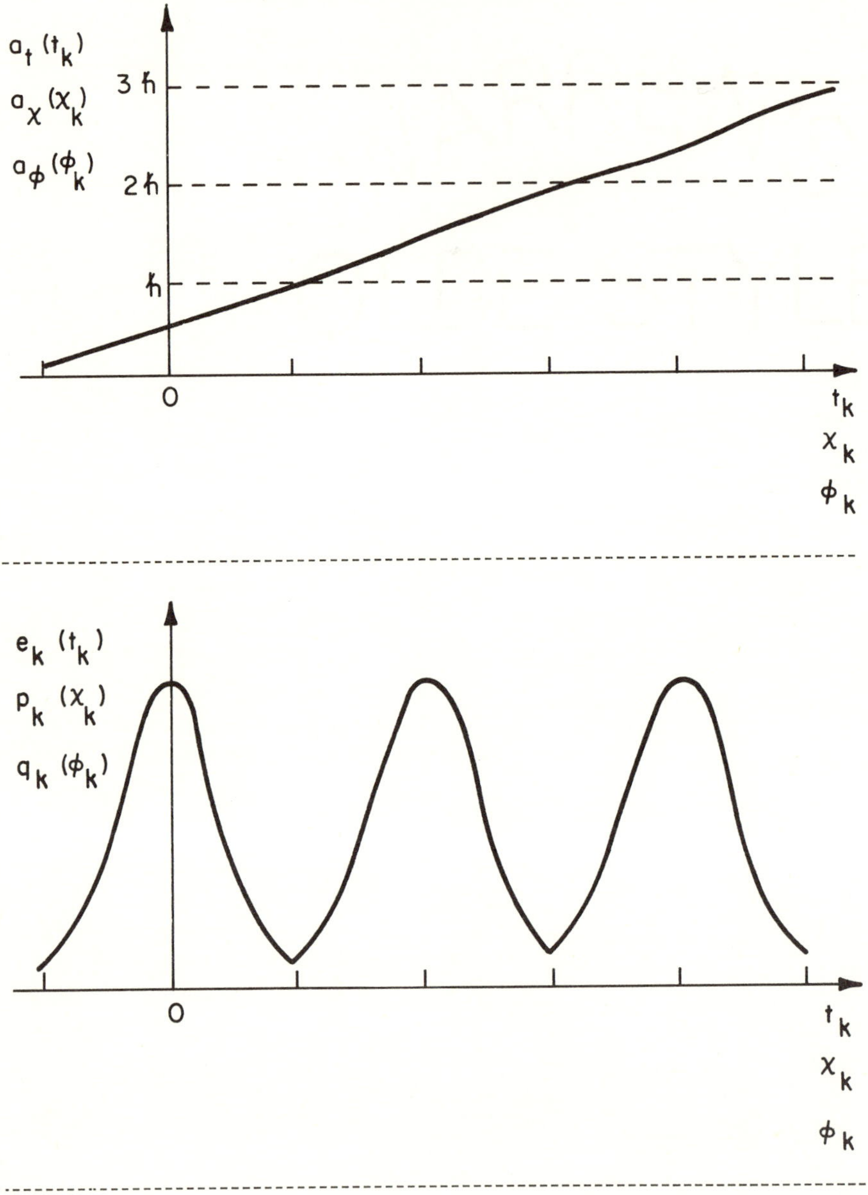

AND CHARGE FUNCTIONS, AND THEIR IMAGES IN THE TRANSFORM DOMAIN

Transform Domain

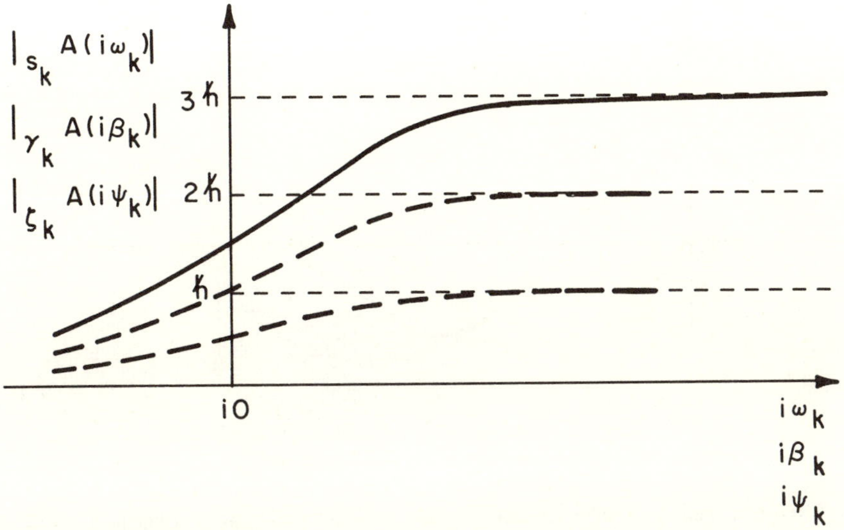

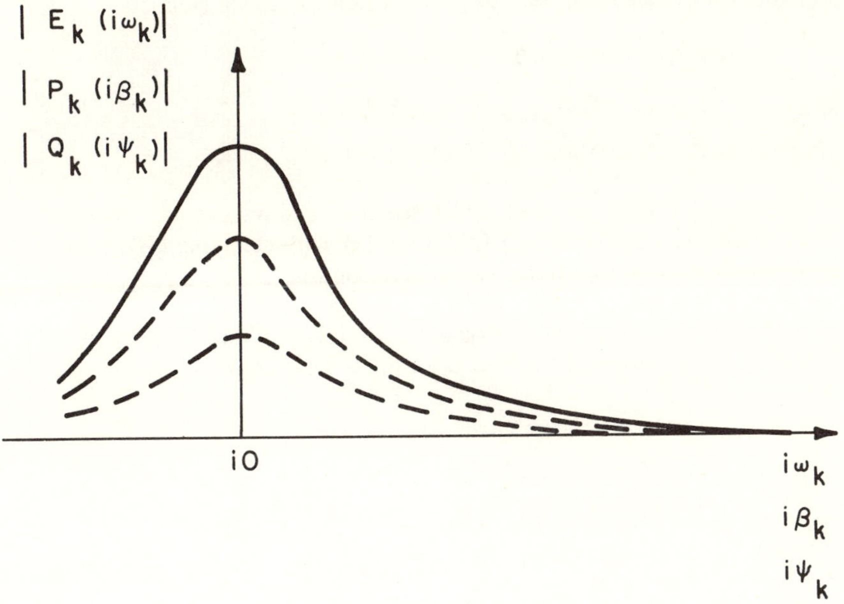

FIG. 3 ACTION FUNCTIONS FROM ENERGY, LINEAR MOMENTUM, LAPLACE–FOURIER

Original Domain

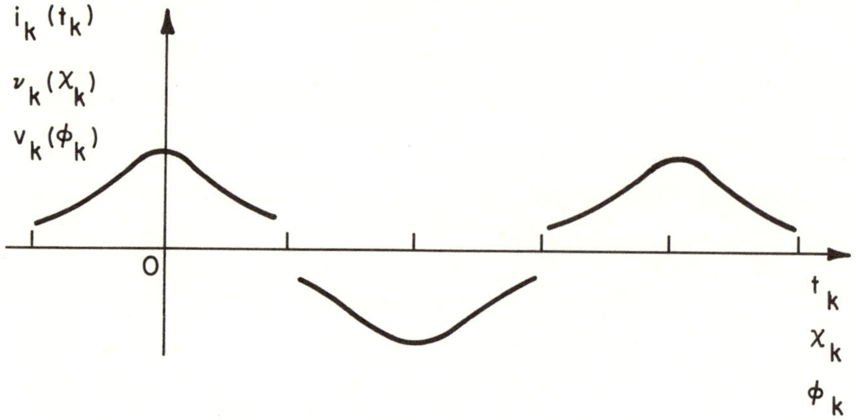

In the second representation the replacements are: magnetic flux coordinate φ_k for x_k, $k = 1, 2, 3$, magnetic flux–angular velocity $i\psi_k$ for $i\beta_k$, and electric charge $q_k(\varphi_k)$ for $p_k(x_k)$. Then the conclusion is:

(a″) $F_k(i\psi_k) \in L_2$ on **I**,

(U″₁) (b″) $1 \leqq \Delta\varphi_k \cdot \Delta|i\psi_k|, \qquad k = 1, 2, 3,$

(E″₁) (c″) $1 = \Delta\varphi_k \cdot \Delta|i\psi_k|$ if $f_k(\varphi_k)$ is a Gaussian pulse.

The partial ordering relation (U_1) provides a foundation for quantum choice-information theory (see [2]). Together with (U_1') and (U_1'') it leads to three quantum uncertainty theorems as follows.

QUANTUM UNCERTAINTY THEOREMS. Multiply (U_1), (U_1'), (U_1'') by $n\hbar$, $n = 0, 1, 2, \ldots$, to obtain

$$(U_2) \qquad \begin{aligned} n\hbar &\leqq \Delta t_k \cdot n\hbar\,\Delta|i\omega_k|, \\ n\hbar &\leqq \Delta x_k \cdot n\hbar\,\Delta|i\beta_k|, \qquad k = 1, 2, 3, \\ n\hbar &\leqq \Delta\varphi_k \cdot n\hbar\,\Delta|i\psi_k|, \qquad n = 0, 1, 2, \ldots. \end{aligned}$$

Then, by Theorem (Th_1), restrictions (**I**), and physically based assumptions (A), there result equations (L). Substituting from (L) in (U_2), under the

AND CHARGE FUNCTIONS, AND THEIR IMAGES IN THE TRANSFORM DOMAIN (*continued*)

Transform Domain

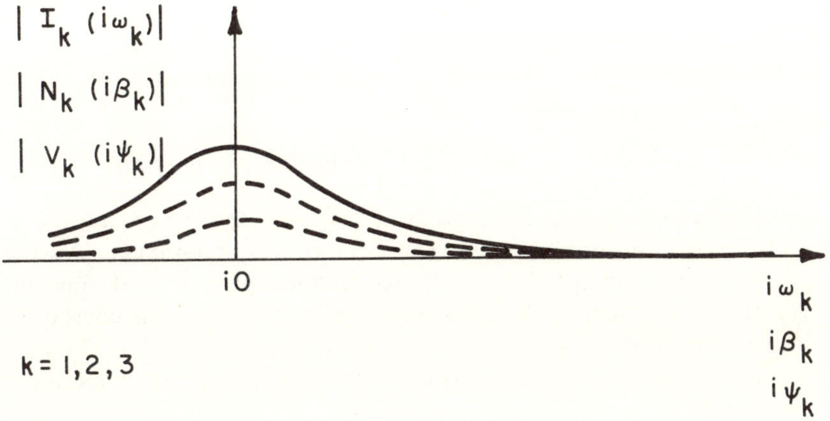

hypotheses of Wiener's Uncertainty Theorem, Theorem (Th_1), (I), and (A), the conclusions are:

$$(Th_2)\quad \begin{aligned} &(H_1) && n\hbar \leqq \Delta t_k \cdot \Delta|E_k(i\omega_k)|, \\ &(H_2) && n\hbar \leqq \Delta x_k \cdot \Delta|P_k(i\beta_k)|, \qquad k = 1, 2, 3, \\ &(D) && n\hbar \leqq \Delta \varphi_k \cdot \Delta|Q_k(i\psi_k)|. \end{aligned}$$

In conclusions (H_1) and (H_2), Theorem (Th_2) gives a precise meaning to Heisenberg's (1927) Uncertainty Principles [14], and, in conclusion (D), to Dirac's (1931) Uncertainty Principle [12]. In [3] the simple deduction of Planck's and de Broglie's Laws was given.

ASYMPTOTIC UNCERTAINTY THEOREM. An example of using Wiener's Basic Uncertainty Theorem for the asymptotic case, $n \to \infty$, i.e., for classical physics and engineering, is the following. Multiply (U_1) by the velocity of electromagnetic waves in free distance-space, c, to obtain

$$(U_3) \qquad c \leqq c\,\Delta t \cdot \Delta|i\omega|. \quad \text{But } r = ct,$$

$$(R) \qquad \Delta r = c\,\Delta t,$$

in which Δr is the $2\sigma_r$ uncertainty in r, the radar pulse distance traveled in time t in free distance-space. Then, under the hypotheses of Wiener's Uncertainty Theorem, using (R) in (U_3), the conclusion is

$$(Th_3) \quad (W) \qquad c \leqq \Delta r \cdot \Delta|i\omega|.$$

Theorem (Th_3)'s conclusion (W) gives a precise meaning to the Radar Uncertainty Principle of Woodward (1953) [18].

INVARIANT METRICS UNDER ANALYTIC COORDINATE TRANSFORMATIONS. 3-D complex (see [8]) vectors and their $\mathscr{L}$–$\mathscr{F}$ transforms will be used in the following model for (special) relativistic physics (see [1]).

ORIGINAL DOMAIN. The distance-space-time displacement *vector* in $\mathbf{C}^3$ is of the *across* type, and is

$$(\mathrm{V}^z) \qquad \overrightarrow{dz} \triangleq dz^k \overrightarrow{1}_k, \qquad k = 1, 2, 3 \text{ (summation convention)},$$

in which complex distance-displacement $z^k \triangleq x^k + ict^k$ with $\overrightarrow{1}_k$ a real unit vector in the k-direction; x^k is a real distance-space coordinate; c is the velocity of electromagnetic waves in free distance-space, ct^k is the imaginary part of the complex distance-displacement, and t^k is time coordinate in the k-direction.

Then the complex distance-displacement *metric* is given by the inner product as

$$(\mathrm{M}^z) \qquad \overrightarrow{dz} \cdot \overrightarrow{dz} = g_{kl}\, dz^k\, dz^l, \qquad k, l = 1, 2, 3,$$

in which g_{kl} are covariant coordinates of a real second-order metric tensor in distance space. The real part of this complex metric includes the Minkowsky hyperbolic metric.

TRANSFORM DOMAIN. Using the complex-to-complex Laplace–Fourier transformation gives for the distance-space-time displacement vector

$$(\mathrm{V}^\gamma) \qquad \overrightarrow{d\gamma} \triangleq d\gamma^k \overrightarrow{1}_k, \qquad k = 1, 2, 3,$$

in which $\gamma^k \triangleq \alpha^k + i\beta^k$. In the $\mathscr{L}$–$\mathscr{F}$ transform domain the inner product gives the metric

$$(\mathrm{M}^\gamma) \qquad \overrightarrow{d\gamma} \cdot \overrightarrow{d\gamma} = g_{kl}\, d\gamma^k\, d\gamma^l, \qquad k, l = 1, 2, 3.$$

ORIGINAL DOMAIN. The *dual* (see [5]), i.e., the Hamiltonian conjugate, of the distance-space-time displacement vector in (V^z) which is of the *through* type and is

$$(\mathrm{V}_{\tilde{p}}) \qquad \overrightarrow{d\tilde{p}} \triangleq d\tilde{p}_k \overrightarrow{1}^k, \qquad k = 1, 2, 3,$$

in which complex linear momentum $\tilde{p}_k \triangleq p_k + icm_{0k}$ with $\overrightarrow{1}^k$ a real unit vector in the k-direction; p_k is a real momentum coordinate; c is the velocity of electromagnetic waves in free distance-space; cm_{0k} is the imaginary part of the complex kth momentum coordinate; and m_{0k} is the rest-mass coordinate in the k-direction.

Then the complex momentum metric is

$$(\mathrm{M}_{\tilde{p}}) \qquad \overrightarrow{d\tilde{p}} \cdot \overrightarrow{d\tilde{p}} = g^{kl}\, d\tilde{p}_k\, d\tilde{p}_l, \qquad k, l = 1, 2, 3,$$

in which g^{kl} are contravariant coordinates of a real second-order metric tensor in momentum space. This metric is the dual of ($\mathrm{M}^{\tilde{z}}$).

TRANSFORM DOMAIN. Using the complex $\mathscr{L}$–$\mathscr{F}$ transformation gives for the linear-momentum vector

$$(\mathrm{V}_{\tilde{P}}) \qquad \overrightarrow{d\tilde{P}} \triangleq d\tilde{P}_k \overrightarrow{1}^k, \qquad k = 1, 2, 3,$$

in which $\tilde{P}_k \triangleq P_{kr} + iP_{ki}$. Here the inner product gives the metric

$$(\mathrm{M}_{\tilde{P}}) \qquad \overrightarrow{d\tilde{P}} \cdot \overrightarrow{d\tilde{P}} = g^{kl}\, d\tilde{P}_k\, d\tilde{P}_l, \qquad k, l = 1, 2, 3.$$

ORIGINAL DOMAIN. The electric *analog* (see [13]), of the linear-graph-preserving type, of the distance-space-time displacement vector ($\mathrm{V}^{\tilde{z}}$) is the magnetic flux-linkage displacement vector in C^3 in (V^{φ}) which is of the *across* type and is

$$(\mathrm{V}^{\tilde{\varphi}}) \qquad \overrightarrow{d\tilde{\varphi}} \triangleq d\tilde{\varphi}^k \overrightarrow{1}_k, \qquad k = 1, 2, 3,$$

in which complex flux displacement $\tilde{\varphi}^k \triangleq \varphi^k + iv_0 t^k$ with $\overrightarrow{1}_k$ a real unit vector in the k-direction; φ^k is a real magnetic flux-linkage coordinate; v_0 is the voltage or time rate of change of magnetic flux of an electromagnetic wave in free complex-magnetic-flux space; $v_0 t^k$ is the imaginary part of the complex-magnetic-flux displacement; and t^k is the time coordinate in the k-direction.

Then the complex flux-displacement metric is

$$(\mathrm{M}^{\tilde{\varphi}}) \qquad \overrightarrow{d\tilde{\varphi}} \cdot \overrightarrow{d\tilde{\varphi}} = g_{kl}\, d\tilde{\varphi}^k\, d\tilde{\varphi}^l, \qquad k, l = 1, 2, 3,$$

in which g_{kl} are covariant coordinates of a real second-order metric tensor in flux space. The real part of this complex metric is the linear-graph-preserving analogue of the extended Minkowski metric and is also hyperbolic.

TRANSFORM DOMAIN. Using the complex $\mathscr{L}$–$\mathscr{F}$ transformation for the complex magnetic-flux-displacement vector

$$(\mathrm{V}^{\tilde{\zeta}}) \qquad \overrightarrow{d\tilde{\zeta}} \triangleq d\tilde{\zeta}^k \overrightarrow{1}_k, \qquad k = 1, 2, 3,$$

in which $\tilde{\zeta}^k \triangleq \zeta^k + i\psi^k$. In the $\mathscr{L}$–$\mathscr{F}$ transform domain the inner product gives the metric

$$(\mathrm{M}^{\tilde{\zeta}}) \qquad \overrightarrow{d\tilde{\zeta}} \cdot \overrightarrow{d\tilde{\zeta}} = g_{kl}\, d\tilde{\zeta}^k\, d\tilde{\zeta}^l, \qquad k, l = 1, 2, 3.$$

ORIGINAL DOMAIN. The dual (see [5]), i.e., Hamiltonian, conjugate of the magnetic-flux-displacement vector (V^{Φ}) is the charge vector in C^3 which is of the *through* type, and is

$$(V_{\tilde{q}}) \qquad \overrightarrow{d\tilde{q}} \triangleq d\tilde{q}_k \overrightarrow{1}^k, \qquad k = 1, 2, 3,$$

in which complex charge $\tilde{q}_k \triangleq q_k + i i_0 t_k$ with $\overrightarrow{1}^k$ a real unit vector in the k-direction and q_k a real charge coordinate; i_0 is the current or time rate of change of charge of an electromagnetic wave in free complex-magnetic-flux space; $i_0 t_k$ is the imaginary part of the complex charge; and t_k is the time coordinate in the k-direction.

Then the charge metric is

$$(M_{\tilde{q}}) \qquad \overrightarrow{d\tilde{q}} \cdot \overrightarrow{d\tilde{q}} = g^{kl}\, d\tilde{q}_k\, d\tilde{q}_l, \qquad k, l = 1, 2, 3,$$

in which g^{kl} are contravariant coordinates of a real second-order metric tensor in charge space. This metric is the dual of (M^{Φ}) and the linear-graph-preserving analogue of ($M_{\tilde{p}}$).

TRANSFORM DOMAIN. Using the complex $\mathscr{L}$–$\mathscr{F}$ transformation gives for the charge vector

$$(V_{\tilde{Q}}) \qquad \overrightarrow{d\tilde{Q}} \triangleq d\tilde{Q}_k \overrightarrow{1}^k, \qquad k = 1, 2, 3,$$

in which $\tilde{Q}_k \triangleq Q_{kr} + iQ_{ki}$. Here the inner product gives the metric

$$(M_{\tilde{Q}}) \qquad \overrightarrow{d\tilde{Q}} \cdot \overrightarrow{d\tilde{Q}} = g^{kl}\, d\tilde{Q}_k\, d\tilde{Q}_l, \qquad k, l = 1, 2, 3.$$

4. Messages, choice information, and probability in the analytic complex domain

Considering the discrete finite examples in Figs. 1a, 1b, and 2 and the discussion which accompanies these it is seen that more generally a function of the form

$$m(x, y) \triangleq y^x$$

defined for real domain x as position and real domain y as height is evidently a function of two independent variables $m(x, y)$ in the message range. Hence it carries both abscissa (position) and ordinate (height) information. Examination of the lowest set of three identically shaped pulses in the original domain on Fig. 3 shows that these carry position as well as height information. The unilateral Laplace–Fourier transforms of such functions are analytic throughout a half plane in their complex domain of definition (see [7]). Only the amplitude function of frequency is shown in the bottom right-hand part of Fig. 3. However, along the imaginary axis in the, say, $s \triangleq \sigma + i\omega$ plane, the complex transform of such a function in polar form can be shown as two real functions, namely, an amplitude function of frequency which carries height information from the original function and a phase-angle function of frequency which carries position information from the original function. Translating a pulse, while leaving its height and hence shape invariant in the original domain with respect to, say, t, changes its phase spectrum only, leaving its amplitude spectrum invariant in the transform domain. Thus a single real positive pulse in the original domain normalized to have its integral equal to one may be interpreted as a real probability density function for a random variable. If it is a Gaussian function (in height), then its $\mathscr{L}$–$\mathscr{F}$ amplitude spectrum is also Gaussian (see [13]), while a change in its horizontal position would show up as a change in its phase spectrum. For a pair of pulses the difference in phase would be of interest. Consequently it can be said that the $\mathscr{L}$–$\mathscr{F}$ transform of a real p.d.f. is an analytic complex probability density function with amplitude and phase spectra which are conditioned by the height and position of the original real p.d.f. This idea has been known at least roughly by the quantum physicists since it was introduced by the Copenhagen School in the late 1920's. It has also been known by the communication engineers concerned with picture (facsimile) transmission since the same time. However, even today it seems not to appear in the literature of strict probabilists and statisticians, although it is strongly appreciated by those who make extensive use of transform methods in stochastics (see [9], [4], and [6]).

The transition from flat photography to 3-D holography can be made in statistics, probabilistics, and stochastics by a transition from the real to the complex domain. The transition can be made from Kolmogorov's axioms for real probability [15] by a corresponding set of axioms for complex probability. A possible set of such axioms for the finite case is the following.

Given a complex space $(\tilde{\mathscr{S}}, \tilde{\mathscr{E}}, \tilde{P})$ in which space $\tilde{\mathscr{S}}$ is a complex set of contingencies, $\tilde{\mathscr{E}}$ the set of subsets, i.e., the events of space $\tilde{\mathscr{S}}$, and $\tilde{P} \triangleq |\tilde{P}|e^{i\angle\tilde{P}}$, a complex probability measure for $\tilde{\mathscr{E}}$, i.e., an appropriate set of complex numbers, which satisfy the following three axioms:

1. If $\tilde{\mathscr{E}} \subset \tilde{\mathscr{S}}$, then $0 \leqq |\tilde{P}(\tilde{\mathscr{E}})|$, $0 \leqq \angle\tilde{P}(\tilde{\mathscr{E}}) < 2\pi$.
2. If $\mathscr{A} \triangleq \tilde{\mathscr{S}}$, then $|\tilde{P}(\mathscr{A})| = 1$.
3. If $\tilde{\mathscr{E}}_k$, $\tilde{\mathscr{E}}_1 \cup \tilde{\mathscr{E}}_2 \subset \tilde{\mathscr{S}}$, $k = 1, 2$, and $\tilde{\mathscr{E}}_1 \cap \tilde{\mathscr{E}}_2 = \tilde{\phi}$, then
   $\tilde{P}(\tilde{\mathscr{E}}_1 \cup \tilde{\mathscr{E}}_2) = \tilde{P}(\tilde{\mathscr{E}}_1) + \tilde{P}(\tilde{\mathscr{E}}_2)$.

Of course, such a triple of axioms as in the case of real probability extends to a countable set of events and to a set of events defined on the complex plane. Analytic complex functions are of interest in models for cumulative distribution functions and probability density functions associated with the physical world.

With *whole probability* it is the difference in position of the modes in, say, a bimodal probability density function of real type which carries the abscissa information. This 1-D information is in addition to the 1-D height information carried by the ordinates. Naturally the real p.d.f. summarizes the information on the behavior of a random variable of real type which carries information on both its height and position.

Next compare $z \triangleq |z|e^{i\angle z}$ in modulus and angle form with $e^s \equiv e^{\sigma+i\omega} \equiv e^\sigma \cdot e^{i\omega}$ which is the complex base in the inverse $\mathscr{L}$–$\mathscr{F}$ transformation (see [13]) kernel $e^{ts} \equiv (e^s)^t$. In turn compare $e^s \equiv e^\sigma \cdot e^{i\omega}$ with the real discrete message function in Figs. 1 and 2, $m \triangleq n_s^{n_c} \equiv n_{sH}^{n_{cH}} \cdot n_{sP}^{n_{cP}}$. It is seen that if the polar-coordinate form of z is looked at in rectangular coordinates $(|z|, i \angle z)$ as in radio-antenna engineering, or $(e^\sigma, i\omega)$ or in the complex-base exponent plane $(\sigma, i\omega)$ that *complex information* results. Here the real part, I_r, is the height information, while the imaginary part, I_i, is the position information in $I \triangleq I_r + iI_i$. Or, alternatively, for the direct $\mathscr{L}$–$\mathscr{F}$ transformation the complex base is e^{-s} in the kernel $(e^{-s})^t \equiv e^{-st}$ which corresponds to $z^{-1} \equiv |z|^{-1} \cdot e^{-i<z}$ and to our message probability $m^{-1} \equiv n_s^{-n_c} \equiv n_{SH}^{-n_{cH}} \cdot n_{SP}^{-n_{cP}}$. Thus damping factor, $e^{-\sigma}$, and decrement, $-\sigma$, are connected with height information, while phase-angle factor, $e^{-i\omega}$, and phase-angle coefficient, $-i\omega$, correspond to position information. Finally, real position cells may be replaced by imaginary position cells to have *complex messages* and their associated complex probability. Thus

$$\tilde{m}_{H+iP} \triangleq n_{SH}^{n_{cH}} \cdot n_{SP}^{in_{cP}} \triangleq n_s^{\tilde{I}}, \quad \text{if } n_{SH} = n_{SP} \triangleq n_S.$$

Finally, as a matter of approximation to the physical world, the asymptotic version of quantum message, choice information, and probability theory leads to the choice-information theory of Hartley and Shannon and to "continuous" probability theory. Nevertheless, in a more precise sense, uncertainty theorems such as those discussed in the present linear context prevent an "exact" continuum model since they place lower bounds on physical resolution due to the existence of upper bounds on the dual-transform quantities, here momentum (including energy) and charge, available in our part of the physical world.

UNIVERSITY OF CALIFORNIA
LOS ANGELES

REFERENCES

1. ATKIN, R. H., *Mathematics and Wave Mechanics*. New York: John Wiley & Sons, Inc., 1957.
2. BARNES, JOHN L., "Information theoretic aspects of feedback control systems (Revised version)," *Automatica*, *4* (1968), 165–185.
3. ———, "Laplace–Fourier transformation as the basis of uncertainty principles, Planck's and de Broglie's Laws, and information theory," paper presented at the International Congress of Mathematicians, Moscow, U.S.S.R., August 1966.
4. BARTLETT, M. S., *Stochastic Processes*. London: Cambridge University Press, 1956.
5. BECKENBACH, E. F., ed., *Modern Mathematics for the Engineer*, Series 1, Chap. 14, "Functional transformations for engineering design," by John L. Barnes. New York: John Wiley & Sons, Inc., 1956.
6. BLANC-LAPIERRE, A., and R. FORTET, *Theorie des Fonctions Aleatoires*. Paris: Masson, 1953.
7. BOCHNER, S., *Vorlesungen über Fouriersche Integrale*. Leipzig: Akademische Verlagsgesellschaft M.B.H., 1932.
8. ———, and W. T. MARTIN, *Several Complex Variables*. Princeton, N.J.: Princeton University Press, 1948.
9. ———, *Harmonic Analysis and the Theory of Probability*. Berkeley, Cal.: University of California Press, 1960.
10. ———, *The Role of Mathematics in the Rise of Science*. Princeton, N.J.: Princeton University Press, 1966.
11. DE BROGLIE, L., *Nature*, *112* (1923), 540; Thesis, Paris, 1924, *Ann. Physique*, *2* (10), 1925.
12. DIRAC, P. A. M., "Quantized singularities in the electromagnetic field," *Royal Society* (*London*), *Proc.*, *133* (1931), 60–72.
13. GARDNER, M. F., and J. L. BARNES, *Transients in Linear Systems Studied by the Laplace Transformation*. New York: John Wiley & Sons, Inc., 1942.
14. HEISENBERG, W., *Zeits. f. Phys.*, *43* (1927), 172.
15. KOLMOGOROV, A. N., *Grundbegriffe der Wahrscheinlichkeitsrechnung*. Berlin: Springer-Verlag, 1933.
16. MESSIAH, A., *Quantum Mechanics*, Vol. 1. Amsterdam: North-Holland Publishing Co., 1961.
17. WIENER, N., *Extrapolation, Interpolation and Smoothing of Stationary Time Series*. New York: John Wiley & Sons, Inc., 1949.
18. WOODWARD, P. M., *Probability and Information Theory, with Application to Radar*. London: Pergamon Press, Inc., 1953.

An Integral Equation Related to the Schroedinger Equation with an Application to Integration in Function Space

R. H. CAMERON AND D. A. STORVICK[1]

1. Introduction

In a previous paper [1][2] the authors proved the existence of a solution of the equation[3]

$$\Gamma(t, \xi, q) = \left(\frac{q}{2\pi it}\right)^{1/2} \overset{(\xi)}{\int_{-\infty}^{\infty}} \psi(u) e^{iq(\xi-u)^2/2t}\, du$$

$$(1.1) \qquad + \left(\frac{q}{2\pi i}\right)^{1/2} \int_0^t \frac{1}{(t-s)^{1/2}} \overset{(\xi)}{\int_{-\infty}^{\infty}} \theta(s, u)\Gamma(s, u, q) e^{iq(\xi-u)^2/2(t-s)}\, du\, ds$$

for almost every real q under the following conditions. It was assumed that $\theta(t, u)$ is continuous almost everywhere in the strip $R: 0 \leqq t \leqq t_0$, $-\infty < u < \infty$, $|\theta(t, u)| \leqq M$ for (t, u) in R and that $\psi(u) \in L_2(-\infty, \infty)$.

Moreover it was shown[4] that for each $t \in (0, t_0]$ and for almost every real q, the solution satisfies the inequality

$$(1.2) \qquad \int_{-\infty}^{\infty} |\Gamma(t, \xi, q)|^2\, d\xi \leqq \|\psi\|^2 e^{2Mt_0}.$$

[1] Research sponsored by the Air Force Office of Scientific Research, Office of Aerospace Research, United States Air Force, Grant No. AF-AFOSR 381–66.

[2] See Theorem 9 and the Corollary to Theorem 5 of [1]. In choosing the value of the square root in the last term of (1.1) we assume that $|\mathrm{Arg}\,(q/i)| = \pi/2$.

[3] We have found it to be convenient to introduce a notation for limits in the mean as $A \to \infty$ of integrals over $(-A, A)$ such as occur in Plancherel's theorem, namely

$$\overset{(\xi)}{\int_{-\infty}^{\infty}} f(u, \xi)\, du \equiv \underset{\substack{A\to\infty \\ (\xi)}}{\mathrm{l.i.m.}} \int_{-A}^{A} f(u, \xi)\, du.$$

[4] See Theorem 9 and the Corollary to Theorem 5 of [1].

It is the purpose of this paper to obtain a corresponding uniqueness theorem and then to strengthen both the existence and the uniqueness theorems. In doing this we will obtain new existence theorems for operator-valued integrals in function space.

Let $\lambda > 0$. Consider the operator $I_\lambda(F)$ which maps the function $\psi \in L_2(-\infty, \infty)$ into the function $I_\lambda(F)\psi$ whose value at the point $\xi \in (-\infty, \infty)$ is given by the Wiener integral

$$(I_\lambda(F)\psi)(\xi) = \int_{C_0[a,b]} F(\lambda^{-1/2}x + \xi)\psi(\lambda^{-1/2}x(b) + \xi)\, dx. \tag{1.3}$$

Thus $I_\lambda(F)$ may be considered as a family of operators depending on the positive parameter λ.

In [1] this family of operators was studied, and formulas were given for its analytic extension to the half-plane, $\operatorname{Re} \lambda > 0$. By the use of weak limits as λ approached the imaginary axis, we were able to extend the definition continuously to almost all points of the imaginary axis. In the present paper we prove a theorem which removes the uncertainty of the expression "almost all."

2. The uniqueness theorem

THEOREM 1. *Let q be any fixed real number, $\psi \in L_2(-\infty, \infty)$, $\theta(t, u)$ continuous almost everywhere in R and $|\theta(t, u)| \leqq M$ in R. Then any two solutions of* (1.1) *for which the left member of* (1.2) *is finite and bounded in t are equal in R except on a set of measure zero; in fact, for each t they are equal for almost all ξ.*

PROOF. Let Γ_1 and Γ_2 be two solutions of (1.1) (for the same fixed q), let

$$G(t, \xi) = \Gamma_1(t, \xi, q) - \Gamma_2(t, \xi, q), \tag{2.1}$$

and assume that the left member of (1.2) is finite and bounded in t for both Γ_1 and Γ_2. Then it follows from (1.1) that

$$G(t, \xi) = \left(\frac{q}{2\pi i}\right)^{1/2} \int_0^t \frac{1}{(t-s)^{1/2}} \int_{-\infty}^{\infty} {}^{(\xi)} \theta(s, u) G(s, u) e^{iq(\xi-u)^2/2(t-s)}\, du\, ds. \tag{2.2}$$

Moreover, by the Minkowski inequality there exists a finite constant B such that

$$\|G(t, \cdot)\| \equiv \left|\int_{-\infty}^{\infty} |G(t, \xi)|^2\, d\xi\right|^{1/2} \leqq B \quad \text{for } 0 < t \leqq t_0. \tag{2.3}$$

We shall show by induction that

$$\|G(t,\cdot)\| \leqq \frac{BM^n t^n}{n!} \quad \text{for } 0 < t \leqq t_0 \text{ and } n = 0, 1, 2, \ldots, \tag{2.4}$$

and hence that

$$\|G(t,\cdot)\| = 0 \quad \text{for } 0 < t \leqq t_0. \tag{2.5}$$

The initial step of the induction (for $n = 0$) follows from (2.3) and we therefore assume that (2.4) holds for $n = N - 1$ and seek to show that it holds for $n = N$.

From Lemma 1 of [1] we have

$$\left\| \left(\frac{q}{2\pi i(t-s)} \right)^{1/2\,(\cdot)} \int_{-\infty}^{\infty} \theta(s,u)G(s,u)e^{iq((\cdot)-u)^2/2(t-s)}\, du \right\| \leqq \|\theta(s,\cdot)G(s,\cdot)\| \leqq M\frac{BM^{N-1}s^{N-1}}{(N-1)!}.$$

Thus we have from (2.2)

$$\|G(t,\cdot)\| \leqq \frac{BM^N}{(N-1)!}\int_0^t S^{N-1}\, ds = \frac{BM^N t^N}{N!}$$

and (2.4) holds when $n = N$ and thus holds for all non-negative integers n. Thus (2.5) holds and (2.1) shows that the solution is essentially unique.

It is clear that a corresponding existence theorem could be established by the techniques of successive substitutions. However we shall obtain the existence theorem in a different way by giving a specific formula for the solution in terms of our operator-valued function space integral. This will yield a stronger existence theorem than that obtained in Theorem 9 of [1]. Our new result establishes the existence of the operator-valued function space integral and hence of the solution of the integral equation for all real q (rather than for almost all real q).

3. Cluster sets

The notion of cluster set has been an extremely useful one in many branches of analysis in describing the behavior of a function near the boundary of its domain of definition. In this section, for the sake of completeness, we shall establish a few supporting facts. We begin with a definition.

DEFINITION. Let $H(z)$ be a function defined throughout a domain Ω of the complex z-plane whose values lie in a separable Hilbert space $\mathscr{H}$. For any point $z_0 \in \bar{\Omega}$ we define the cluster set, $\mathscr{C}_\Omega(H, z_0)$, of $H(z)$ at z_0 to be the set of points $\alpha \in \mathscr{H}$ such that there exists a sequence $\{z_n\} \subset \Omega - \{z_0\}$ such that $\lim_{n\to\infty} z_n = z_0$ and $w \lim_{n\to\infty} H(z_n) = \alpha$. (The last limit shall be interpreted as the weak limit in the Hilbert space $\mathscr{H}$.)

We shall begin by proving that $\mathscr{C}_\Omega(H, z_0)$ is a nonempty closed set whenever $H(z)$ is bounded. The first lemma is analogous to the Bolzano–Weirstrass theorem and is equivalent to the statement and the proof of the weak compactness of the unit ball.

LEMMA 1. *If $\{\alpha_n\}$ is a bounded sequence, $\|\alpha_n\| \leqq B$, in $\mathscr{H}$, then there exist an element $\alpha \in \mathscr{H}$ and a subsequence $\{\alpha_{n_\nu}\}$ such that $\omega \lim_{\nu\to\infty} \alpha_{n_\nu} = \alpha$, i.e., $\lim_{\nu\to\infty} (\alpha_{n_\nu}, \varphi) = (\alpha, \varphi)$ for all $\varphi \in \mathscr{H}$.*

PROOF. Let $\{\varphi_k\}$ be a complete orthonormal sequence in $\mathscr{H}$, and let $(\alpha_n, \varphi_1) = z_{n,1}$. Extract a subsequence $\{\alpha_{n_j}\}$ such that $(\alpha_{n_j}, \varphi_1) = z_{n_j,1} \to z_1^*$. By repeating this procedure of extracting a subsequence from the previous subsequence and then applying the diagonal process, we obtain a subsequence $\{\alpha_\nu'\}$ of the original sequence such that $(\alpha_\nu', \varphi_k) \to z_k^*$ as $\nu \to \infty$ for every k. Now

$$\sum_{k=1}^{\infty} |(\alpha_\nu', \varphi_k)|^2 = \|\alpha_\nu'\|^2 \leqq B^2,$$

and in particular

$$\sum_{k=1}^{N} |(\alpha_\nu', \varphi_k)|^2 \leqq B^2 \quad \text{for each } N.$$

Taking the limit $\nu \to \infty$, we obtain

$$\sum_{k=1}^{N} |z_k^*|^2 \leqq B^2$$

and hence

$$\sum_{k=1}^{\infty} |z_k^*|^2 \leqq B^2.$$

Then $\alpha \equiv \sum_{k=1}^{\infty} z_k^* \varphi_k$ exists and $(\alpha, \varphi_k) = z_k^*$ and $(\alpha_\nu', \varphi_k) \to (\alpha, \varphi_k)$ as $\nu \to \infty$. Because the $\{\varphi_k\}$ forms a complete orthonormal sequence and the $\{\alpha_\nu'\}$ is bounded, we have $(\alpha_\nu', \varphi) \to (\alpha, \varphi)$ for all $\varphi \in \mathscr{H}$. This completes the proof of the lemma.

LEMMA 2. *The cluster set $\mathscr{C}_\Omega(H, z_0)$ is nonempty and weakly closed whenever $H(z)$ is a bounded function.*

PROOF. The fact that $\mathscr{C}_\Omega(H, z_0)$ is nonempty follows from Lemma 1. To show that the cluster set $\mathscr{C}_\Omega(H, z_0)$ is a closed set, let $\{\alpha^k\}$ be any sequence contained in $\mathscr{C} = \mathscr{C}_\Omega(H, z_0)$ which converges weakly to an element $\alpha \in \mathscr{H}$, and we shall prove that $\alpha \in \mathscr{C}$.

We shall assume without loss of generality that $H(z)$ is bounded by one and consider $\{\varphi_n\}$ a complete orthonormal sequence in $\mathscr{H}$. We form the metric $\Delta(\alpha, \beta) = \sum_{n=1}^{n=\infty} \frac{1}{2^n} |(\alpha, \varphi_n) - (\beta, \varphi_n)|$ which gives a topology equivalent to the weak topology inside the unit ball.

Since $\alpha^k \to \alpha$ weakly, given any $\varepsilon > 0$, there exists K such that for $k \geqq K > 0$, $\Delta(\alpha^k, \alpha) < \varepsilon/2$. Because each α^k is an element of $\mathscr{C}$, for each α^k there exists a sequence $\{z_j^k\}$, $z_j^k \in \Omega$, $\lim_{j\to\infty} z_j^k = z_0$ and $H(z_j^k) \underset{j\to\infty}{\to} \alpha^k$ weakly. For each index k, we choose an index j_k such that $|z_0 - z_{j_k}^k| < 1/k$ and $\Delta(H(z_{j_k}^k), \alpha^k) < 1/k$. The sequence $\{z_k^*\}$ defined by $z_k^* = z_{j_k}^k$ has the property that $z_k^* \underset{k\to\infty}{\to} z_0$ and because $\Delta(H(z_k^*), \alpha) \leqq \Delta(H(z_k^*), \alpha^k) + \Delta(\alpha^k, \alpha)$ we have $\Delta(H(z_k^*), \alpha) < \varepsilon$ whenever $k \geqq K + 2/\varepsilon$ and so $H(z_k^*) \to \alpha$ weakly and consequently $\mathscr{C}$ is closed.

LEMMA 3. *If the cluster set consists of only one element α, the function possesses α as a weak limit, i.e., if $z_n \to z_0$ and $\mathscr{C}_\Omega(H(z), z_0) = \{\alpha\}$ then $\lim_{n\to\infty} (H(z_n) - \alpha, \varphi) = 0$ for every $\varphi \in \mathscr{H}$.*

PROOF. This follows from Lemma 1 by a reductio ad absurdum argument.

4. Operator-valued function space integrals

We now introduce the operator-valued function space integral that was defined in [1]. Let λ be a parameter which eventually will assume complex values but for the present is real and positive. Let $C[a, b] = \{x(\cdot) | x(t)$ real and continuous on $[a, b]\}$, and let $C_0[a, b] = \{x(\cdot) | x \in C[a, b], x(a) = 0\}$. Let F be a complex-valued functional defined on $C[a, b]$ such that the following Wiener integral exists:

$$(4.1) \qquad (I_\lambda(F)\psi)(\xi) = \int_{C_0[a,b]} F(\lambda^{-1/2}x + \xi)\psi(\lambda^{-1/2}x(b) + \xi)\, dx$$

for $\psi \in L_2(-\infty, \infty)$ and real ξ. Let $I_\lambda(F)\psi$ denote the function which maps ξ into $(I_\lambda(F)\psi)(\xi)$ for almost all values of ξ in $(-\infty, \infty)$. Let $I_\lambda(F)$ denote the operator that maps ψ into $I_\lambda(F)\psi$.

We extend the definition of $I_\lambda(F)$ to complex values of the parameter λ in two different ways. Using analytic extension, we define $I_\lambda^{\text{an}}(F)$ to be the operator-valued analytic function of λ which agrees with $I_\lambda(F)$ for real λ and is analytic through $\operatorname{Re} \lambda > 0$. It is clear that if $I_\lambda^{\text{an}}(F)$ exists it is unique.

We also make a sequential definition which in a large class of cases coincides with $I_\lambda^{\mathrm{an}}(F)$. Let us consider functionals defined on the larger space $B[a, b]$:

$B[a, b] = \{x(\cdot)|x(t)$ defined and continuous on $[a, b]$ except for a finite number of finite jump discontinuities$\}$.

Let σ be any subdivision of $[a, b]$, $\sigma\colon a = t_0 < t_1 < t_2 < \cdots < t_n = b$, and let norm $\sigma = \max\,(t_j - t_{j-1})$. For any $x \in B[a, b]$, let

$$x_\sigma(t) = \begin{cases} x(t_{j-1}) & \text{if } t_{j-1} \leqq t < t_j \text{ for } j = 1, \ldots, n \\ x(b) & \text{if } t = b. \end{cases} \tag{4.2}$$

Let

$$F_\sigma(x) = F(x_\sigma) \tag{4.3}$$

for $x \in B[a, b]$. Let us define the function $f_\sigma(v_0, \ldots, v_n)$ by the equation $f_\sigma[x(t_0), \ldots, x(t_n)] = F(x_\sigma)$ which holds for all $x \in B[a, b]$. For $\operatorname{Re}(\lambda) > 0$ let

$$\begin{aligned}(I_\lambda^\sigma(F)\psi)(\xi) &= \frac{\lambda^{n/2}}{(2\pi)^{n/2}(t_1 - a)^{1/2}\cdots(t_n - t_{n-1})^{1/2}} \int_{-\infty}^{\infty} \overset{(n)}{\cdots} \int_{-\infty}^{\infty} f_\sigma(\xi, v_1, \ldots, v_n) \\ &\quad \cdot \psi(v_n) \exp\left\{-\sum_{j=1}^{n} \frac{\lambda(v_j - v_{j-1})^2}{2(t_j - t_{j-1})}\right\} dv_1 \cdots dv_n,\end{aligned} \tag{4.4}$$

where $\psi \in L_2(-\infty, \infty)$ and $v_0 \equiv \xi$. We now define

$$I_\lambda^{\mathrm{seq}}(F) = \operatorname*{w\,lim}_{\sigma\to 0} I_\lambda^\sigma(F), \tag{4.5}$$

where w lim denotes weak limit.

In Theorem 1 of [1] we have shown that if $F(x)$ is bounded and continuous in the uniform topology on $B[a, b]$, $I_\lambda^{\mathrm{seq}}(F)$ exists and equals $I_\lambda(F)$ for real positive λ. We have shown in [1] that for a large class of functionals F, $I_\lambda^{\mathrm{an}}(F)$ and $I_\lambda^{\mathrm{seq}}(F)$ exist and are equal for $\operatorname{Re}(\lambda) > 0$. When $I_\lambda^{\mathrm{an}}(F)$ and $I_\lambda^{\mathrm{seq}}(F)$ exist and are equal for $\operatorname{Re}\lambda > 0$, we shall call their common value $I_\lambda(F)$.

Let F be a functional for which $I_\lambda^{\mathrm{seq}}(F)$ and $I_\lambda^{\mathrm{an}}(F)$ exist and are equal for $\operatorname{Re}\lambda > 0$. Then if the following weak limit exists, we define

$$J_q(F) = \operatorname*{w\,lim}_{\lambda\to iq,\ \operatorname{Re}\lambda>0} I_\lambda(F). \tag{4.6}$$

In order to obtain the existence of the weak limit $J_q(F)$, we shall first prove a theorem showing that certain functions defined by cluster sets satisfy our integral equation.

5. The integral equation for cluster values

THEOREM 2. *Let $\theta(t, u)$ be continuous almost everywhere in the strip $R\colon 0 \leqq t \leqq t_0$, $-\infty < u < \infty$, let $|\theta(t, u)| \leqq M$ for $(t, u) \in R$, and let q be any real number, $q \neq 0$. Let $\psi \in L_2(-\infty, \infty)$ and let*

$$F_t(x) = \exp\left\{\int_0^t \theta(t - s, x(s))\, ds\right\}. \tag{5.1}$$

Let $G(\zeta, \lambda) \equiv G(t, \xi, \lambda) \equiv (I_\lambda(F_t)\psi)(\xi)$, where $\zeta = (t, \xi)$. Then every element $\Gamma(\zeta) \equiv \Gamma(t, \xi)$ of $\mathscr{C}(G(\cdot, \lambda), -iq)$ satisfies the integral equation:

$$\begin{aligned}\Gamma(t, \xi) = {} & \left(\frac{q}{2\pi it}\right)^{1/2} \overset{(\xi)}{\int_{-\infty}^{\infty}} \psi(u) \exp\left(\frac{iq(\xi - u)^2}{2t}\right) du \\ & + \left(\frac{q}{2\pi i}\right)^{1/2} \overset{(\xi)}{\int_0^t} \int_{-\infty}^{\infty} (t - s)^{-1/2} \theta(s, u) \Gamma(s, u) \exp\left(\frac{iq(\xi - u)^2}{2(t - s)}\right) du\, ds\end{aligned} \tag{5.2}$$

for almost all $(t, \xi) \in R$.

The proof of this theorem parallels the proof of Theorem 9 of [1], but it differs at significant points. For the convenience of the reader the entire proof is presented.

PROOF. Let t and q be given real numbers, $t \in (0, t_0)$ and $q \neq 0$. By applying Lemmas 4 and 5 of [1, p. 527] to $\theta^*(s, x(s)) = \theta(t - s, x(s))$ we observe that $\theta(t - s, x(s))$ is measurable for almost every $x \in C_0[0, t]$, and is measurable in the product space. Thus $F_t(x)$ is Wiener integrable on $C_0[0, t]$ and hence also on $C_0[0, t_0]$. By Theorem 4 of [1], $I_\lambda(F_t)$ exists and is bounded for $\operatorname{Re} \lambda > 0$, $\|I_\lambda(F_t)\| \leqq e^{Mt_0} = M'$. Hence we obtain

$$\|I_\lambda(F_t)\psi\| \leqq M'\|\psi\|$$

and

$$\int_{-\infty}^{\infty} |(I_\lambda(F_t)\psi)(\xi)|^2\, d\xi \leqq M'^2\|\psi\|^2.$$

Upon integration with respect to t we obtain

$$\int_0^{t_0} \int_{-\infty}^{\infty} |(I_\lambda(F_t)\psi)(\xi)|^2\, d\xi\, dt \leqq M'^2 t_0 \|\psi\|^2.$$

If we set $(I_\lambda(F_t)\psi)(\xi) \equiv G(t, \xi, \lambda) \equiv G(\zeta, \lambda)$ where $\zeta = (t, \xi)$ we have by the Fubini theorem

$$\int_R |G(\zeta, \lambda)|^2\, d\zeta \leqq M'^2 t_0 \|\psi\|^2$$

and

$$\|G(\cdot, \lambda)\| \leqq M'(t_0)^{1/2}\|\psi\|.$$

From Lemma 2 of Section 2, we conclude that the cluster set $C = \mathscr{C}_\Omega(G(\cdot, \lambda), -iq)$ is a nonempty set. Here the Hilbert space H is understood to be $L_2(R)$, and the domain Ω is the right half plane, $\operatorname{Re} \lambda > 0$.

Let λ_n be a sequence such that $\operatorname{Re} \lambda_n > 0$, $\lambda_n \to -iq$, and $G(\zeta, \lambda_n) \to \Gamma(\zeta)$ weakly, i.e.,

$$\lim_{n\to\infty} \int_R G(\zeta, \lambda_n)\varphi(\zeta)\, d\zeta = \int_R \Gamma(\zeta)\varphi(\zeta)\, d\zeta \tag{5.3}$$

for all $\varphi \in L_2(R)$. Our function $G(t, \xi, \lambda_n)$ satisfies the integral equation of Theorem 8 of [1]:

$$\begin{aligned} G(t, \xi, \lambda_n) = {} & \left(\frac{\lambda_n}{2\pi t}\right)^{1/2} \int_{-\infty}^{\infty} \psi(u) e^{-\lambda_n(\xi-u)^2/2t}\, du \\ & + \left(\frac{\lambda_n}{2\pi}\right)^{1/2} \int_0^t \frac{1}{(t-s)^{1/2}} \int_{-\infty}^{\infty} \theta(s,u) G(s,u,\lambda_n) e^{-\lambda_n(\xi-u)^2/2(t-s)}\, du\, ds. \end{aligned} \tag{5.4}$$

We now multiply this equation by $\varphi_k(\xi) \equiv H_k(\xi)e^{-\xi^2/2}$ (where $H_k(\xi)$ is essentially the Hermite polynomial of degree k, i.e., $H_k(\xi) = (-1)^k e^{\xi^2/2} \cdot (d^k/d\xi^k)e^{-\xi^2/2}$), and integrate with respect to ξ over $(-\infty, \infty)$:

$$\begin{aligned} \int_{-\infty}^{\infty} \varphi_k(\xi) G(t, \xi, \lambda_n)\, d\xi = {} & \int_{-\infty}^{\infty} \varphi_k(\xi) \left(\frac{\lambda_n}{2\pi t}\right)^{1/2} \int_{-\infty}^{\infty} \psi(u) e^{-\lambda_n(\xi-u)^2/2t}\, du\, d\xi \\ & + \int_{-\infty}^{\infty} \left(\frac{\lambda_n}{2\pi}\right)^{1/2} \varphi_k(\xi) \int_0^t \frac{1}{(t-s)^{1/2}} \\ & \cdot \int_{-\infty}^{\infty} \theta(s, u) G(s, u, \lambda_n) e^{-\lambda_n(\xi-u)^2/2(t-s)}\, du\, ds\, d\xi. \end{aligned} \tag{5.5}$$

Let us begin by considering the second term on the right-hand side of the integral equation (5.5) which we write as follows:

$$\int_{-\infty}^{\infty} \varphi_k(\xi) \int_0^t g(s, \xi, \lambda)\, ds\, d\xi \tag{5.6}$$

where

$$g(s, \xi, \lambda) = \left(\frac{\lambda}{2\pi(t-s)}\right)^{1/2} \int_{-\infty}^{\infty} \theta(s, u) G(s, u, \lambda) e^{-(\lambda/2)[(\xi-u)^2/(t-s)]}\, du. \tag{5.7}$$

Then by Lemma 1 and Theorem 4 of [1], the following estimates hold:

$$\|G(s, \cdot, \lambda)\| \leqq \|\psi\| e^{Mt_0},$$

and

$$\begin{aligned} \|g(s, \cdot, \lambda)\| & \leqq \|\theta(s, u) G(s, \cdot, \lambda)\| \\ & \leqq M \|G(s, \cdot, \lambda)\| \\ & \leqq M \|\psi\| e^{Mt_0}, \end{aligned} \tag{5.8}$$

and

$$\begin{aligned}\int_{-\infty}^{\infty}\left(\int_0^t |g(s,\xi,\lambda)|\,ds\right)^2 d\xi &\leqq \int_{-\infty}^{\infty}\int_0^t |g(s,\xi,\lambda)|\,ds\int_0^t |g(s',\xi,\lambda)|\,ds'\,d\xi\\ &= \int_0^t ds\int_0^t ds'\int_{-\infty}^{\infty} |g(s,\xi,\lambda)|\,|g(s',\xi,\lambda)|\,d\xi\\ &\leqq \int_0^t ds\int_0^t \|g(s,\cdot,\lambda)\|\cdot\|g(s',\cdot,\lambda)\|\,ds'\\ &\leqq \int_0^t ds\int_0^t M^2\|\psi\|^2 e^{2Mt_0}\,ds'\\ &\leqq t^2M^2\|\psi\|^2e^{2Mt_0}.\end{aligned}$$

This enables us to conclude that

$$\int_{-\infty}^{\infty}\left(\int_0^t |g(s,\xi,\lambda)|\,ds\right)^2 d\xi \leqq t^2M^2\|\psi\|^2e^{2Mt_0},$$

and thus by applying the Schwarz inequality we obtain for an arbitrary function $\varphi\in L_2(-\infty,\infty)$,

$$\int_{-\infty}^{\infty}|\varphi(\xi)|\int_0^t |g(s,\xi,\lambda)|\,ds\,d\xi \leqq \|\varphi\|\cdot|t|\cdot M\,\|\psi\|e^{Mt_0}.$$

We now apply the Fubini theorem to interchange the order of integration

$$\begin{aligned}\int_{-\infty}^{\infty}\varphi(\xi)\int_0^t g(s,\xi,\lambda)\,ds\,d\xi\\ &= \int_0^t\int_{-\infty}^{\infty}\varphi(\xi)g(s,\xi,\lambda)\,d\xi\,ds\\ &= \left(\frac{\lambda}{2\pi}\right)^{1/2}\int_0^t\frac{1}{(t-s)^{1/2}}\\ &\quad\cdot\int_{-\infty}^{\infty}\int_{-\infty}^{\infty}\varphi(\xi)\theta(s,u)G(s,u,\lambda)e^{-(\lambda/2)[(\xi-u)^2/(t-s)]}\,du\,d\xi\,ds.\end{aligned}\tag{5.9}$$

We now consider the double integral:

$$\int_{-\infty}^{\infty}\int_{-\infty}^{\infty}\varphi(\xi)h(u)e^{-(\Lambda/2)(\xi-u)^2}\,du\,d\xi,$$

where

$$h(u) = \theta(s,u)G(s,u,\lambda)\quad\text{and}\quad \Lambda = \frac{\lambda}{t-s}\cdot$$

By Lemma 1 of [1], since $\|h\|<\infty$, we may observe that

$$\left\|\left(\frac{\operatorname{Re}\Lambda}{2\pi}\right)^{1/2}\int_{-\infty}^{\infty}|h(u)|e^{-(\operatorname{Re}\Lambda/2)(\xi-u)^2}\,du\right\| \leqq \|h\|;$$

hence by the Schwarz inequality

$$\left(\frac{\operatorname{Re}\Lambda}{2\pi}\right)^{1/2}\int_{-\infty}^{\infty}\int_{-\infty}^{\infty}|\varphi(\xi)|\,|h(u)|e^{-(\operatorname{Re}\Lambda/2)(\xi-u)^2}\,du\,d\xi \leqq \|\varphi\|\cdot\|h\|.$$

Therefore by the Fubini theorem we may conclude that

$$\int_{-\infty}^{\infty}\int_{-\infty}^{\infty}\varphi(\xi)h(u)e^{-(\Lambda/2)(\xi-u)^2}\,du\,d\xi = \int_{-\infty}^{\infty}\int_{-\infty}^{\infty}\varphi(\xi)h(u)e^{-(\Lambda/2)(\xi-u)^2}\,d\xi\,du, \tag{5.10}$$

and therefore

$$\left(\frac{\lambda}{2\pi}\right)^{1/2}\int_0^t\frac{1}{(t-s)^{1/2}}\int_{-\infty}^{\infty}\int_{-\infty}^{\infty}\varphi(\xi)\theta(s,u)G(s,u,\lambda)\cdot e^{-(\lambda/2)[(\xi-u)^2/(t-s)]}\,du\,d\xi\,ds$$
$$= \left(\frac{\lambda}{2\pi}\right)^{1/2}\int_0^t\frac{1}{(t-s)^{1/2}}\int_{-\infty}^{\infty}\int_{-\infty}^{\infty}\varphi(\xi)\theta(s,u)G(s,u,\lambda)\cdot e^{-(\lambda/2)[(\xi-u)^2/(t-s)]}\,d\xi\,du\,ds. \tag{5.11}$$

We have established that for an arbitrary function $\varphi\in L_2(-\infty,\infty)$, by (5.7), (5.9), and (5.10)

$$\int_{-\infty}^{\infty}\varphi(\xi)\int_0^t\frac{1}{(t-s)^{1/2}}\int_{-\infty}^{\infty}\theta(s,u)G(s,u,\lambda)e^{-(\lambda/2)[(\xi-u)^2/(t-s)]}\,du\,ds\,d\xi$$
$$= \int_0^t\frac{1}{(t-s)^{1/2}}\int_{-\infty}^{\infty}\theta(s,u)G(s,u,\lambda)\int_{-\infty}^{\infty}\varphi(\xi)\cdot e^{-(\lambda/2)[(\xi-u)^2/(t-s)]}\,d\xi\,du\,ds. \tag{5.12}$$

We now consider the Hermite polynomials $H_k(\xi)$ and write

$$\int_{-\infty}^{\infty}H_k(\xi)e^{-\xi^2/2}\int_0^t\frac{1}{(t-s)^{1/2}}\int_{-\infty}^{\infty}\theta(s,u)G(s,u,\lambda)e^{-(\lambda/2)[(\xi-u)^2/t-s)]}\,du\,ds\,d\xi$$
$$= \int_0^t\frac{ds}{(t-s)^{1/2}}\int_{-\infty}^{\infty}\theta(s,u)G(s,u,\lambda)\Phi_k\left(u,\frac{\lambda}{t-s}\right)du,$$

where

$$\Phi_k(u,\Lambda) = \int_{-\infty}^{\infty}H_k(\xi)e^{-\xi^2/2}e^{-\Lambda(\xi-u)^2/2}\,d\xi \quad\text{and}\quad \Lambda = \frac{\lambda}{t-s}.$$

Now we consider the Fourier transform of Φ_k,

$$\mathscr{F}(\Phi_k) = \frac{1}{(2\pi)^{1/2}}\int_{-\infty}^{\infty}\Phi_k(u,\Lambda)e^{iu\rho}\,du,$$

and observe that

$$\mathscr{F}(H_k(\xi)e^{-\xi^2/2}) = (i\rho)^k e^{-\rho^2/2}.$$

Because the Fourier transform of a convolution is the product of Fourier transforms we have

$$\frac{1}{(2\pi)^{1/2}}\int_{-\infty}^{\infty}\Phi_k(u,\Lambda)e^{iu\rho}\,du = \mathscr{F}(\Phi_k) = (2\pi)^{1/2}(i\rho)^k e^{-\rho^2/2}\frac{1}{\Lambda^{1/2}}\,e^{-\rho^2/2\Lambda}$$

$$= \sqrt{\frac{2\pi(t-s)}{\lambda}}\,(i\rho)^k e^{-(\lambda+t-s)\rho^2/2\lambda}.$$

Because the Fourier transform is an isometric transform, we have

$$\|\Phi_k\| = \|\mathscr{F}(\Phi_k)\| = \left[2\int_0^{\infty}\frac{2\pi(t-s)}{|\lambda|}\,\rho^{2k}e^{-(|\lambda|^2+(t-s)\,\mathrm{Re}\,\lambda)\rho^2/|\lambda|^2}\cdot d\rho\right]^{1/2}.$$

If we let

$$\sigma = \left(\frac{|\lambda|^2+(t-s)\,\mathrm{Re}\,\lambda}{|\lambda|^2}\right)\rho^2,$$

then

$$\|\Phi_k\| = \left[\frac{2\pi(t-s)|\lambda|^{2k+1}}{|\lambda|(|\lambda|^2+(t-s)\,\mathrm{Re}\,\lambda)^{k+1/2}}\int_0^{\infty}\sigma^{k-1/2}e^{-\sigma}\,d\sigma\right]^{1/2},$$

and using the gamma function we have

$$\|\Phi_k\| = \left[\frac{2\pi(t-s)|\lambda|^{2k}}{(|\lambda|^2+(t-s)\,\mathrm{Re}\,\lambda)^{k+1/2}}\,\Gamma(k+\tfrac{1}{2})\right]^{1/2}.$$

We shall later perform a similar analysis for differences of Φ_k's.

We shall undertake to establish both the pointwise limit and the limit in the mean for Φ_k. We begin with the pointwise limit. Set $\Lambda_\mu = \lambda_\mu/(t-s)$. For a fixed value of u,

$$\lim_{\mu\to\infty}\int_{-\infty}^{\infty}H_k(\xi)e^{-\xi^2/2}e^{-\Lambda_\mu(\xi-u)^2/2}\,d\xi = \int_{-\infty}^{\infty}H_k(\xi)e^{-\xi^2/2}\exp\left(\frac{i(\xi-u)^2}{2(t-s)}\right)d\xi.$$

This follows from Lebesgue's convergence theorem because

$$\left|e^{-(\Lambda/2)(\xi-u)^2}\right| = e^{-\mathrm{Re}\,\Lambda(\xi-u)^2/2} \leqq 1 \quad \text{if } \mathrm{Re}\,\Lambda > 0,$$

and we have

$$\left|H_k(\xi)e^{-\xi^2/2}e^{-(\Lambda/2)(\xi-u)^2}\right| \leqq |H_k(\xi)|e^{-\xi^2/2}.$$

Thus we have the pointwise limit for $\mathrm{Re}\,\Lambda > 0$ and for each fixed u,

$$\lim_{\mu\to\infty}\Phi_k(u,\Lambda_\mu) = \Phi_k\left(u,\frac{-iq}{t-s}\right).$$

If we now apply Lemma 1 of [1] to the function Φ_k,

$$\Phi_k(u,\Lambda) = \int_{-\infty}^{\infty}H_k(\xi)e^{-\xi^2/2}e^{-(\Lambda/2)(\xi-u)^2}\,d\xi,$$

we obtain the inequality

$$\|\Phi_k(\cdot, \Lambda)\| \leqq \left(\frac{2\pi}{|\Lambda|}\right)^{1/2} \|H_k(\cdot)e^{-(\cdot)^2}\|.$$

In order to establish the limit in the mean for the Φ_k's, we observe the general property; if a sequence of functions f_n possesses the property that $\lim_{n\to\infty} f_n(x) = g(x)$ and if $\lim_{n\to\infty} f_n(x) = h(x)$ almost everywhere, then $g(x) = h(x)$ almost everywhere.

Now if μ, ν are distinct positive integers,

$$\left\|\Phi_k\left(\cdot, \frac{\lambda_\mu}{t-s}\right) - \Phi_k\left(\cdot, \frac{\lambda_\nu}{t-s}\right)\right\|^2 = \left\|\mathscr{F}\Phi_k\left(\cdot, \frac{\lambda_\mu}{t-s}\right) - \mathscr{F}\Phi_k\left(\cdot, \frac{\lambda_\nu}{t-s}\right)\right\|^2$$

$$= 2\pi(t-s)\int_{-\infty}^{\infty}\left[\frac{(i\rho)^k e^{-(\lambda_\mu+t-s)/(2\lambda_\mu)\rho^2}}{(\lambda_\mu)^{1/2}} - \frac{(i\rho)^k e^{-(\lambda_\nu-t-s)/(2\lambda_\nu)\rho^2}}{(\lambda_\nu)^{1/2}}\right]$$

$$\cdot\left[\frac{(-i\rho)^k e^{-(\bar\lambda_\mu+t-s)/(2\bar\lambda_\mu)\rho^2}}{(\bar\lambda_\mu)^{1/2}} - \frac{(-i\rho)^k e^{-(\bar\lambda_\nu+t-s)/(2\bar\lambda_\nu)\rho^2}}{(\bar\lambda_\nu)^{1/2}}\right] d\rho.$$

We have established that the integral tends to 0 as $\mu, \nu \to \infty$, since the integrand tends to zero as $\mu, \nu \to \infty$ and is dominated by a function integrable with respect to ρ and independent of μ and ν, namely $(8e^{-\rho^2/2}\rho^{2k})/|q|$ for $|\lambda_\mu| > \frac{1}{2}|q|$, $|\lambda_\nu| > \frac{1}{2}|q|$. Thus by dominated convergence, we have established that

$$\left\|\Phi_k\left(\cdot, \frac{\lambda_\mu}{t-s}\right) - \Phi_k\left(\cdot, \frac{\lambda_\nu}{t-s}\right)\right\| \to 0$$

as $\mu, \nu \to \infty$, and by the general property indicated above, we have a.e.

$$\lim_{n\to\infty} \Phi_k\left(u, \frac{\lambda_n}{t-s}\right) = \Phi_k\left(u, \frac{-iq}{t-s}\right).$$

Let us consider the functions

$$I_k(\lambda) \equiv \int_0^t \frac{ds}{(t-s)^{1/2}} \int_{-\infty}^{\infty} \theta(s,u)G(s,u,\lambda)\Phi_k\left(u, \frac{\lambda}{t-s}\right) du$$

and

$$J_k(q) \equiv \int_0^t \frac{ds}{(t-s)^{1/2}} \int_{-\infty}^{\infty} \theta(s,u)\Gamma(s,u)\Phi_k\left(u, \frac{-iq}{t-s}\right) du.$$

Then

$$I_k(\lambda_n) - J_k(q) = \mathscr{I}_1 + \mathscr{I}_2,$$

where

$$\mathscr{I}_1 = \int_0^t \frac{1}{(t-s)^{1/2}} \int_{-\infty}^{\infty} \theta(s, u)G(s, u, \lambda_n) \cdot \left[\Phi_k\left(u, \frac{\lambda_n}{t-s}\right) - \Phi_k\left(u, \frac{-iq}{t-s}\right)\right] du\, ds$$

and

$$\mathscr{I}_2 = \int_0^t \frac{1}{(t-s)^{1/2}} \int_{-\infty}^{\infty} \theta(s, u)\Phi_k\left(u, \frac{-iq}{t-s}\right)[G(s, u, \lambda_n) - \Gamma(s, u)]\, du\, ds.$$

By equations (5.5) and (5.12), we have

$$\begin{aligned}\int_{-\infty}^{\infty} & \varphi_k(\xi)G(t, \xi, \lambda_n)\, d\xi \\ &= \int_{-\infty}^{\infty} \varphi_k(\xi)\left(\frac{\lambda_n}{2\pi t}\right)^{1/2} \int_{-\infty}^{\infty} \psi(u) \exp\left(\frac{-\lambda_n(\xi-u)^2}{2t}\right) du\, d\xi \\ &\quad + \left(\frac{\lambda_n}{2\pi}\right)^{1/2} \int_0^t \frac{1}{(t-s)^{1/2}} \int_{-\infty}^{\infty} \theta(s, u)G(s, u, \lambda_n) \\ &\qquad \cdot \int_{-\infty}^{\infty} \varphi_k(\xi) \exp\left(\frac{-\lambda_n(\xi-u)^2}{2(t-s)}\right) d\xi\, du\, ds \\ &= \int_{-\infty}^{\infty} \varphi_k(\xi)\left(\frac{\lambda_n}{2\pi t}\right)^{1/2} \int_{-\infty}^{\infty} \psi(u) \exp\left(\frac{-\lambda_n(\xi-u)^2}{2t}\right) du\, d\xi + \left(\frac{\lambda_n}{2\pi}\right)^{1/2} I_k(\lambda_n).\end{aligned}$$

We next note that if $R_t \equiv (0, t) \times (-\infty, \infty)$

$$\frac{1}{(t-s)^{1/2}}\, \theta(s, u)\Phi_k\left(u, \frac{-iq}{t-s}\right) \in L_2(R_t);$$

for, if we consider

$$\int_0^t \int_{-\infty}^{\infty} \left|\frac{1}{(t-s)^{1/2}}\, \theta(s, u)\Phi_k\left(u, \frac{-iq}{t-s}\right)\right|^2 du\, ds$$

and apply Lemma 1 of [1] to the function Φ_k, we obtain

$$\int_{-\infty}^{\infty} |\Phi_k(u, \Lambda)|^2\, du \leqq \frac{2\pi}{|\Lambda|}\, \|\varphi_k\|^2$$

and

$$\int_{-\infty}^{\infty} \left|\Phi_k\left(u, \frac{-iq}{t-s}\right)\right|^2 du \leqq \frac{2\pi(t-s)}{|q|}\, \|\varphi_k\|^2.$$

Thus we have established that

$$\int_0^t \int_{-\infty}^{\infty} \left|\frac{1}{(t-s)^{1/2}}\, \theta(s, u)\Phi_k\left(u, \frac{-iq}{t-s}\right)\right|^2 du\, ds \leqq \frac{M^2 t_0 2\pi}{|q|}\, \|\varphi_k\|^2$$

and have proved that

$$\frac{1}{(t-s)^{1/2}}\,\theta(s,u)\Phi_k\left(u,\frac{-iq}{t-s}\right)\in L_2(R_t).$$

Thus by equation (5.3), we obtain $\mathscr{I}_2 \to 0$ as $n \to \infty$.

The integral

$$\mathscr{I}_1 \equiv \int_0^t \frac{1}{(t-s)^{1/2}}\int_{-\infty}^{\infty}\theta(s,u)G(s,u,\lambda_n)$$
$$\cdot\left[\Phi_k\left(u,\frac{\lambda_n}{t-s}\right)-\Phi_k\left(u,\frac{-iq}{t-s}\right)\right]du\,ds,$$

and by an application of the Schwarz inequality, we have

$$|\mathscr{I}_1| \leqq \int_0^t \frac{1}{(t-s)^{1/2}}\,M\,\|G(s,\cdot,\lambda_n)\|\,\left\|\Phi_k\left(\cdot,\frac{\lambda_n}{t-s}\right)-\Phi_k\left(\cdot,\frac{-iq}{t-s}\right)\right\|ds$$
$$\leqq Me^{Mt_0}\int_0^t \|\psi\|\cdot\left\|\Phi_k\left(\cdot,\frac{\lambda_n}{t-s}\right)-\Phi_k\left(\cdot,\frac{-iq}{t-s}\right)\right\|\frac{1}{(t-s)^{1/2}}\,ds.$$

We have already established that for each fixed s, this integrand tends to zero as $n \to \infty$. We also have

$$\left\|\Phi_k\left(\cdot,\frac{\lambda_n}{t-s}\right)-\Phi_k\left(\cdot,\frac{-iq}{t-s}\right)\right\|$$
$$\leqq \left(\frac{2\pi(t-s)}{|\lambda_n|}\right)^{1/2}\|\varphi_k\| + \left(\frac{2\pi(t-s)}{|q|}\right)^{1/2}\|\varphi_k\|.$$

Since we may assume that $|\lambda| \geqq |q|/2$, by bounded convergence we have $\mathscr{I}_1 \to 0$. Consequently $I_k(\lambda_n) - J_k(q) \to 0$ as $n \to \infty$, i.e.,

$$\int_{-\infty}^{\infty} H_k(\xi)e^{-\xi^2/2}\int_0^t \frac{1}{(t-s)^{1/2}}\int_{-\infty}^{\infty}\theta(s,u)G(s,u,\lambda_n)$$
$$\cdot e^{-(\lambda_n/2)[(\xi-u)^2/(t-s)]}\,du\,ds\,d\xi \to J_k(q) \tag{5.13}$$

as $n \to \infty$.

By applying Lemma 10 of [1] to the inner two integrals of

$$J_k(q) \equiv \int_0^t \frac{1}{(t-s)^{1/2}}\int_{-\infty}^{\infty}\theta(s,u)\Gamma(s,u)\int_{-\infty}^{\infty}\varphi_k(\xi)e^{(iq/2)[(\xi-u)^2/(t-s)]}\,d\xi\,du\,ds$$

and using $f(u) = \theta(s,u)\Gamma(s,u)$ we have

$$J_k(q) = \int_0^t \frac{1}{(t-s)^{1/2}}\int_{-\infty}^{\infty}\varphi_k(\xi)\overset{(\xi)}{\int_{-\infty}^{\infty}}\theta(s,u)\Gamma(s,u)e^{+iq(\xi-u)^2/2(t-s)}\,du\,d\xi\,ds$$
$$= \left(\frac{2\pi i}{q}\right)^{1/2}\int_0^t\int_{-\infty}^{\infty}\varphi_k(\xi)g(s,\xi)\,d\xi\,ds$$

where

$$g(s, \xi) = \left(\frac{q}{2\pi i(t-s)}\right)^{1/2} {}^{(\xi)}\!\!\int_{-\infty}^{\infty} \theta(s, u)\Gamma(s, u)e^{+iq(\xi-u)^2/2(t-s)}\, du,$$

and by Lemma 1 of [1] and (5.3)

$$\|g(s, \cdot)\| \leqq M\|\Gamma(s, \cdot)\|$$
$$\|g(\cdot, \cdot)\| \leqq M\|\Gamma(\cdot, \cdot)\| \leqq \liminf_{n\to\infty} M\|G(\cdot, \cdot, \lambda_n)\| \leqq MM'(t_0)^{1/2}\|\psi\|.$$

Thus by the Fubini theorem

$$J_k(q) = \left(\frac{2\pi i}{q}\right)^{1/2} \int_{-\infty}^{\infty} \varphi_k(\xi)g(s, \xi)\, ds\, d\xi$$

$$= \int_{-\infty}^{\infty} \varphi_k(\xi) \int_0^t \frac{1}{(t-s)^{1/2}} {}^{(\xi)}\!\!\int_{-\infty}^{\infty} \theta(s, u)\Gamma(s, u) \cdot e^{+iq(\xi-u)^2/2(t-s)}\, du\, ds\, d\xi. \tag{5.14}$$

Thus we have proved by equations (5.13) and (5.14) that

$$\lim_{n\to\infty} \left(\frac{\lambda_n}{2\pi}\right)^{1/2} \int_{-\infty}^{\infty} \varphi_k(\xi) \int_0^t \int_{-\infty}^{\infty} \frac{\theta(s, u)G(s, u, \lambda_n)}{(t-s)^{1/2}} \exp\left(\frac{-\lambda_n(\xi-u)^2}{2(t-s)}\right) du\, ds\, d\xi$$

$$= \left(\frac{q}{2\pi i}\right)^{1/2} \int_{-\infty}^{\infty} \varphi_k(\xi) \int_{-\infty}^{\infty} \frac{\theta(s, u)\Gamma(s, u)}{(t-s)^{1/2}} \exp\left(\frac{iq(\xi-u)^2}{2(t-s)}\right) du\, ds\, d\xi. \tag{5.15}$$

We now turn our attention to the first integral on the right-hand side of the equation. By dominated convergence, since $\varphi_k \in L_1$,

$$\lim_{n\to\infty} \int_{-\infty}^{\infty} \varphi_k(\xi)e^{-(\lambda_n/2)[(\xi-u)^2/t]}\, d\xi = \int_{-\infty}^{\infty} \varphi_k(\xi)e^{iq(\xi-u)^2/2t}\, d\xi$$

for all fixed u. By Lemma 1 of [1] we see that

$$\left\| \int_{-\infty}^{\infty} \varphi_k(\xi)e^{-(\lambda_n/2)[(\xi-u)^2/2t]}\, d\xi \right\| \leqq c\|\varphi_k\| \quad \text{for some constant } c.$$

Hence the pointwise limit can be replaced by the weak limit

$$\lim_{n\to\infty} \left(\frac{\lambda_n}{2\pi t}\right)^{1/2} \int_{-\infty}^{\infty} \psi(u) \int_{-\infty}^{\infty} \varphi_k(\xi)e^{-(\lambda_n/2t)[(\xi-u)^2]}\, d\xi\, du$$

$$= \left(\frac{q}{2\pi it}\right)^{1/2} \int_{-\infty}^{\infty} \psi(u) \int_{-\infty}^{\infty} \varphi_k(\xi)e^{iq(\xi-u)^2/2t}\, d\xi\, du.$$

We interchange the order of integration on both sides, using the Fubini theorem for the limitand on the left and Lemma 10 of [1, p. 542] on the right. We can do the latter because $\varphi_k \in L_1 \cap L_2$. We obtain

$$\lim_{\lambda\to iq} \left(\frac{\lambda}{2\pi t}\right)^{1/2} \int_{-\infty}^{\infty} \varphi_k(\xi) \int_{-\infty}^{\infty} \psi(u)e^{-(\lambda/2)[(\xi-u)^2/t]}\, du\, d\xi$$

$$= \left(\frac{q}{2\pi it}\right)^{1/2} \int_{-\infty}^{\infty} \varphi_k(\xi) \int_{-\infty}^{\infty} \psi(u)e^{iq(\xi-u)^2/2t}\, du\, d\xi. \tag{5.16}$$

Let us multiply the following integral equation of Theorem 8 of [1], valid for Re $\lambda > 0$,

$$G(t, \xi, \lambda) = \left(\frac{\lambda}{2\pi t}\right)^{1/2} \int_{-\infty}^{\infty} \psi(u) e^{-\lambda(\xi - u)^2/2t}\, du$$

$$(5.17) \qquad + \left(\frac{\lambda}{2\pi}\right)^{1/2} \int_0^t \frac{ds}{(t-s)^{1/2}} \int_{-\infty}^{\infty} \theta(s, u) G(s, u, \lambda) e^{-(\lambda/2)[(\xi-u)^2(t-s)]}\, du,$$

by $\varphi_k(\xi) = H_k(\xi) e^{-\xi^2/2}$ and integrate both sides with respect to ξ from $-\infty$ to ∞ and pass to the limit as $n \to \infty$ with λ replaced by λ_n.

Since we know that the limit on the right-hand side exists, it follows that the limit on the left-hand side exists also, and we have by (5.15), (5.16), and (5.17)

$$(5.18) \qquad \lim_{n \to \infty} \int_{-\infty}^{\infty} \varphi_k(\xi) G(t, \xi, \lambda_n)\, d\xi = \int_{-\infty}^{\infty} \varphi_k(\xi) \Gamma^*(t, \xi)\, d\xi$$

where

$$\Gamma^*(t, \xi) = \left(\frac{q}{2\pi i t}\right)^{1/2} \overset{(\xi)}{\int_{-\infty}^{\infty}} \psi(u) \exp\left(\frac{iq(\xi - u)^2}{2t}\right) du$$

$$(5.19) \qquad + \left(\frac{q}{2\pi i}\right)^{1/2} \overset{(\xi)}{\int_0^t} \int_{-\infty}^{\infty} \frac{\theta(s, u)\Gamma(s, u)}{(t-s)^{1/2}} \exp\left(\frac{iq(\xi - u)^2}{2(t-s)}\right) du\, ds$$

for each $t \in (0, t_0]$ and almost all $\xi \in (-\infty, \infty)$. Thus for each fixed t, $0 < t \leqq t_0$, the weak limit of $G(t, \xi, \lambda_n)$ is $\Gamma^*(t, \xi)$,

$$(5.20) \qquad \underset{n \to \infty}{\text{w lim}}\ G(t, \xi, \lambda_n) = \Gamma^*(t, \xi).$$

Letting $\beta(t) \in L_2(0, t_0)$ we write

$$\int_0^{t_0} \beta(t) \lim_{n \to \infty} \int_{-\infty}^{\infty} \varphi_k(\xi) G(t, \xi, \lambda_n)\, d\xi\, dt$$

$$(5.21) \qquad = \int_0^{t_0} \beta(t) \int_{-\infty}^{\infty} \varphi_k(\xi) \Gamma^*(t, \xi)\, d\xi\, dt.$$

By dominated convergence we have

$$\lim_{n \to \infty} \int_0^{t_0} \beta(t) \int_{-\infty}^{\infty} \varphi_k(\xi) G(t, \xi, \lambda_n)\, d\xi\, dt$$

$$(5.22) \qquad = \int_0^{t_0} \beta(t) \int_{-\infty}^{\infty} \varphi_k(\xi) \Gamma^*(t, \xi)\, d\xi\, dt.$$

By the definition of $\Gamma(t, \xi)$ we have

$$\lim_{n \to \infty} \int_0^{t_0} \beta(t) \int_{-\infty}^{\infty} \varphi_k(\xi) G(t, \xi, \lambda_n)\, d\xi\, dt = \int_0^{t_0} \beta(t) \int_{-\infty}^{\infty} \varphi_k(\xi) \Gamma(t, \xi)\, d\xi\, dt,$$

and therefore $\Gamma(t, \xi) = \Gamma^*(t, \xi)$ almost everywhere in R, and the proof of Theorem 2 is complete.

COROLLARY 1. *Under the hypotheses of Theorem 2, for each fixed* $t \in (0, t_0]$, *the right-hand side of equation* (5.2) *as a function of* ξ *is an element of* $L_2(-\infty, \infty)$ *and its norm is bounded by* $\|\psi\| + MA_0(t_0)^{1/2}$, *where* $A_0 = \|\Gamma(\cdot, \cdot)\|$.

PROOF. Let t be fixed, $0 < t \leqq t_0$. By Lemma 1 of [1], the first term of the right-hand side as a function of ξ is an element of $L_2(-\infty, \infty)$. Indeed,

$$\left\| \left(\frac{q}{2\pi i t}\right)^{1/2} {}^{(\xi)}\!\!\int_{-\infty}^{\infty} \psi(u) \exp\left(\frac{iq(\xi - u)^2}{2t}\right) du \right\| \leqq \|\psi\|.$$

Using the notation of Theorem 2 we have

$$\Gamma(\cdot, \cdot) \in L_2(R),$$

and thus there exists a null set S_0, such that if $s \notin S_0$,

$$\Gamma(s, \cdot) \in L_2(-\infty, \infty).$$

Now

$$A_0^2 = \int_0^{t_0} \int_{-\infty}^{\infty} |\Gamma(s, u)|^2 \, ds \, du.$$

Let $A(s) = \|\Gamma(s, \cdot)\|$, then $A^2(s) = \int_{-\infty}^{\infty} |\Gamma(s, u)|^2 \, du$; thus $A_0^2 = \int_0^{t_0} [A(s)]^2 \, ds$, and $A(s) \in L_2(0, t_0)$. We set

$$Q(s, t, \xi) = \left(\frac{q}{2\pi i}\right)^{1/2} {}^{(\xi)}\!\!\int_{-\infty}^{\infty} \frac{\theta(s, u)\Gamma(s, u)}{(t - s)^{1/2}} \exp\left(\frac{iq(\xi - u)^2}{2(t - s)}\right) du.$$

By Lemma 1 of [1], for any fixed $t \in (0, t_0]$ and for $s \notin S_0$ and $0 < s < t$, $Q(s, t, \xi)$ is of $L_2(-\infty, \infty)$ in ξ, and $\|Q(s, t, \cdot)\| \leqq A(s)M$. Then we have by the Schwarz inequality that

$$\begin{aligned}
\int_{-\infty}^{\infty} \left| \int_0^t Q(s, t, \xi) \, ds \right|^2 d\xi &= \int_{-\infty}^{\infty} \int_0^t Q(s, t, \xi) \, ds \int_0^t \bar{Q}(s', t, \xi) \, ds' \, d\xi \\
&= \int_{-\infty}^{\infty} \int_0^t \int_0^t Q(s, t, \xi) \bar{Q}(s', t, \xi) \, ds \, ds' \, d\xi \\
&= \int_0^t \int_0^t \int_{-\infty}^{\infty} Q(s, t, \xi) \bar{Q}(s', t, \xi) \, d\xi \, ds \, ds' \\
&\leqq \int_0^t \int_0^t A(s) \, A(s') \, M^2 \, ds \, ds' \leqq M^2 A_0^2 t.
\end{aligned}$$

Thus we have proved that for each fixed $t \in (0, t_0]$, $\int_0^t Q(s, t, \xi) \, ds \in L_2(-\infty, \infty)$ as a function of ξ.

COROLLARY 2. *Let $G(t, \xi, \lambda)$ satisfy the hypotheses of Theorem 2. Let $\lambda_1, \lambda_2, \ldots$ be a sequence such that* $\operatorname{Re} \lambda_n > 0$, $\lim_{n\to\infty} \lambda_n = -iq$, *and* $\mathrm{w}\lim_{n\to\infty} G(t, \xi, \lambda_n) = \Gamma(t, \xi)$ *over R. Then for each $t \in (0, t_0]$, it follows that* $\mathrm{w}\lim_{n\to\infty} G(t, \xi, \lambda_n) \equiv \Gamma^*(t, \xi)$ *exists over $(-\infty, \infty)$. Moreover $\Gamma^* = \Gamma$ almost everywhere on R, and Γ^* is given by* (5.19) *for each $t \in (0, t_0]$ and almost all real ξ.*

PROOF. This follows from equation (5.18).

6. Existence theorem

THEOREM 3. *Let $\theta(t, u)$ be continuous almost everywhere in the strip $R: 0 \leqq t \leqq t_0$, $-\infty < u < \infty$, let $|\theta(t, u)| \leqq M$ for $(t, u) \in R$, and let q be any real number, $q \neq 0$. Let $\psi \in L_2(-\infty, \infty)$ and let*

$$F_t(x) = \exp\left\{\int_0^t \theta(t - s, x(s))\, ds\right\}. \tag{6.1}$$

Then for each fixed $t \in (0, t_0]$, $J_q(F_t)\psi$ exists and is an element of $L_2(-\infty, \infty)$. Moreover, if $\Gamma(t, \xi) = (J_q(F_t)\psi)(\xi)$ for $(t, \xi) \in R$, then for each $t \in (0, t_0]$ we have

$$\begin{aligned}\Gamma(t, \xi) = {} & \left(\frac{q}{2\pi i t}\right)^{1/2} \overset{(\xi)}{\int_{-\infty}^{\infty}} \psi(u) \exp\left(\frac{iq(\xi - u)^2}{2t}\right) du \\ & + \left(\frac{q}{2\pi i}\right)^{1/2} \int_0^t \frac{ds}{(t - s)^{1/2}} \overset{(\xi)}{\int_{-\infty}^{\infty}} \theta(s, u)\Gamma(s, u) \\ & \qquad \cdot \exp\left(\frac{iq(\xi - u)^2}{2(t - s)}\right) du\end{aligned} \tag{6.2}$$

for almost all $\xi \in (-\infty, \infty)$. Moreover Γ satisfies (1.2).

PROOF. In Theorem 2 we proved that every element $\tilde{\Gamma}$ of the cluster set $\mathscr{C}(G(\cdot, \lambda), -iq)$ satisfied the integral equation (5.2). By Lemma 2, this cluster set is not empty. If we denote the right-hand side of equation (5.2) (with Γ replaced by $\tilde{\Gamma}$) by Γ^* as we did in (5.19), by Corollary 1 to Theorem 2 we have

$$\|\Gamma^*\| \leqq \|\psi\| + MA_0(t_0)^{1/2}. \tag{6.3}$$

Moreover $\tilde{\Gamma}$ and Γ^* differ at most on set of measure zero in R. Thus Γ^* is a representation of our cluster point $\tilde{\Gamma}$ and by our uniqueness theorem, the cluster set at $-iq$ consists of a single point in $L_2(R)$. Thus our function $\tilde{\Gamma}(t, \xi)$ is unique in this sense.

By Lemma 3 of Section 2, we have the existence of the weak limit over R:

$$\underset{\substack{\lambda \to iq,\ \mathrm{Re}\,\lambda > 0 \\ (t,\xi)}}{\mathrm{w\,lim}} G(t, \xi, \lambda) = \tilde{\Gamma}(t, \xi). \tag{6.4}$$

By Corollary 2 of Theorem 2, for any sequence $\{\lambda_n\}$, $\mathrm{Re}\,\lambda_n > 0$, $\lambda_n \to -iq$, we have for each $t \in (0, t_0]$, $\underset{n\to\infty}{\mathrm{w\,lim}}\ G(t, \xi, \lambda_n) = \Gamma^*(t, \xi)$ over $(-\infty, \infty)$. Thus we have for each fixed $t \in (0, t_0]$

$$\underset{n\to\infty}{\mathrm{w\,lim}}\,(I_{\lambda_n}(F_t)\psi)(\xi) = \Gamma^*(t, \xi) \tag{6.5}$$

for almost all $\xi \in (-\infty, \infty)$. Since the sequential limit exists for all sequences $\{\lambda_n\}$ and Γ^* is unique by (5.19), we have

$$(J_q(F_t)\psi)(\xi) \equiv \underset{\lambda \to iq,\ \mathrm{Re}\,\lambda > 0}{\mathrm{w\,lim}} (I_\lambda(F_t)\psi)(\xi) = \Gamma^*(t, \xi). \tag{6.6}$$

Thus Γ as defined in the hypotheses of the theorem exists, equals Γ^* and satisfies the integral equation (6.2).

7. Deterministic theorem for $J_q(F)$

We now show that with a slightly stronger hypothesis, Theorem 5 of [1] can be made deterministic, i.e., its conclusion can be shown to hold for all nonzero q instead of merely almost all q.

THEOREM 4. *Let $\theta(s, u)$ be bounded and almost everywhere continuous in R. Let $F(x) = \exp\left\{\int_0^{t_0} \theta(s, x(s))\,ds\right\}$. Then if q is real and $q \neq 0$, $J_q(F)$ exists as a bounded linear operator taking $L_2(-\infty, \infty)$ into itself.*

PROOF. Let $\theta^*(s, u) = \theta(t_0 - s, u)$ and $F_t^*(x) = \exp\left\{\int_0^t \theta^*(t - s, x(s))\,ds\right\}$ $= \exp\left\{\int_0^t \theta(t_0 - t + s, x(s))\,ds\right\}$, for $0 < t \leqq t_0$. Then by Theorem 3, $J_q(F_t^*)$ exists as a bounded linear operator taking $L_2(-\infty, \infty)$ into itself. In particular, this is true when $t = t_0$. However,

$$F_{t_0}^*(x) = \exp\left\{\int_0^{t_0} \theta(s, x(s))\,ds\right\} = F(x).$$

Hence $J_q(F)$ exists and is a bounded linear operator taking $L_2(-\infty, \infty)$ into itself.

REMARK. Clearly the interval $[0, t_0]$ can be changed into the interval $[a, b]$ used in the definition of $I_\lambda(F)$ and $J_q(F)$.

REFERENCE

1. CAMERON, R. H., and D. A. STORVICK, "An operator valued function space integral and a related integral equation," *J. Math. Mech.*, *18* (1968), 517–552.

A Lower Bound for the Smallest Eigenvalue of the Laplacian

JEFF CHEEGER

Various authors have studied the geometrical and topological significance of the spectrum of the Laplacian Δ^2, on a Riemannian manifold. (The excellent survey article of Berger [2] contains background, references, and open problems.) The purpose of this note is to give a lower bound for the smallest eigenvalue $\lambda > 0$ of Δ^2 applied to functions. The bound is in terms of a certain global geometric invariant, essentially the constant in the isoperimetric inequality. The technique works for compact manifolds of arbitrary dimension with or without boundary.

The author wishes to thank J. Simons for helpful conversations and in particular for suggesting the importance of understanding the following example of E. Calabi. Consider the "dumbbell" manifold homeomorphic to S^2, shown in Fig. 1. The pipe connecting the two halves is to be thought of as having fixed length l and variable radius r.

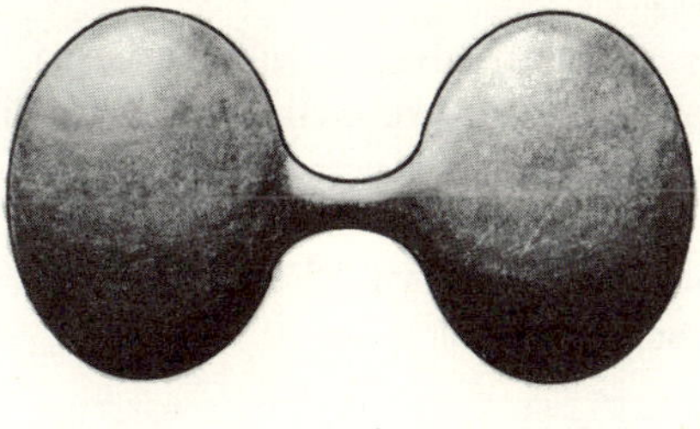

Fig. 1

One sees that $\lambda \to 0$ as $r \to 0$. Calabi's original argument involved consideration of the heat equation, $\frac{-\partial T}{\partial t} = \Delta^2 T$.

A somewhat more direct argument is as follows: Let f be a function which is equal to c on the right-hand bulb, $-c$ on the left-hand bulb and changes linearly from c to $-c$ across the pipe. (c chosen so that $\int_M f^2 = 1$.) Then $\int_M f = 0$ and

$$\|\operatorname{grad} f\| \approx \begin{cases} 0 & \text{bulbs,} \\ \dfrac{2c}{l} & \text{pipe.} \end{cases}$$

One has by Stokes' theorem

$$\lambda \leqq \int \Delta^2 f \cdot f = \int \|\operatorname{grad}\|^2 \approx \frac{2c}{l} \cdot 2\pi \cdot r \cdot l. \tag{1}$$

Clearly $\lambda \to 0$ as $r \to 0$. The Calabi example makes it evident that in bounding λ from below, it is not enough to consider just the diameter or volume of M. It also suggests

DEFINITION 1.

(a) Let M be a compact n-dimensional Riemannian manifold, $\partial M = \phi$. Set $h = \inf \dfrac{A(S)}{\min V(M_i)}$, where $A(\ )$ denotes $(n-1)$-dimensional area, $V(\ )$ denotes volume, and the inf is taken over all compact $(n-1)$-dimensional submanifolds S, dividing M into submanifolds with boundary M_1, M_2, with $M = M_1 \cup M_2$, and $\partial M_i = S$.

(b) If $\partial M \neq \phi$, set

$$h = \inf_S \frac{A(S)}{V(M_1)},$$

where we stipulate $S \cap \partial M = \phi$, and there is a submanifold with boundary M_1 such that $S = \partial M_1$. M_1 is necessarily unique.

In the preceding definition ∂M, M_1, M_2, S are not assumed to be connected.

THEOREM. *In the situation just described* $\lambda \geqq \frac{1}{4}h^2$. (*If* $\partial M \neq \phi$ *we assume* $f * df|\partial M = 0$.)

PROOF. If M is not orientable, it will suffice to look at its 2-fold orientable cover. Let f be the eigenfunction corresponding to λ. We make the assumption that f has nondegenerate (and therefore isolated) critical points. If this is not the case we use an obvious approximation argument based on Sard's theorem, which will be left to the reader. First note that for any region R, such that $f * df|\partial R = 0$,

$$\lambda = \frac{\int_R \Delta^2 f \cdot f}{\int_R f^2} = \frac{\int_R \|\operatorname{grad} f\|^2}{\int_R f^2} \tag{2}$$

$$= \frac{\left(\int_R \|\operatorname{grad} f\|^2\right)}{\left(\int_R f^2\right)^2} \cdot \left(\int_R f^2\right) \geqq \frac{\left(\int_R |f| \cdot \|\operatorname{grad} f\|\right)^2}{\left(\int_R f^2\right)^2} \tag{3}$$

$$= \frac{1}{4} \frac{\left(\int_R \|\operatorname{grad} f^2\|\right)^2}{\left(\int_R f^2\right)^2}, \tag{4}$$

where the inequality in (3) is obtained by squaring the Schwarz inequality.

We now assume that zero is not a critical value of f. (Again if this is not the case the argument undergoes a trivial modification.) Now the submanifold $Z = \{x|f(x) = 0\}$ divides M into n-dimensional submanifolds with boundary $M_1 = \{x|f(x) \geqq 0\}$ and $M_2 = \{x|f(x) \leqq 0\}$. It is in asserting that Z, M_1, and M_2 exist that we are using the information that f is a nonconstant eigenfunction ($\lambda \neq 0$) and hence must take on positive and negative values. Let h, h_1, and h_2 be the constants corresponding to M, M_1, and M_2. Clearly if, say, $V(M_1) \leqq V(M_2)$, then $h_1 \geqq h$. It will then suffice to prove the estimate for the submanifold with boundary M_1 and, moreover, the same argument will work for any manifold with boundary. Now the regions of M_1 lying between the critical levels of f^2 have a natural product structure $L \times I$ given by the level surfaces and their orthogonal trajectories. We introduce product coordinates (x, t) by choosing local coordinates $\{x_i\}$ on some L and setting $t = f^2$. Since dt is orthogonal to dx_i, the volume element dv may be written in coordinates as

$$dV = v_1(t, x)\, dt \times v_2(t, x)\, dx. \tag{5}$$

Since $f^2 = dt$, $v_1(t, x) = \left\|\frac{\partial}{\partial t}\right\|$, we have

$$\|\operatorname{grad} f^2\| \cdot v_1(t, x) = \left\langle \operatorname{grad} t, \frac{\partial}{\partial t}\Big/\left\|\frac{\partial}{\partial t}\right\|\right\rangle \cdot \left\|\frac{\partial}{\partial t}\right\| \tag{6}$$

$$= dt\left(\frac{\partial}{\partial t}\right) = 1. \tag{7}$$

Let $V(t)$ denote the volume of the set $\{x \in M_1 | f^2(x) \geqq t\}$. $V(t)$ is continuous and differentiable.

$$\int_{M_1} \|\operatorname{grad} f^2\| \cdot dv = \int_L \left(\int_0^\infty \|\operatorname{grad} f^2\| \cdot v_1 \cdot v_2 \cdot dt\right) dx. \tag{8}$$

By (7) this is equal to

$$\int_L \left(\int_0^\infty v_2 \cdot dt\right) dx = \int_0^\infty \left(\int_L v_2 \cdot dx\right) dt \tag{9}$$

$$= \int_0^\infty A(L_t)\, dt \geqq h_1 \int_0^\infty V(t) \cdot dt \tag{10}$$

$$= -h_1 \int_0^\infty t \cdot \frac{dV(t)}{dt} \cdot dt. \tag{11}$$

Moreover,

$$V(t) = V(M_1) - \int_0^t \left\{\int_L v_1(x, t) \cdot v_2(x, t) \cdot dx\right\} dt, \tag{12}$$

and $t = f^2$. Thus (11) becomes

$$h_1 \int_0^\infty t\left\{\int_L v_1(t, x)\cdot v_2(t, x)\,dx\right\} dt$$

$$= h_1 \int_0^\infty \left\{\int_L t\cdot v_1(t, x)\cdot v_2(t, x)\,dx\right\} dt \tag{13}$$

$$= h_1 \int_{M_1} f^2\cdot dV. \tag{14}$$

Squaring the inequality (8)–(14) and dividing through by $\left(\int_{M_1} f^2\right)^2$ yields

$$\frac{\left(\int_M \|\operatorname{grad} f^2\|\right)^2}{\left(\int_{M_1} f^2\right)^2} \geqq h_1^2 \geqq h^2. \tag{15}$$

Combining (15) with (2)–(4) completes the proof.[1]

In dimension 2, it is relatively easy to see that h is always strictly greater than zero. In fact, let $V(M) = V$, and let c be such that a metric ball of radius $r < c$ is always convex. Then, if

$$\frac{A(S)}{\min V(M_i)} \leqq \frac{c}{V}, \tag{16}$$

it follows that each component of S must lie in a convex ball. On such balls the metric g satisfies $k\cdot E \geqq g \geqq \frac{1}{k} E$ where E is the Euclidean metric in normal coordinates. Hence $h > 0$ is implied by the usual isoperimetric inequality in the plane. Now, according to a theorem of the author (see [3]), c may be estimated from below by knowing a bound on the absolute value of the sectional curvature s_M, an upper bound for the diameter $d(M)$, and a lower bound for the volume. Once this is done, it is elementary that k may be estimated from $|s_M|$. This yields

COROLLARY. *If* $\dim M = 2$, $\partial M = \phi$, *then given* δ *there exists* ε *such that if* $\frac{1}{V} + d(M) + |s_M| < \delta$, *then* $\lambda > \varepsilon$.

In case $\dim M > 2$ the situation is not so elementary but 6.1 and 6.2 of [4], or [5], will still imply that $h > 0$.[2] Actually the results of [4] and [5] show the existence of an integral current T whose boundary in S, such that $A(S)$ divided by the mass of the current is always bounded away from zero independent of S. However, since T is of top dimension, it is known that T

[1] The equality $\int_M \|\operatorname{grad} f^2\|\,dV = \int_0^\infty A(L_t)\,dt$ is actually a special case of the "co-area formula" (see [6]).

[2] Thanks are due to F. Almgren for supplying these references.

may be taken to be either M_1 or M_2. If $\partial M \neq \phi$, the fact that $h > 0$ may also be deduced from Theorem 1 of [1] without too much difficulty.

It would be of interest to generalize the argument given here to Δ^2 acting on k-forms. Singer has pointed out that this would give a new proof of the fact that the dimension of the space of harmonic forms is independent of the metric and that the techniques might be applicable to other situations. To date, we have not been able to accomplish this except in the case dim $M = 2$. The essential point here is that for any eigenvalue of Δ^2 on 1-forms one can find an eigenform of the form df, where f is an eigenfunction corresponding to λ. This observation is probably of little help if $n > 2$.

UNIVERSITY OF MICHIGAN
and
SUNY, Stony Brook

REFERENCES

1. ALMGREN, F., "Three Theorems on Manifolds with Bounded Mean Curvature," *B.A.M.S.*, *71*, No. 5 (1965), 755–756.
2. BERGER, M., Lecture notes, Berkeley Conference on Global Analysis, 1968.
3. CHEEGER, J., "Finiteness Theorems for Riemannian Manifolds," *Am. J. Math.* (in press).
4. FEDERER, H., and W. FLEMING, "Normal Integral Currents," *Ann. of Math*, *72* (1960).
5. FEDERER, H., "Approximating Integral Currents by Cycles," *AMS Proceedings*, *12* (1961).
6. ———, "Curvature Measures," *Trans. A.M.S.*, *93* (1959), 418–491.

The Integral Equation Method in Scattering Theory

C. L. DOLPH[1]

1. Introduction

The subject of this essay seems singularly appropriate to this occasion for several reasons. Much of the material on which it depends stems from the Berlin and Munich schools where Salomon Bochner spent many of his early mathematical years. The foundations of the theory are perhaps best expressed via the Bochner integral (see Wilcox [1] and Dolph [2]). Finally, the subject has reached some sort of culmination for the direct problems with the as yet unpublished work of Shenk and Thoe [3], while the inverse problem has achieved a significant new impetus from the recently published work of Lax and Phillips [4] and that of Ludwig and Morawetz [5].

In this review I hope to give some indication of the history and status of this subject in the hope that perhaps a new generation of mathematicians can be interested. For this reason very few theorems will be proved in detail, and the discussion will be limited to the appropriate operators with independent variables restricted mainly to two- and three-dimensional spaces, although almost all that will be said will be valid, with modifications, for $n \geqq 2$.

At the outset it is necessary to make a clear distinction between the basic physical problems which will be treated and the method of the title, since,

[1] The author wishes to express his great gratitude to P. Lax and R. S. Phillips for making their new unpublished work [4] available to him as soon as possible; to S. Sternberg [42] for the same reason; to P. Werner for making the dissertations of all of his recent students including those of Kussmaul [52] and Haf [63] available; to C. H. Wilcox, who furnished me a Xerox copy of his Edinburgh lecture notes; and in particular to N. A. Shenk II, and D. Thoe for their unremitting efforts both by mail and personal contacts to keep me informed of their progress. I would also like to express my thanks to H. Kramer, R. Lane, and R. R. Rutherford for the stimulating summer they provided at G. E. Tempo 1968. I have taken the liberty to reproduce some things from those in Section 7; see also [82].

Sponsored in part by NSF grant GP-6600.

in many cases, the physical problems can be treated by other methods, the most extensive such treatment of which is to be found in the work of Lax and Phillips [6].

The mathematical problem is as follows: Let Ω in R^n with $n = 2$ or 3 be an unbounded domain with a compact C^2 hypersurface Γ as boundary, where Γ may have several components or be empty. In the former case Γ is considered as the disjoint union $\Gamma_1 \cup \Gamma_2$ with each Γ_j either empty or consisting of some of the components of Γ. Let ν be the unit normal to Γ which points into Ω, Γ^i the interior bounded by Γ, and σ a real-valued Hölder continuous function on Γ_1, and let $q(x)$ be a real-valued Hölder continuous function on $\Omega \cup \Gamma$ which satisfies

$$q(x) = 0[\exp(-2a|x|)] \quad \text{as } |x| \to \infty \tag{1.0}$$

for some $a > 0$. It should be noted that this latter assumption is more general than that used by Lax and Phillips [6] who require a q of compact support and a domain which is star-shaped.

Further let K denote the subset $\{k; \operatorname{Im} k > -a\}$ of the complex plane if n is odd and the portion

$$\{k \neq 0 |\operatorname{Im} k| < a \text{ and } -\infty < \arg k < \infty\} \cup \{k; \operatorname{Im} K > 0 \text{ and } 0 < \arg k < \pi\}$$

of the logarithmic Riemann surface $\{k \neq 0, -\infty < \arg k < \infty\}$ if n is even. This is a type of normalization and, to obtain that used by Lax and Phillips [6] and Newton [7], one should reflect about the real k-axis.

One now considers the reduced wave equation

$$[-\Delta + q(x) - k^2]\phi(x) = f_1 \quad \text{in } \Omega \tag{1.1}$$

and we desire outgoing solutions satisfying

$$\left(\frac{\partial}{\partial \nu} - \sigma\right)\phi(x + o_x) = f_2 \quad \text{on } \Gamma_1 \tag{1.2}$$

$$\phi(x + o_x) = f_3 \quad \text{on } \Gamma_2 \tag{1.3}$$

for $k \in K$ and with function $[f_j]$ which satisfy for some $0 < \beta < 1$, $f_i = 0(\bar{e}^{a|x|})$ as $|x| \to \infty$, $f_1 \in C^\beta(\Omega \cup \Gamma)$, $f_2 \in C(\Gamma_1)$, $f_3 \in C^1 + (\Gamma_2)$ and $f_i = 0$.

In addition one needs an appropriate outgoing radiation condition whose history and various formulations will be commented upon subsequently.

There are four main areas of interest for problems of this type, all mathematical, but all strongly influenced by both classical and modern physics: (i) to establish a suitable Hilbert space, define an appropriate self-adjoint operator in it and, in terms of "distorted plane waves," develop a spectral theory for real k by considering the overall problem as a perturbation on those which can be solved by Fourier analysis; (ii) to define the so-called wave and scattering operators, decide when the latter is unitary and relate it to the asymptotic form of the solution for the so-called integral equation of scattering (for the Schrödinger equation it is usually called the Lippman–Schwinger equation by physicists); (iii) to discuss the analytic properties of the scattering operator and its analytic continuation as well as that of, it turns out, not the resolvent but the resolvent kernel (or its transform, for $n > 3$); and (iv) to use the knowledge gained by some or all of this to obtain asymptotic solutions in time which, while they are the cause of the so-called exponential catastrophe, can be discussed in suitably enlarged spaces. Once this has been accomplished the inverse problem can be faced. A most promising beginning on it has been made by Lax and Phillips [4].[2] The areas of (i) and (ii) are eminently associated with Carleman [8], Kato [9], Pozvner [10], Ikeba [11], Lax and Phillips [6], Shenk [17], Thoe [18], and Kuroda [19]. An excellent summary of results up to the recent collaboration of Shenk and Thoe can be found in Shenk [20]. Thus we shall limit our remarks concerning them to a brief updating of the spectral results achieved by Shenk and Thoe [3]. Area (iii) is the one the author has been most interested in and the only one to which he has occasionally contributed. It will be the one reported most fully here. While Lax and Phillips [4] and Ludwig and Morawetz [5] have made recent contributions to area (iv), these, unlike the recent results of Shenk and Thoe [3], have just been published, so that only hints of them will be reported here. To date, the integral equation method in this area has produced only fragmentary results (compare Dolph [21], Thoe [22]), although it is clear that it will eventually succeed and is part of the overall plan of Shenk and Thoe. To orient the reader, a semichronological approach, in which at the end several difficulties of the method of this title will be indicated, will be used.

Physical problems of the type described, with some or most of the conditions (1.1)–(1.3) present, occur frequently in acoustics and electromagnetic and quantum scattering, as reference to Frank von Mises [23] or Morse and Feshbach [24] clearly show. Initially work was limited to the case where the variables separate and where q is identically zero, and if condition (1.2) occurs it is most often with σ = identically zero. Most

[2] This approach to the inverse problem differs from the method of Gelfand–Levitan [12] and Faddeev [13], [14], and [15] in that the hope is to use knowledge of the poles in $\text{Im}\, k < 0$ to deduce the perturbation for a body. For an introduction, see Dolph [16].

importantly of all, initially k was assumed *real.* The earliest (1881) such solution known to me is due to Lord Rayleigh [25] for the case of a cylinder or circle of radius $r = a$ for $n = 2$ with Dirichlet boundary conditions. The total field takes the form

$$\phi = \sum_{-\infty}^{\infty} \exp\left[in\left(\theta - \frac{\pi}{2}\right)\right]\left\{J_n(kr) - \frac{J_n(ka)}{H_n^{(1)}(ka)} H_n^{(1)}(kr)\right\}$$

in which the first terms are the well-known expansion of an incident plane wave while the second terms represent the scattered field. All subsequent remarks will be built around this solution since it explicitly illustrates all that is known about the problem formulated for the case at hand.

The first thing to be noted is the type of condition the scattered field satisfies as $r \to \infty$. Abstracted, it leads to the justly celebrated Sommerfeld [26] radiation condition introduction by him in 1912, which states that for uniqueness with the exterior problem for real k one must require that the limit $\lim_{r\to\infty} r^{(n-1)/2}\left[\frac{\partial}{\partial r} - ik\right]\phi \to 0$. It was not until 1943, however, that Rellich [27] showed that this was all that was necessary (Sommerfeld had required more) and also demonstrated that any solution of the reduced wave equation with $q = 0$ that went to zero faster than $1/r$, $r = |x|$ must vanish identically. This last result has played a vital role in many subsequent developments (see, for example, Kato [28], Ikebe [11], and others). Loosely speaking, the Sommerfeld radiation condition says that there can be no sources at infinity and it is sufficient to make a certain surface integral vanish on a large sphere as $r \to \infty$. For the case $n = 3$, Stratton [29, p. 486] gives a heuristic development on which I cannot improve.

Returning to the Rayleigh solution, one notices that there are "eigensolutions" of the exterior problem at the zeros of $H_n^{(1)}(ka)$ (the Hankel function of the first kind) which occur for discrete values of k with $\operatorname{Im} k < 0$. These are the so-called "complex eigenvalues" first introduced by Lord Kelvin [30] in 1884 for the case of a perfectly conducting sphere and are similar to those introduced by Gamow [31] in 1928 for the case of α-decay. Their introduction was criticized and improved by Lamb [32] in 1900 and Love [33] in 1904, and in my judgment their best interpretation in the Gamow case was by van Kampen [34] in 1953. Perhaps the most informative historical discussion of them is to be found in Beck and Nussenzveig [35], Nussenzweig [36], [37], and Petzold [38] for simple cases. They are the subject of the book by Lax and Phillips [6] referred to earlier and will play a vital role in our subsequent discussion. With these preliminary remarks, attention will now be directed to the integral equation method in scattering theory itself.

2. The integral equation method for obstacle scattering

While I cannot be absolutely certain, it appears that Kupradze [39], [40] in the period 1934–1936 was the first to apply the Fredholm integral equation method to an arbitrary exterior problem under Dirichlet boundary conditions, and he appears to have been the first to state a radiation condition valid for all k (see, for instance, Schwarz [41]). As in the Sternberg paper [42] for the similar problem with complex k, there were lacunae, at least according to Freundenthal [43].

In 1947 Maue [44] applied the same method and actually treated the Rayleigh case described. In the simplest case for that problem, his integral equation took the following form for the cylinder:

$$\phi(\theta) - \frac{a}{2i}\int_0^{2\pi} \frac{\partial}{\partial a} H_0^{(1)}\left[ka\left|\sin\frac{(\theta-\theta')}{2}\right|\right]\phi(\theta')\,d\theta' = 2e^{-ika\cos\theta}. \tag{2.1}$$

Using expansion techniques from special function theory, it can readily be shown that the corresponding homogeneous integral equation has nontrivial solution only at the zeros of $J_n^1(ka)$ and of $H_n^{(1)}(ka)$, i.e., at the zeros of the derivative of a regular Bessel function or (in the normalization used here) at the zeros of a Hankel function of the first kind. The first type of zeros correspond to eigenvalues of the associated interior Neumann problem for the cylinder in a way completely analogous to that which occurs in ordinary potential theory. Examination of Rayleigh's solution shows that they do not, however, affect either the scattered or the total field and thus in some sense they are spurious, because of the method, and should be proven so. This will be done for a quite general case in Section 4. The second set of zeros all occur for $\operatorname{Im} k < 0$ and imply that the exterior problem for the circle has so-called "complex eigenvalues" which are not removable. This is indeed the case in general and is the subject of areas (iii) and (iv) defined above.

Weyl [45] considered the Dirichlet problem quite generally via a double-layer assumption, the same one that leads to the Maue equation. The difficulty of the interior eigenvalues of the Neumann problem he overcame by the addition of a suitably chosen single layer, mimicking the methods of potential theory, even under the assumptions that nonsimple elementary divisors occurred and, as I have remarked before (see Dolph [21]), he thought they might. Müller [41] soon gave an argument which showed that they did not and by so doing considerably simplified Weyl's argument.

The next step which yielded analyticity in $\operatorname{Im} k \geqq 0$ was taken by Müller's student Werner [47], who replaced the simple double-layer Ansatz by one of the form

$$\phi(x, k) = k^{n-2}\left\{\int_\Gamma \frac{\partial}{\partial v_y} F_k[(x - s)\mu(s)\, dS_s] + \int_{\Gamma'} F_k[(x - y)\tau(y)\, dV_y]\right\}, \tag{2.2}$$

where, in n-dimensional Euclidean space,

$$F^{\pm}[k|x|] = \pm\frac{i}{4}\left(\frac{k}{2\pi|x|}\right)^p H_p^{(i)}[k|x|]; \qquad p = \frac{n-2}{2}; \qquad + \leftrightarrow H_p^{(1)}, \ - \leftrightarrow H_p^{(2)} \tag{2.3}$$

and in which τ is an unknown volume potential. Under this assumption the difficulties associated with the interior Neumann problem disappear and one has merely to deal with the first part of the Fredholm alternative to get existence and uniqueness since one can set up a vector integral equation involving a compact operator in a suitable Banach space. One obtains, in addition, analyticity in k for the solution for all of k, $\operatorname{Im} k \geqq 0$ and this method was used subsequently by Shenk [17] and is part of the technique in the current version of Shenk and Thoe [3] which will be discussed in Section 6. From a practical point of view the presence of an additional volume integral makes numerical calculations immensely more difficult. As a result several people—Leis, another student of Müller [48], Werner and Birkhage [49], and Pannic [50]—replaced the double-layer assumption by one which reads, for both the Dirichlet and Neumann problems, as follows:

$$\phi(x, k) = \int_\Gamma \left(\frac{\partial}{\partial v_s} - i\eta\right) F_k^+(x - s)\mu(s)\, dS_s,$$

where

$$\begin{aligned} \eta = \eta(k) &= 1 \quad \text{for } \operatorname{Re} k \geqq 0 \\ &= -1 \quad \text{for } \operatorname{Re} k < 0. \end{aligned}$$

Under this assumption one needs only the first part of the Fredholm alternative. Pannič also went on to treat the corresponding exterior electromagnetic problems although here I find the treatment somewhat obscure. As Werner and Greenspen [51] have shown for the Dirichlet problem and Kussmaul [52] for the Neumann problem, it is possible to obtain excellent results numerically for at least the cylinder problem and certainly the absence of the volume integral is of tremendous help. I fully anticipate that a subsequent version of Shenk and Thoe [3] will employ it also.

As sometimes happens in mathematics, good mathematics results, ironically enough, come from a method failure rather than from the true physical situation. To return to the Maue integral equation (2.1), I conjectured that what was true there was true in general. This led Wilcox and me to suspect that a technique of Tamarkin's [53], combined with a theorem of Rellich [27], could be used to prove that this was always the case. This we were able to establish. Subsequently, for his Edinburgh address of 1967, Wilcox developed an extensive Fredholm-type treatment valid for the entire complex plane. Since it seems unlikely that these results will ever be published I shall use this occasion to state them in some detail.

Before doing this, a few comments on the radiation condition are in order.

3. The radiation condition

There are many possible forms for a radiation condition and even today new ones are being devised for special purposes (compare Ludwig and Morawetz [54]). In 1956, Wilcox [56] gave a particularly elegant reformulation of that of Sommerfeld, in which for any solution $\phi(x, k)$ in C^2 he showed that it was sufficient to require that

$$\lim_{R\to\infty} \int_{r=R} \left|\left(\frac{\partial}{\partial r} - ik\right)\phi\right|^2 dS \to 0.$$

Earlier, in 1950, Kupradze [56] gave the following conditions (compare the 1956 German translation [57]):

$$r\left(\frac{\partial}{\partial r} - ik\right)\phi = e^{\pm ikr}o(1)$$

$$r\phi(r) = e^{\pm ikr}o(1),$$

where the plus sign holds for Im $k < 0$ and the minus sign for Im $k \geqq 0$. Subsequently, most workers, including Shenk and Thoe [3], have used a 1960 reformulation due to Reichardt [58] for the plane and a 1965 generalization of Schwartz [41]. If $F_k^+(x - y)$ is any fundamental solution of (1.1), these conditions are merely the requirement that the surface integral arising from Green's formula, namely:

$$\int_{R=r} \left[\phi(s)\frac{\partial}{\partial v_s} F_k^+(x - s) - F_k^+(x - s)\frac{\partial}{\partial v_s}\phi(s)\right] dS_s$$

converge to zero as $r \to \infty$. For $k \neq 0$ and $0 \leqq \arg k \leqq \pi$ this condition is equivalent to the Sommerfeld radiation condition if one takes $F_k^+(x - y)$ to be that given in (2.3).

The elegant formulation of a radiation condition due to Wilcox [59] in 1959 seems somehow to have been overlooked, probably since he did not explicitly state its validity for the entire complex plane at the time of its introduction. To state it, let $k \in [K - [0]]$ and let v be a solution of

$$\Delta v + k^2 v = 0 \tag{3.1}$$

for

$$|x| \geqq c.$$

Then he observes that the mean,

$$M_{x,r}[v] = \frac{1}{4\pi} \int_{|\omega|=1} v(x + r\omega)\, d\omega,$$

is in fact equal to

$$M_{x,r}[v] = v_+(x) \frac{e^{ikr}}{r} + v_-(x) \frac{e^{-ikr}}{r}, \qquad r = r(x),$$

where

$$v_\pm(x) = \pm \frac{e^{ikr}}{8\pi ik} \int_{|\omega|=1} \left(\frac{\partial}{\partial r} + ik\right) rv(x + r\omega)\, d\omega$$

is defined for all $x \in R^3$ (independently of $r \geqq r_0(x)$) and $v_\pm(x)$ are entire solutions of (k). v is a k-radiation function for, if $v \in C^2[\Omega]$, v is a solution of (3.1) for x such that

$$\int_{|\omega|=1} \left(\frac{\partial}{\partial r} - ik\right) rv(x + r\omega)\, d\omega = 0.$$

For $\operatorname{Im} k \geqq 0$ these conditions are equivalent to those of Sommerfeld. With this result we can return to the Kupradze–Maue–Weyl integral equation.

4. The Fredholm integral for the exterior Dirichlet problem

Following Wilcox [60], we assume that Γ is a closed bounded surface in R^3 of class C^2, that Ω is the connected exterior of Γ, and that Γ' is the interior of Γ. The problem to be solved is that of finding a solution of

$$\Delta\phi + k^2\phi = \rho(x) \quad \text{for } x \in \Omega$$

and

$$\phi(x) = 0 \qquad \text{for } x \in \Gamma.$$

Here $\rho(x)$ is a prescribed function—the source—and k is the wave number in the set $\{K - (0)\}$ but is otherwise an arbitrary complex number. Let

$$\phi_0(x, k) = -\frac{1}{4\pi^3} \int \frac{e^{ik|x-y|}}{|x - y|} \rho(y) dy$$

be the solution for the case in which $\Omega = R^3$ and $\Gamma = 0$. In the general case, let

$$\phi(x, k) = \phi_0(x, k) + v(x, k) \quad \text{for } x \in \Omega,$$

where $v(x, k)$ is a k-radiation function for Ω in the sense of Wilcox such that

$$v(x, k) \in C^1(\bar{\Omega})$$

and

$$v(s, k) = -\phi_0(s, k) \quad \text{for } x \in \Gamma.$$

As in the Kupradze–Maue–Weyl equations, the double-layer assumption implies a representation for $v(x, k)$ in the form

$$v(x, k) = \int_\Gamma \frac{\partial}{\partial \nu_s} \frac{e^{i|k|\,|x-s|}}{2\pi|x - s|} \mu(s)\, dS_s, \tag{4.1}$$

where ν is a unit exterior-directed normal on Γ. This leads to the following integral equation (in view of the well-known jump relation):

$$\mu(s) = \int_\Gamma K(s, s'; k)\mu(s')\, dS_{s'} = \phi_0(s, k) \qquad \text{for } s, s' \in \Gamma, \tag{4.2}$$

when

$$K(s, s'; k) = \frac{1}{2\pi} \frac{\partial}{\partial \nu_{s'}} \frac{e^{i|k|\,|s-s'|}}{|s - s'|}.$$

Then in terms of iterates of the kernels, one can represent the resolvent kernel of (4.2) in the form

$$R(s, s'; k) = \frac{N(s, s'; k)}{D(k)}, \qquad D(k) \neq 0,$$

where

$$R(s, s'; k) = K(s, s'; k') + \int_\Gamma K(s, s''; k)R(s'', s'; k)\, dS_{s''}$$

and

$$R(s, s'; k) = K(s, s'; k) + \int_\Gamma R(s, s''; k)K(s'', s'; k)\, dS_{s''}.$$

for all k such that $D(k) \neq 0$. This in turn implies that the double-layer density admits the representation

$$\mu(s, k) = v(s, k) + \int_\Gamma R(s, s'; k)v(s', k)\, dS_{s'}$$

and that the total field admits the representation

$$\phi(x, k) = \int_{\Omega} G(x, y; k)\rho(y)\, dy.$$

Here the resolvent Green's function has the following explicit expression:

$$G(x, y; k) = -\frac{1}{4\pi}\frac{e^{i|k|\,|x-y|}}{|x-y|} + \int_{\Gamma} K(x, s; k) \cdot \left\{\frac{e^{i|k|\,|s-y|}}{|s-y|} + \int_{\Gamma} R(s, s'; k)\frac{e^{i|k|\,|s'-y|}}{|s'-y|}\, dS_{s'}\right\} dS_s.$$

As such it is defined for all k except for those values of $D(k) = 0$. Let us examine the nature of these zeros. First of all $D(k_0) = 0$ if and only if the homogeneous integral equation

$$\mu(s, k) - \int_{\Gamma} K(s, s; k)\mu(s')\, dS_{s'} = 0. \tag{4.3}$$

If $\mu(s, k_0) \neq 0$ is a solution of (4.3), consider

$$v(x) = \int_{\Gamma} K(x, s'; k_0)\mu(s', k_0)\, dS_{s'},$$

so that v is a k-radiation function, $v \in C^2[\Omega] \cap C^1[\bar{\Omega}]$ and v vanishes on Γ. Thus if $\operatorname{Im} k_0 \geqq 0$, then $v(x) \equiv 0$ in Ω.

Hence

$$\left.\frac{\partial v}{\partial \nu_{x+}}\right|_{\Gamma} = \left.\frac{\partial v}{\partial \nu_{x-}}\right|_{\Gamma} = 0, \qquad \Delta v + k_0^2 v = 0;\ x \in \Gamma'$$

and

$$v(s, k_0) = \mu(s, k_0) + \int_{\Gamma} K(s, s'; k)\mu(s', k_0)\, dS_{s'}$$
$$= 2\mu(s, k_0) \neq 0,$$

so that v is an eigenfunction of the interior Neumann problem for k_0^2. It is not difficult to show that there are no zeros in $\operatorname{Im} k > 0$ and that the only zeros in $\operatorname{Im} k \geqq 0$ are of the form $k_0 = [\lambda_n]^{1/2}$, where $0 \leqq \lambda_0 \leqq \lambda_1 \leqq \lambda_2 \cdots$, the eigenvalues of the interior Neumann problem. The case solved by Maue shows that there can be zeros in $\operatorname{Im} k < 0$. Consideration of the integral equation adjoint to (4.3) in the usual Fredholm sense shows that the latter occur also if and only if $D(k) = 0$. Returning to the representation (4.1), one has the following Dolph–Wilcox theorem referred to in the Introduction:

THEOREM 4.1. *The poles of the total field* $\phi(x, k)$ *at* $k = [\lambda_n]^{1/2}$ *are all removable. Hence the only poles of the scattered field come from the region* $\operatorname{Im} k < 0$ *at those values for which* $D(k) = 0$.

To prove this, suppose for simplicity that k_0 is a simple zero of $D(k)$ with $\operatorname{Im} k = 0$. Then one has

$$D(k) = (k - k_0)D_1(k); \qquad D_1(k_0) \neq 0.$$

Then

$$\phi(x, k) = \phi_0(x, k) + \int_\Gamma K(x, s'; k)\mu(s', k)\, dS_{s'}, \tag{4.4}$$

where

$$\mu(s, k) = \phi_0(s, k) + \int_\Gamma R(s, s'; k)\phi_0(s', k)\, dS_{s'}$$

$$= \frac{\phi_{-1}(s)}{k - k_0} + \phi_0(s) + \phi_1(s)(k - k_0) + \cdots. \tag{4.5}$$

Moreover,

$$\mu(s, k) - \int_\Gamma K(s, s'; k)\mu(s; k)\, dS_{s'} = \phi_0(s, k). \tag{4.6}$$

The substitution of (4.5) into (4.6) and the use of a series expansion in powers of $(k - k_0)$ implies, since $\phi_0(x, k)$ is analytic at $k = k_0$, that

$$\phi_{-1}(s) - \int_\Gamma K(s, s'; k_0)\phi_{-1}(s')\, dS_{s'} = 0. \tag{4.7}$$

On the other hand, the substitution of (4.5) into (4.4) followed by expansion yields

$$\phi(x, k) = \frac{\phi_{-1}(x)}{k - k_0} + \phi_0(x) + \phi_1(x)(k - k_0), + \cdots$$

where

$$\phi_{-1}(x) = \int_\Gamma K(x, s', k_0)\phi_{-1}(s')\, dS'$$

is obviously a radiation function for k_0, $\phi_{-1} \in C^2(\Omega) \cap C^1(\overline{\Omega})$ and

$$\phi_{-1}(s) = -\phi_{-1}(s) + \int_\Gamma K(s, s', k_0)\phi_{-1}(s')\, dS_{s'} = 0$$

by (4.7). Thus $\phi_{-1}(x) \equiv 0$ in Ω, and hence, by Rellich's uniqueness theorem, or that of ordinary potential theory if $k_0 = (0)$, the pole is removable and does not affect the scattered field. A similar proof can be given for higher-order poles.

If one considers the spaces

$$L_2^m(\Omega) = \{\phi,\ D^\alpha\phi \in L_2(\Omega),\ 0 \leqq |\alpha| \leqq m\},$$

one can show that $L_2(\Omega)$ is the closure of $C_0(\Omega)$ in $L_2^1(\Omega)$ and one can introduce a self-adjoint operator A in $L_2(\Omega)$ whose domain is

$$L_2(\Omega) \cap \{\phi,\ \Delta\phi \in L_2(\Omega)\}$$

and whose spectrum is the positive real line, as was first shown by Rellich [27]. Moreover, the associated resolvent operator

$$R_\lambda = (A - \lambda I)^{-1}$$

is defined on the entire complex plane deleted by the positive-real axis. As such, for this cut plane it is a bounded operator which maps $L_2(\Omega)$ into itself and is also a strongly analytic function of λ there. If $\operatorname{Im} k > 0$, then the resolvent is an integral operator and then $G(x, y, k)$ as given earlier is its kernel with $\lambda_1 = k^2$. Moreover, the relation

$$[G(k)\rho](x) = \int_\Omega G(x, y; k)\rho(y')\,dy'$$

defines a mapping of $C(\Omega)$ into $C^1(\bar{\Omega})$ which is meromorphic in the strong sense. The associated resolvent having $G(x, y, k)$ as its kernel defines a bounded operator of $C_0(\Omega)$ into $C^1(\bar{\Omega})$ for $\operatorname{Im} k > 0$ which is strongly analytic in the latter Frechet space. For $\operatorname{Im} k < 0$, the resolvent operator possesses an analytic continuation which is meromorphic with all its poles in $\operatorname{Im} k < 0$. A different method was first used by Dolph, McLeod, and Thoe [61] to obtain the meromorphic character of the resolvent kernel for the situation to be discussed in the next section, and general theorems have been obtained subsequently by Steinberg [62] and by Haf [63]. We shall return to this subject in Section 5. Finally, to conclude this section, we should note that McLeod [64] treated the two-dimensional problem and obtained results similar to those of Wilcox [60] by a method closer to the spirit of Dolph et al. [61].

5. The Schrödinger equation and the meromorphic "spin off"

Physicists over a long period of years have studied the analytic continuations of the spherically symmetric Schrödinger equation to the region $\operatorname{Im} k < 0$. This history can be found either in the work of Wu and Ohmura [6] or in the review paper by Newton [7]. It was from the latter that McLeod, Thoe, and I initiated an attempt to achieve similar results without the restriction of spherical symmetry on the potential. In this we succeeded and our results are published in Dolph et al. [61] but not without the help of MacCamby and Werner who directed our attention to the possibility of adapting a result of J. D. Tamarkin's [66] to the case at hand. The widely known Dunford–Schwarz result [67] did not imply that the continuation was meromorphic.[3]

In brief, we first introduced the enlarged Hilbert space $H_1 = L_2[d\mu(x)]$ where $d\mu(x) = \exp[-a|x|]\,dx$ with an inner product

$$\langle \phi, \psi \rangle = \int_\Omega \phi(y)\overline{\psi(y)}\, e^{-2a|y|}\, dy$$

and a norm

$$\|\phi\|_{H_1} = \langle \phi, \psi \rangle.$$

The introduction of this space avoids the so-called "exponential catastrophe" which results if one works in a time-independent way in the usual Hilbert space appropriate for the region $\operatorname{Im} k \geqq 0$. When, in this enlarged Hilbert space, under the assumption (1.0), k is restricted to the region $\operatorname{Im} k \geqq -a$, the integral operator

$$T_k\phi = -\frac{1}{4\pi}\int_\Omega \frac{e^{i|k|\,|x-y|}}{|x-y|}\, p(y)\phi(y)\, d\mu(y), \tag{5.1}$$

where $p(y) = e^{2a|x|}q(x)$ is an analytic compact family of operators of the Hilbert–Schmidt class with the property that

$$\|T(k)\|_{H_1} = 0[(\operatorname{Im} k)^{1/2}] \quad \text{as } \operatorname{Im} k \to \infty.$$

Moreover, $[I - T(k)]^{-1}$ is a meromorphic operator-valued function for $\operatorname{Im} k \geqq -a$ and has a pole at k_0 if and only if there exists a ϕ such that $\phi = T(k)\phi$ in H_1. Explicitly, it was the resolvent kernel of the scattering integral equation (see Dolph et al. [61] or Ikebe [11]) which could be continued since it is not apparently possible to continue the resolvent relations of Hilbert. We also showed that the poles were symmetrically placed with respect to the imaginary k-axis and that those in the region $\operatorname{Im} k \geqq 0$ all occurred on $\operatorname{Re} k = 0$. We also discussed the scattering operator in the form given by Ikebe [70] and showed (i) that under these same conditions it could be continued analytically through the region $|\operatorname{Im} k| < a$; (ii) that it was meromorphic in this region with poles confined to the subregion $-a < \operatorname{Im} k < 0$ and to the non-negative imaginary axis; and (iii) that the poles in the next-to-last region were also symmetrically placed with respect to the imaginary axis. We did *not*, however, relate the *singularities* of the *integral operator* and/or resolvent Green's kernel to

[3] Similar results were announced by Ramm [68]. I have just received a reprint of his paper [69] containing the full details.

those of the *scattering operator*. However we did give an example which indicated that the region of continuation to $\operatorname{Im} k > -a$ was in agreement with the known results of radially symmetric potentials and was the best possible. This example led us to suspect, but this has not been verified as yet, that under the hypothesis (1.0), in contrast to that of compact support for $q(x)$, there will always be a contribution from infinity in the lower half-plane due to the occurrence of a branch cut along $\operatorname{Im} k = -a$.

Subsequent to our joint work, Thoe [22] gave a proof under a differentiability condition on $q(x)$ that there were only a finite number of poles in any strip in the region and made a beginning on the associated asymptotic series for the wave equation $\operatorname{Im} k > 0$. (There is a known simple relation between the reduced wave equation with potential when derived from the wave equation and when derived from Schrödinger's equation.) This relation is, for example, explicitly spelled out in Shenk [20]. Independently, McLeod [71] introduced ellipsoidal coordinates and in a manner reminiscent of the way physicists prove the convergence of the Born series (see Newton [72], Zemach and Klein [73], and Khuri [74]), he was able to obtain Thoe's result without the additional differentiability condition. One essentially integrates by parts in this new coordinate system and then makes appropriate estimates.

The results on the meromorphic nature of a family of compact operators have subsequently been considerably generalized by students of Phillips and Werner, respectively, namely Steinberg [62] and Haf [63]. Steinberg obtains his results by working with the projection operator defined by the contour integral involving the resolvent (see, for example, Riesz–Nagy [75]) while Haf bases his on the Schmidt procedure of approximation by degenerate kernels. Since Steinberg's results are the most general and will be used in the next section, we shall state them explicitly. From the viewpoint of one who teaches applied analysis, however, the method used by Haf is probably more easily understood *today* by graduate engineers and physicists. Haf's thesis also contains several detailed applications of his results to various scattering problems.

Following Steinberg, we let B be a Banach space and $0(B)$ be the bounded operators on B. Let K_1 be a subset of a complex plane which is open and connected. A family of operators $T(k)$ is said to be meromorphic in K_1 if it is defined and analytic in K_1 except for a discrete set of points where it is assumed that in the neighborhood of one such point k_0

$$T(k) = \sum_{j=-N}^{\infty} T_j(k - k_0)^j, \tag{5.2}$$

where the operator $T_j \in 0(B)$, $N \geqq 0$, and where the series converges in the uniform operator topology in some deleted neighborhood of k_0. The family $T(k)$ is said to be analytic at k_0 if $N = 0$ in (5.2). Then one has the following theorems:

THEOREM 5.1. *If $T(k)$ is an analytic family of compact operators for $k \in K_1$, then either $[I - T(k)]$ is nowhere invertible in K_1 or $[I - T(k)]^{-1}$ is meromorphic in K_1.*

THEOREM 5.2. *Suppose $T(k, x)$ is a family of compact operators analytic in k and jointly continuous in (k, x) for each $(k, x) \in \Omega \times R$ (R is the set of real numbers) then if $[I - T(k, x)]$ is somewhere invertible for each x, then its inverse is meromorphic in k for each x. Moreover, if k_0 is not a pole of $[I - T(k, x_0)]^{-1}$, then the inverse is jointly continuous in (k, x) at (k_0, x_0) and the poles of the inverse depend continuously on x and can appear and disappear only at the boundary of Ω, including the point at infinity.*

In addition to these results, Steinberg also gives sufficient conditions for poles to be of order one.

Again these results illustrate the usefulness of the interplay between mathematics and physics.

6. Intimations concerning the unified Shenk–Thoe theory

All of what precedes has been unified for the problem given by the relations (1.0)–(1.3) under the hypotheses stated below them. To state some of these results we rewrite the Werner relation in the form

$$\phi(x, k) = S\eta(x) + D\mu(x) + V\tau(x),$$

where the first term stands for a single layer, explicitly

$$S\eta(x) = -2 \int_{\Gamma_1} F_k(x - s)\eta(s)\, dS_s,$$

the second term stands for the double layer of (4.1), and the third term is a slightly modified volume potential of (2.2). One next defines a matrix operator

$$M = M_{\pm}\begin{Bmatrix} \lambda S & \lambda D & \lambda V \\ S & D & V \\ \tilde{S} & \tilde{D} & \tilde{V} \end{Bmatrix}.$$

In this matrix, the third row is obtained by applying the operator

$$\frac{\partial}{\partial \nu} - \sigma(x)$$

under the integral signs in the definitions of S, D, V and

$$\lambda = \lambda(k) = q(x)e^{2a|x|} \qquad \text{for } x \in \Omega$$
$$= k^2 + i^{2a|x|} \quad \text{for } x \in \Gamma.$$

Let B_i, $i = 1, 2, 3$, denote three Banach spaces where $B_1 = C(\Gamma_1)$, $B_2 = C^\alpha(\Gamma_2)$ for a fixed α, $0 < \alpha < 1$, and $B_3 = [\tau \in C(\Omega \cup \Gamma)]$, where τ has continuous extension to $\Omega \cup \Gamma$ and to $\Gamma' \cup \Gamma$, and where under the hypothesis (1.0) the norm in B_3 is

$$\|\tau\|_{B_3} = \sup \exp\,[-a|x|][|\tau(x)|].$$

The following theorem is now established:

THEOREM 6.1. *The operator M maps $B_1 \times B_2 \times B_3$ into itself and this mapping is completely continuous for each k and analytic in k for* $\operatorname{Im} k > -a$ (*with a logarithmic singularity at $k = 0$ if n is even*).

It is next shown that every outgoing solution $\phi(x, k)$ can be obtained by solving

$$[I + M(k)][\tau e^{-2a|x|}, \mu, \eta] = [f, e^{2a|x|}, f_2, f_3] \tag{6.1}$$

in the product Banach spaces. In connection with Steinberg's results discussed in Section 5, it also follows that $[I + M(k)]^{-1}$ is a meromorphic function of k in all $\operatorname{Im} k > -a$ and this in turn yields meromorphic continuation of the outgoing solutions from $\operatorname{Im} k > 0$ to all of $\operatorname{Im} k > -a$, the only poles occurring at the poles of the last inverse. Except for these poles, these analytically continued solutions are unique and satisfy the outgoing radiation condition. At the poles the "complex eigenvalues" occur and the outgoing solutions are not unique. However, as in the usual Riesz–Fredholm theory, a solution will exist if and only if the $[f_i]$ involved are in the null space of the "adjoint" of $[I + M(k)]$. As in Werner's work [47] and that discussed in Section 4, this approach eliminates the (removable) singularities of the corresponding interior problems. The resolvent Green's operator can be constructed and used to build a spectral theory for real k completely analogous to that of Ikebe [11] and Shenk [17] for this general problem, but its description is too difficult and lengthy to present here. Perhaps more importantly, the poles of the operator $[I + M(k)]^{-1}$ can be related to those of the scattering operator in a very precise sense, thus answering the question left open by Dolph et al. [61] about the relationship of singularities of the scattering operator and the nontrivial solution of the extended integral operator discussed there. If one lets A denote the self-adjoint extension of the operator defined by equations (1.1)–(1.3) in $L_2[\Omega]$, then the scattering operator $S(k)$ for the "Schrödinger wave equation" applicable to this problem, taken in the form

$$i\frac{dU(t)}{dt} = AU(t),$$

is a unitary operator on $L_2[S^{n-1}]$ for each positive k, where S^{n-1} is the unit hypersphere in n dimensions. As such, $S(k)$ has a meromorphic extension (with a branch point at $k = 0$) if n is even to the entire complex k-plane and has no nonzero real poles. The poles of $S(k)$ in the upper half-plane are simple, lie on the imaginary axis, and have as their square the nonzero eigenvalues of A. Shenk and Thoe now define a resonant state as a nontrivial solution of

$$u = T_k^+ u = \int_\Gamma u(s)\frac{\partial}{\partial v_s} F_k^+(x - s) - F_k^+(x - s)\frac{\partial}{\partial v_s} u(s)\, dS_s$$
$$- \int_\Omega F_k^+(x - y)q(y)u(y)\, dV_y$$

in $H_2^{\text{loc}}[\bar{\Omega}]$ satisfying the boundary condition (1.2) and (1.3). Such a function will automatically satisfy

$$[-\Delta + q - k^2]u = 0$$

as a distribution in Ω, since for $H_2^{\text{loc}}[\bar{\Omega}]$ one has

$$[-\Delta - k^2]T_k^{\pm} v = 0$$

as distributions in Ω. Suppose now for simplicity that one considers the poles of $S(k)$ with $\operatorname{Im} k < 0$ and k^2 not an eigenvalue of A. (Shenk and Thoe do succeed in eliminating this last hypothesis but at a considerable cost in complexity.) They then proceed to establish the following characterization:

THEOREM 6.2. *For k with $\operatorname{Im} k < 0$ and with k^2 not an eigenvalue of A, we have the following:*

(i) *If $S(\bar{k})^*h = 0$ with $h \neq 0$ and if we define*

$$U_k h(x) = \int h(\omega)\phi_-(x, k, \omega)\, d\omega, \qquad x \in \bar{\Omega},$$

where the distorted plane wave ϕ_- is the solution of

$$[-\Delta + q(x) - k^2]\phi_- = 0$$

in Ω which satisfies the boundary condition of (1.2)–(1.3) *and which differs from the plane wave $e^{ikx\cdot\omega}$ by a function satisfying the classical incoming radiation condition*

$$\phi_-(x, k, \omega) - e^{ikx\cdot\omega} = 0[\exp[\operatorname{Im} k|x|]]$$

as $|x| \to \infty$, *then* $u(x) = U_k h(x)$ *is a resonant state at* k. *Furthermore,*

$$u(s\theta) \sim \left(\frac{2\pi}{iks}\right)^{(n-1)/2} h(\theta)e^{iks}.$$

(ii) *Conversely, if* $u(x)$ *is a resonant state at* k *and if we define* $h(\theta)$ *by the last (asymptotic) relation, then*

$$S(\bar{k})^*h = 0, \qquad u = U_k h \quad \textit{and} \quad h \neq 0.$$

(iii) *Moreover*

$$(I - T_k^+)U_k h(x) = \int e^{ikx\cdot\omega} S(\bar{k})^* h(\omega)\, d\omega.$$

With the additional remark that, since each $\phi_-(x, k, \omega)$ is in $C^1(\bar{\Omega}) \cap C^2(\Omega)$, the resonant states $u = U_k h$ will also be, space limitation forces me to conclude this discussion.

7. Open question of the integral equation method versus the physical problem

1. When are there only a finite number of poles in any strip parallel to the axis $\operatorname{Im} k = 0$?

As the Thoe–McLeod results show, this is the case for the Schrödinger equation under the hypothesis used here. McLeod has told me that the same is true in two dimensions for the Dirichlet problem. It certainly cannot be true in general as the famous counterexample of my late friend B. Friedman clearly shows. Since he never sketched it in detail except in an unpublished lecture, let me take this opportunity to do so. Consider the problem

$$\phi_{xx} - \frac{1}{c^2}\phi_{tt} = 0 \qquad \begin{aligned} \phi(0, t) &= 0 \\ \phi(x, 0) &= 0 \\ \phi_t(x, 0) &= f(x) \text{ for } 0 \leqq x < \alpha \\ &= 0 \text{ for } x > \alpha, \end{aligned}$$

where

$$\begin{aligned} c^2 &= c_1^2 \qquad 0 \leqq x < \alpha \\ &= c_2^2 \qquad \alpha \leqq x < \infty \end{aligned}$$

and take its Laplace transform via the usual relation:

$$\bar{\phi}(x, s) = \int_0^\infty e^{-st}\phi(x, t)\, dt.$$

This of course yields

$$\bar{\phi}_{xx} - \frac{s^2}{c^2}\bar{\phi} = -\frac{f}{c_1^2} \qquad 0 \leqq x < \alpha$$
$$= 0 \qquad \alpha \leqq x < \infty,$$

whose solution is given by

$$\bar{\phi}(x, s) = -\int_0^\alpha G(x, y; s)\frac{f(y)}{c_1^2}\, dy$$

Employing here the usual continuity and jump relations for G, one obtains that for large x,

$$\bar{\phi}(x, s) = \frac{e^{-s\alpha/c_2}}{s\Delta c_1^2}\int_0^\alpha \left[2 \sinh \frac{s}{c_1} y - \cosh \frac{s}{c_1} y \sinh \frac{s}{c_1}(y - \alpha)\right] f(y)\, dy,$$

where

$$\Delta = \Delta(s) = -e^{-sx/c_2}\left[\frac{\cosh s\alpha/c_1}{c_1} + \frac{\sinh s\alpha/c_1}{c_2}\right].$$

If one now takes $f(x)$ to be $\delta(x - x_0)$, $0 < x_0 < \alpha$, then

$$\bar{\phi}(x, s) = \frac{e^{-s\alpha/c_2}}{s\Delta c_1^2}\left[2 \sinh \frac{s}{c_1} x_0 - \cosh \frac{s}{c_1} x_0 \sinh \frac{s}{c_1}(x_0 - x)\right].$$

The inverse of this can of course be evaluated in several different ways. If one defines for convenience

$$\xi = \frac{1}{c_1} + \frac{1}{c_2}$$

$$\eta = \frac{1}{c_1} - \frac{1}{c_2}$$

and

$$\varepsilon = \frac{\eta}{\xi},$$

one easily sees by elementary manipulation that

$$\bar{\phi}(x, s) = \frac{-2}{c_1^2\xi}\sum_{i=1}^{6} \beta_i \frac{e^{s(\theta_i - \eta\alpha - x/c_2)}}{s(1 + \varepsilon e^{-2s\alpha/c_1})}$$

where

$$\beta_1 = -\beta_2 = 1 \qquad \theta_1 = -\theta_2 = \frac{x_0}{c_1}$$

$$\beta_3 = -\beta_4 = \tfrac{1}{4} \qquad \theta_3 = -\theta_4 = \frac{2x_0}{c_1} - \frac{\alpha}{c_1}$$

$$\beta_5 = -\beta_6 = \tfrac{1}{4} \qquad \theta_5 = -\theta_6 = \frac{\alpha}{c_1}.$$

Now consider the function

$$u(t) = 1 - \varepsilon u(t - m),$$

whose transform is

$$\bar{u}(s) = \frac{1}{s(1 + \varepsilon e^{-ms})}.$$

One then finds that

$$\begin{aligned}\phi(x, t) &= \frac{-2}{c_1^2 \xi} \sum_{i=1}^{6} \beta_i u\left(t + \theta_1 - \eta\alpha - \frac{x}{c_2}\right) \\ &= \frac{-2}{c_1^2 \xi} \sum_{i=1}^{6} \beta_i \left[1 - \varepsilon u\left(t + \theta_i - \eta\alpha - \frac{x}{c_2} - \frac{2\alpha}{c_1}\right)\right],\end{aligned}$$

where $u(\tau) = 0$ for $\tau < 0$, and note that for large t, $\phi(x, t) = 0$ which agrees with the final value theorem of Laplace transform theory. R. R. Rutherford plotted this out for $\varepsilon = 0.5$ and $x_0 = a/2$ and obtained the results shown in Fig. 1.

If, on the other hand, one uses the inversion integral

$$\phi(x, t) = \frac{1}{2\pi i} \int_{\gamma - i\infty}^{\gamma + i\infty} \frac{e^{s(t - x/c_2)}}{s\Delta c_1^2} \left[2 \sinh \frac{s}{c_1} x_0 - \cosh \frac{s}{c_1} x_0 \sinh \frac{s}{c_1}(x_0 - \alpha)\right] ds,$$

the theory of residues involves zero's of Δ, which occur at $s_0 = 0$, and, for $R = \frac{-c_2}{c_1}$, at

$$s_n = \frac{c_1}{2\alpha} \ln \left|\frac{1 - R}{1 + R}\right| \pm \frac{c_1}{\alpha} n\pi i.$$

These all lie parallel and at fixed distance from the axis Re $s = 0$. Their number is *denumerable*, not finite. Closing the contour to the left (there is no contribution as a consequence of Jordan's lemma), one can obtain an explicit asymptotic result.

If one now sets $s = -ik$ (s, k complex) in this example, thus rotating everything so that the domain of analyticity becomes the upper half space, one has the normalization used in the rest of this paper. This type of phenomenon also occurs in what H. Kramer and I like to call the spherical whale model to all orders of angular momentum. This is a generalization of Friedman's example turned inside out. Imagine a spherical interface between two liquids, the inner one being of greater density and the excitation being a point source far away in the exterior one. Assuming continuity of the scalar field and its radial derivatives at the interface, one finds the same type of behavior.

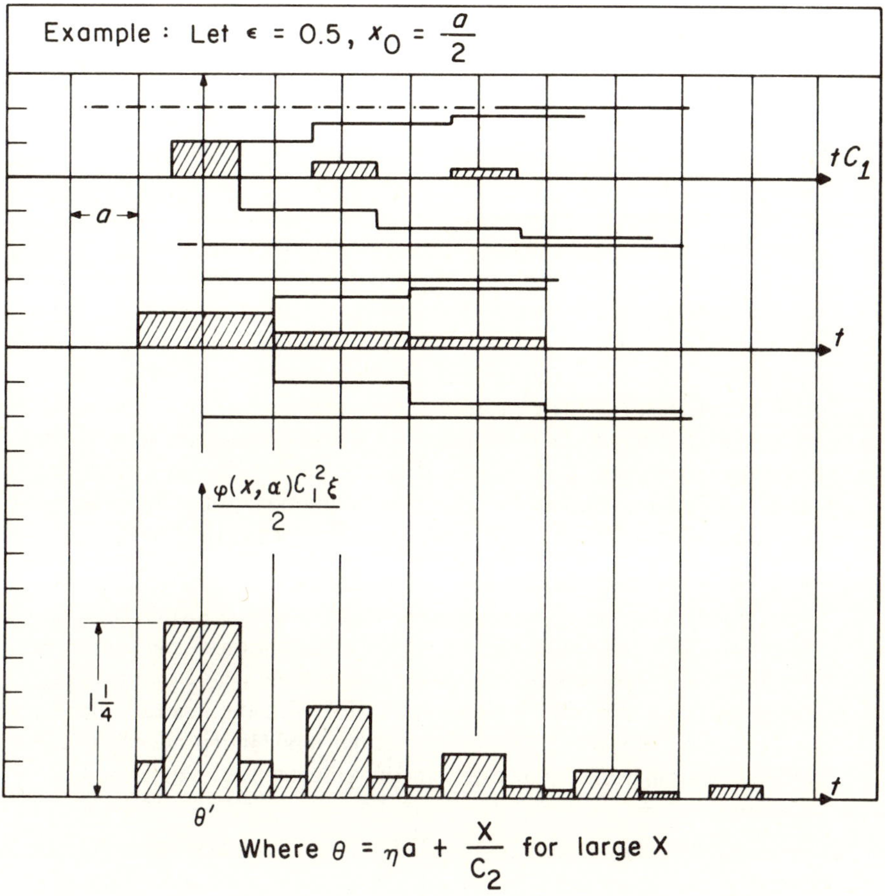

FIG. 1. Rutherford's results

2. When are the poles in Im $k < 0$ simple?

This is at the moment a completely open question. Shenk and Thoe have made a start on settling it and I added to their considerations the conjecture that it must somehow be associated with the *real* symmetry of the resolvent Green's function (see Ikebe [11]) or with the situation that occurs when the Hermitian inner product reduces to the real symmetric product (see Dolph, Marx, and McLaughlin [76], a paper, which due to these recent developments clearly needs extension to, at least, the separable Hilbert space case).

3. What is the most effective way to compute the poles in Im $k < 0$?

Since the integral equations (2.1) in the enlarged Hilbert space do represent compact operators, they could of course be approximated by degenerate kernels and the poles computed approximately by algebraic methods. A question this author hopes to investigate soon is the possibility of developing variational principles to take advantage of the real symmetry of the resolvent Green's function when it exists (see Dolph and Ritt [77] and Dolph [78]).

4. How intrinsically determined are the poles by the scatterer?

Recent work by Lax and Phillips [4] demonstrates that, at least for the Dirichlet and Neumann problem, there is a monotonicity property for star-shaped bodies (at least for those poles, in our normalization, with $\operatorname{Im} k < 0$, $\operatorname{Re} k = 0$), and that one can expect asymptotic formulas reminiscent of those of Carleman and Weyl for the interior problem. Somehow, at least for these problems, the invariance of these under the various Ansatz's has to be established for the integral equation method. Moreover, as pointed out in Beck and Nussenzweig [35] and exemplified by the Friedman example, the associated nonmodal solutions have no separate existence of their own. I quote from the former (p. 22):

> "It must be strongly emphasized, in connection with the physical interpretation of the transient modes, that no special significance can be attached, in general, to the amplitude of excitation of each separate mode. In fact, the transient modes are not orthogonal, so that the total energy is not a sum of the terms associated with the separate modes, and it is not possible to excite one particular mode independently of the others. A similar situation exists in the case of transients in discrete systems."

A related result should also be noted. La Vitta, a student of Lax's, has shown [79] that analytically perturbed bounded obstacles depending on a parameter a lead to algebraic functions of the parameter $k(\varepsilon)$ for sufficiently small ε and that the first variation of $k(\varepsilon)$ depends only on the perturbation of the boundary and solutions to the reduced wave equation for the unperturbed region.

At the time of the Bochner symposium, H. Rauch and I initiated an attack which we hope will throw further light on this subject. We propose to solve the Kupradze–Maue–Weyl integral equations for the Dirichlet and Neumann problems in other separable coordinate systems in both two and three dimensions. In addition, we hope to constructively use the method, based on conformal mapping, due to P. Garabedian [80]. We are of course seeking some type of invariance.

5. Another closely related fundamental question is a detailed examination of the question when should one use the Hermitian inner product and

when should one use the real symmetric inner product and, of course, the associated Green's formulas? An example of the literature shows the value of each in an isolated instance but no overall governing principle has as yet been announced to my knowledge.

6. Is it possible to replace the unified Shenk–Thoe theory by an essentially equivalent one in a Hilbert space? This has been achieved by Kuroda for the quantum scattering theory [19] and I conjecture that it should be possible here. If so, this would remove the, to my taste, somewhat arbitrary choice of the Banach spaces, a choice which seems purely utilitarian and not based on any profound mathematical reasons.

7. What is the relationship between analytic continuation of the resolvent kernel, the scattering operator, and the phenomenon of spectral concentration? In the general case our understanding of the situation appears to be about the same as that reported by McLeod [81]. Knowland [82], [84][4] has had some success for the case of finite rank perturbations. For an overall introduction to the subject see Kato [85, pp. 471–477].

8. Possible generalizations

1. The extension of every known result to the electromagnetic case should certainly be made since the physical phenomena are still linear. A similar case can be made for the Dirac equation and the Klein–Gordon equation.

2. Although the underlying theory is nonlinear, the linear part of it as described for example by Kupradze [86] should be examined in the entire complex plane. A beginning in this direction has been made by Kramer, Lane, and Rutherford [87].

3. One should examine these same problems for general differential equations, much as Vainberg [88] did, to summarize our knowledge of the more general case involving the principles of radiation, limit absorption, and limit amplitude in the general theory of partial differential equations.

4. One should investigate in general what happens on the line Im $k = -a$ and what the true mathematical and physical implications of this limit are.

With these I shall conclude, paraphrasing M. Kac and K. Case, by saying that some day we may be able to answer the question whether we can or cannot see the shape of a cavity.

UNIVERSITY OF MICHIGAN

[4] Reference [83] is added for completeness. It contains simplified proofs of the results in [61].

REFERENCES

1. WILCOX, C. H., "The mathematical foundation of diffraction theory," in *Electromagnetic Theory*, Madison, Wisc.: University of Wisconsin Press, 1962.
2. DOLPH, C. L., "Positive real resolvents and linear passive Hilbert systems," *Ann. Acad. Sci. Fennicae*, Ser. A, #336/9, 1963, pp. 1–39.
3. SHENK, N., II, and D. THOE, "A unified theory for exterior scattering problem." To appear in *J. Math. Anal. Appl.*
4. LAX, P. D., and R. S. PHILLIPS, "Purely decaying modes for the wave equation in the exterior of an obstacle," *Comm. Pure and Appl. Math.*, *22* (1969) pp. 737–787.
5. LUDWIG, D., and C. MORAWETZ, "The generalized Huyghens principle for reflecting bodies," *Comm. Pure and Appl. Math. 22* (1969), pp. 189–205.
6. LAX, P. D., and R. S. PHILLIPS, *Scattering Theory*. New York: Academic Press, 1967.
7. NEWTON, R. G., "Analytic properties of radial wave functions," *J. Math. Phys.*, *1* (1960), 319–347.
8. CARLEMAN, T., "Sur la theorie mathematique de l'equation de Schroedinger," *Arkiv. Mat. Abstr. o. Fys.*, *11* (1934), 1–7.
9. KATO, T., "Fundamental properties of Hamiltonian operators of Schroedinger type," *Trans. Am. Math. Soc.*, *70* (1951), 195–211.
10. POVZNER, A. Ya., "On the expansion of arbitrary functions in terms of the eigenfunctions of the operator $-\Delta u + cu$ (in Russian), *Mat. Sbornik*, *32*, 74 (1953), 109–156.
11. IKEBE, T., "Eigenfunction expansions associated with the Schroedinger operators and their applications to scattering theory," *Arch. for Rat. Mech. and Anal.*, *5* (1960), 1–4.
12. GELFAND, I. M., and B. M. LEVITAN, "On the determination of a differential equation by its spectral function," translated from the original Russian in *AMS Trans.*, *1* (1951), 253.
13. FADDEEV, L. D., "The inverse problem in the quantum theory of scattering," translated from the Russian original and in Math. *Phys.*, *4*, 72 (1963), 72.
14. ———, "Increasing solution of the Schroedinger equation," *Soviet Physics, Doklady*, *10* (1966), 1033–35.
15. ———, "Factorization of the *S*-matrix for the multidimensional Schroedinger operator," *Soviet Physics, Doklady*, *11* (1966), 209–211.
16. DOLPH, C. L., "Motivation and heuristic development of inverse scattering theory," U. of Mich. Rad. Lab., 8579-60-10, Sept. 1967, pp. 1–19.
17. SHENK, N. A., II, "Eigenfunction expansions and scattering theory for the wave equation in an exterior region," *Arch. Rat. Mech. Anal.*, *21* (1966), 120–150.
18. THOE, D., "Spectral theory for the wave equation with a potential term," *Arch. for Rat. Mech. and Anal.*, *22* (1966), 364–406.
19. KURODA, S. T., "An abstract stationary approach to perturbation of continuous spectra and scattering theory, *J. Analyse Math.*, *20* (1967), 57–117.
20. SHENK, N. A., II, "The invariance of wave operators associated with perturbations of $-\Delta$," *J. Math. and Mech.*, *17* (1968), 1005–1022.
21. DOLPH, C. L., "Recent developments in some nonself-adjoint problems of mathematical physics," *Bull. Am. Math. Soc.* (1961), 1–69.
22. THOE, D., "On the exponential decay of solutions of the wave equation," *J. Math. Anal. and Appl.*, *16* (1966), 333–346.
23. FRANK, P., and R. V. MISES, *Die Differential und Integralgleichungen der Mechanik und Physik*, I and II. New York: S. M. Rosenberg, 1943.
24. MORSE, P. M., and H. FESHBACH, *Methods of Theoretical Physics*, I and II. New York: McGraw-Hill Book Co., Inc., 1953.
25. LORD RAYLEIGH, *Phil. Mag.*, *4*, 12 (1881), 86.

26. Sommerfeld, A., "Die Greensche Function der Schwingungsgleichung," *Jahresber. deut. math. Ver.*, *21* (1912), 326.
27. Rellich, F., "Uber das asymptotische Verhalten der Losungen von $\Delta u + k^2 u = 0$ in unendlichen Gebieten," *Jahresber. Deutschen. Math. Vereigigung*, *53* (1943) 57.
28. Kato, T., "Growth properties of solutions of the reduced wave equation with a variable coefficient," *Comm. Pure and Appl. Math.*, *9* (1957), 403–425.
29. Stratton, J. A., *Electromagnetic Theory*. New York: McGraw-Hill Book Co., Inc., 1941.
30. Lord Kelvin, *Proc. Lond. Math. Soc.*, *15*, 1 (1884), 197.
31. Gamow, G., *Z. Phys.*, *51* (1928), 204.
32. Lamb, H., *Proc. Lond. Math. Soc.*, *32*, 1 (1900), 208.
33. Love, A. E. H., *Proc. Lond. Math. Soc.*, *2*, 2 (1904), 88.
34. van Kampfen, N. G., *Phys. Rev.*, *91* (1953), 1267.
35. Beck, G., and H. M. Nussenzveig, "On the physical interpretation of complex poles of the *S*-matrix-I," *Nuovo Cimento X*, *16* (1960), 416–449.
36. Nussenzweig, H. M., "On the physical interpretation of complex poles of the *S*-matrix-II," *Nuovo Cimento X*, *20* (1961), 694–714.
37. ———, *Analytic Properties of Non-Relativistic Scattering Amplitudes*, notes from Escuela Latino Americana de Fisicia, Universidad de Mexico, 1962.
38. Petzold, J., "Wie gut das Exponentialgesetz beim α-Zerfall," *Z. Phys.*, *155* (1959), 422–432.
39. Kupradze, V., "Uber das Austrahlungprinzip von A. Sommerfeld," *C. R. URSS* (n.s.), *1* (1934), 55–58.
40. ———, "Verbreitung der elektromagnetischen Wellen in nicthomogenem Medium," *C. R. URSS* (n.s.), *1* (1936), 7–9.
41. Schwarze, G., "Über die 1., 2., und 3., aussere Randwertaufgabe der Schwingungsgleichung $\Delta F + K^2 F = 0$," *Math. Nach.*, *28* (1965), 337–367.
42. Sternberg, W., "Anwendung der Integralgleichungen in der elektromagnetische Lichttheorie," *Comp. Math.*, *3* (1936), 254–275.
43. Freundenthal, H., "Über ein Beugungsproblem aus der elektromagnetische Lichttheorie," *Comp. Math.*, *6* (1938), 221–227.
44. Maue, A. W., "On the formulation of a general diffraction problem through an integral equation" (in German), *Z. Physik*, *126* (1949), 601–618.
45. Weyl, H., "Kapazität von Strahlungsfeldern," *Math. Z.*, *55* (1952), 187.
46. Müller, C., "Zur Methode der Strahlungskapazität von H. Weyl," *Math. Z.*, *56* (1952), 80.
47. Werner, P., "Randwertprobleme der mathematischen Akustik," *Arch. Rat. Mech. and Anal.*, *10* (1962), 29–66.
48. Leis, R., "Zur Dirichletschen Randwertaufgeben des Aussenraumes der Schwingungsgleichung," *Math. Z.*, *90* (1965), 205–211.
49. Birkhage, H., and P. Werner, "Uber das Dirichletsche Aussenraumproblem für die Helmholtzsche Schwingungsgleichung," *Archiv. d. Math.*, *16* (1965) 325–329.
50. Pannič, I. J., "On the solubility of exterior boundary value problems for the wave equation and for a system of Maxwell's equations" (Russian), *Uspeshi Mat. Nauk*, *20* (1965), 221–226.
51. Greenspen, D., and P. Werner, "A numerical method for the exterior Dirichlet problem for the reduced wave equation," *Archiv. for Rat. Mech. and Anal.*, *23* (1966), 288–316.
52. Kussmaul, R., "Ein numerische Verfahren zur Lösung des Neumannschen Aussenraumproblems für die Helmholtzsche Schwingungsgleichun," dissertation, Tech. Hochschule Stuttgart, 1968.
53. Tamarkin, J. D., and W. Feller, "*Partial Differential Equations*," notes from Brown University by W. Kincaid, 1941, pp. 106–107.

54. Ludwig, D., and C. Morawetz, "An inequality for the reduced wave operator and the justification of geometric optics," *Comm. Pure and Appl. Math.*, *21* (1968), 187–203.
55. Wilcox, C. H., "A generalization of theorems of Rellich and Atkinson," *Proc. AMS*, *7* (1956), 271–276.
56. Kupradze, V., 1950, Russian; see 57.
57. ———, *Randwertprobleme der Schwingungsgleichung*, Berlin, 1956 (German translation of 56).
58. Reichardt, H., "Austrahlungsbedingungen für die Wellengleichung," *Abh. math. Sem. Univ. Hamburg*, *24* (1960), 41–53.
59. Wilcox, C. H., "Spherical means and radiation conditions," *Arch. Rat. Mech and Anal.*, *3* (1959), 133–148.
60. ———, "The analytic continuation of the resolvent kernel in scattering theory," Edinburgh Lecture, Sept. 21, 1967.
61. Dolph, C. L., J. B. McLeod, and D. Thoe, "The analytic continuation of the resolvent kernel and scattering operator associated with the Schroedinger operator," *J. Math. Anal. and Appl.*, *16* (1966), 311–332.
62. Steinberg, S., "Meromorphic families of compact operators," *Arch. for Rat. Mech.*, *31* (1968), 372–380.
63. Haf, H., "Zur Theorie parameterabhängiger Operatorgleichungen," dissertation, Tech. Hochschule Stuttgart, 1968.
64. McLeod, J. B., "The analytic continuation of the Green's function associated with obstacle scattering," *Quart. J. Math. Oxford*, *2*, 18 (1967), 169–180.
65. Wu, T., and T. Ohmura, *Quantum Theory of Scattering*. Englewood Cliffs, N.J.: Prentice-Hall, Inc., 1962.
66. Tamarkin, J. D., "On Fredholm's integral equations whose kernels are analytic in a parameter," *Ann. Math.*, *28* (1927), 127–152.
67. Dunford, N., and J. Schwartz, *Linear Operators*, Vol. I. New York: McGraw-Hill Book Co., Inc., 1958.
68. Ramm, A. G., "Regions free of resonance poles in the scattering problem for a three-dimensional potential," *Soviet Physics, Doklady*, *11* (1966), 114–116.
69. ———, "Some theorems of analytic continuation in the spectral parameter of the resolvent kernel of the Schroedinger operator" (Russian), *Izv. Akad. Nauk Armjan SSR Ser. Math.*, *3*, #6 (1968), 443–464.
70. Ikebe, T., "On the phase-shift formula for the scattering operator," *Pacific Journal of Mathematics*, *15* (1965), 511–523.
71. McLeod, J. B., "The analytic continuation to the unphysical sheet of the resolvent kernel associated with the Schroedinger operator," *Quart. J. Math. Oxford*, *2*, 18 (1967), 219–231.
72. Newton, R. G., *Scattering Theory of Waves and Particles*. New York: McGraw-Hill Book Co., Inc., 1966.
73. Zemach, C., and A. Klein, "The Born expansion in nonrelativistic quantum theory," *Nuovo Cimento X*, *10* (1958), 1078.
74. Khuri, N., "Analyticity of the Schroedinger scattering amplitude and non-relativistic dispersion relations," *Phys. Rev.*, *107* (1957), 1148.
75. Riesz, F., and B. Sz-Nagy, *Functional Analysis*. New York: Frederick Ungar, 1955. See pp. 418–422.
76. Dolph, C. L., J. McLaughlin, and I. Marx, "Symmetric linear transformations and complex quadratic forms," *Comm. Pure Appl. Math.*, *7* (1954), 621.
77. Dolph, C. L., and R. K. Ritt, "The Schwinger variational principle for one-dimensional scattering," *Math. Z.*, *65* (1956), 309–326.
78. Dolph, C. L., "A saddle point characterization of the Schwinger stationary points in exterior scattering problems," *J. Soc. Ind. Appl. Math.*, *5* (1957), 89.

78a. Dolph. C. L., "The mathematician grapples with linear problems associated with the radiation condition," *I.R.E. Trans. on Antennas and Propagation*, AP-4, #3, 1956, p. 302.

79. LaVita, J. A., "Perturbation of the poles of the scattering matrix," Math. Dept., Univ. of Denver, MS-R 6903 (a Courant-Institute dissertation).

80. Garabedian, P., "An integral equation governing electromagnetic waves," *Quart. Appl. Math.*, *16* (1955), 428.

81. McLeod, J. B., "Spectral concentration, I—the one-dimensional Schroedinger operator," in *Perturbation Theory and Its Applications in Quantum Mechanics*, ed. C. H. Wilcox. New York: John Wiley & Sons, 1966, pp. 119–127.

82. Howland, J. S., "Perturbation of embedded eigenvalues by operators of finite rank." To appear in *J. Math. Anal. and Appl.*

83. ———, "Analytic continuation of the *S*-matrix for potential scattering." To appear in *Pac. J. Math.*

84. ———, "Embedded eigenvalues and virtual poles." To appear in *J. Math. Anal. and Appl.*

85. Kato, T., *Perturbation Theory of Linear Operators*. New York: Springer-Verlag, 1966.

86. Kupradze, V., *Potential Methods in the Theory of Elasticity*. Jerusalem: Israel Program for Scientific Translations, 1965, 339 pages.

87. Kramer, H. P., R. N. Lane, and R. R. Rutherford, "Scattering from smooth elastic shells in water," *G.E. Tempo 69*, YMP-17, Santa Barbara, Calif., February 1969.

88. Vainberg, B. R., "Principles of radiation, limit absorption, and limit amplitude in the general theory of partial differential equations," *Russian Math. Surveys*, #3, *21* (1966), 115–193.

Group Algebra Bundles

BERNARD R. GELBAUM[1]

1. Introduction

Let G be a topological space and let A be a Banach space. On $\mathscr{S}$, the σ-ring generated by the compact sets of G, let μ be a measure (such that $\mu(K) < \infty$ for all compact K). Let $\mathscr{E}$ be a Banach algebra bundle (an equivalence class of Banach algebra coordinate bundles) with total space E, base space G, fibre A, projection $p: E \to G$. Let one coordinate bundle $\tilde{\mathscr{B}}$ in $\mathscr{E}$ be specified by an open covering $\tilde{\mathscr{U}}$ of G and for each $U \in \tilde{\mathscr{U}}$ a coordinate map (fibre-preserving homeomorphism) $\varphi_U: U \times A \to p^{-1}(U)$. Furthermore, let the group of the bundle be $\mathscr{A}$, the set of isometric $\mathbf{C}$-automorphisms of A. We assume that $\mathscr{A}$ is appropriately topologized and that the coordinate transformations $g_{UV}: U \cap V \to \mathscr{A}$ are continuous (see [2], [3]). For $y \in p^{-1}(U)$ we write $\varphi_U^{-1}(y) = (p(y), t_U(y))$.

Let $\Gamma_0(\mathscr{E})$ be the set of continuous sections having compact support. On the one hand we shall construct the space $L^1(G, \mathscr{E})$ of Bochner-integrable functions generated by $\Gamma_0(\mathscr{E})$. On the other hand, by imposing mild restrictions on μ we shall show how to define an integral $I: L^1(G, \mathscr{E}) \to A$. The integral I will depend, *a priori*, on the choice of $\tilde{\mathscr{B}}$. If G is a locally compact Abelian group and if μ is Haar measure, then I will serve to define (via convolution) a group algebra bundle. Any two such, corresponding to different choices of $\tilde{\mathscr{B}}$, will be shown to be isomorphic, and indeed isomorphic to $L^1(G, A)$, the generalized group algebra of A-valued functions on G.

It thus appears that the structure of the group algebra bundle $L^1(G, \mathscr{E})$, when G is a locally compact group, is insensitive to the bundle structure. This is in sharp contrast to the phenomena associated with the Banach algebra bundle $\Gamma(\mathscr{E})$ of continuous sections of a fibre bundle over a compact Hausdorff space (see [2]). In both cases the algebra studied is an algebra of sections. The *continuous* sections with pointwise multiplication

[1] Some of this work was supported by National Science Foundation Grant GP 5436, for which the author is grateful.

give rise to a Banach algebra that reflects the underlying bundle structure. The "integrable" sections with pointwise addition and "convolution" as multiplication give rise to a Banach algebra corresponding to a trivial bundle.

We shall assume that the topology of G permits the following constructions. In particular we shall assume that each finite collection $\{U_i\}_{i=1}^n$ of open sets permits a "partition of unity" $\{\psi_i\}_{i=1}^n$ subordinate to the collection: $\psi_i(x) \geqq 0$, $\sum_{i=1}^n \psi_i(x) = 1$ if $x \in \bigcup_{i=1}^n U_i$, $\psi_i(x) = 0$ off U_i, each $\psi_i(x)$ is continuous and has compact support, i.e., $\psi_i \in C_0(G) \equiv$ the set of continuous functions having compact support.

2. Integration

Clearly $\Gamma_0(\mathscr{E})$ is a $C_0(G)$-module (see [2]). For the coordinate bundle $\tilde{\mathscr{B}}$ defined by $\tilde{\mathscr{U}}$ we can define *approximations* to I as follows. Let $\gamma \in \Gamma_0(\mathscr{E})$, and let $K = \text{supp}(\gamma) \equiv$ support of γ. Choose $\{U_i\}_{i=1}^n \subset \tilde{\mathscr{U}}$ such that $\bigcup_{i=1}^n U_i \supset K$. Let $\{\psi_i\}_{i=1}^n$ be a partition of unity subordinate to $\{U_i\}_{i=1}^n$. Then let an approximation in question be

$$I(\gamma; \{U_i\}_{i=1}^n, \{\psi_i\}_{i=1}^n) = \int_G \sum_{i=1}^n t_{U_i}(\psi_i(x)\gamma(x))\, dx$$

$$= \int_G \sum_{i=1}^n \psi_i(x) t_{U_i}(\gamma(x))\, dx,$$

where $\int \cdots dx$ represents Bochner integration with respect to μ.

If $\{V_j\}_{j=1}^m \subset \tilde{\mathscr{U}}$ is a second covering of K and if $\{\xi_j\}_{j=1}^m$ is a corresponding partition of unity subordinate to $\{V_j\}_{j=1}^m$, then we shall compare the preceding approximation with $I(\gamma; \{V_j\}_{j=1}^m, \{\xi_j\}_{j=1}^m)$. To do this we write

$$\sum_{i=1}^n t_{U_i}(\psi_i(x)\gamma(x)) = \sum_{j=1}^m \sum_{i=1}^n \xi_j(x)\psi_i(x) t_{U_i}(\gamma(x))$$

$$= \sum_{j=1}^m \sum_{i=1}^n \xi_j(x)\psi_i(x) g_{U_i V_j}(x) t_{V_j}(\gamma(x)).$$

Let $\tilde{K} \subset K$ be a (possibly empty) compact set such that for $U', U'' \in \tilde{\mathscr{U}}$ and $x \in U' \cap U'' \cap \tilde{K}$, $g_{U'U''}(x) = e =$ identity of the group $\mathscr{A}$.

Then, since $g_{U_iV_j}(x) = e$ on $\tilde{K}$,

$$\begin{aligned}\int_G \sum_{i=1}^{n} t_{U_i}(\psi_i(x)\gamma(x))\,dx &= \int_{\tilde{K}} \sum_{j=1}^{m} \sum_{i=1}^{n} \xi_j(x)\psi_i(x)t_{V_j}(\gamma(x))\,dx \\ &\quad + \int_{K\setminus\tilde{K}} \sum_{j=1}^{m} \sum_{i=1}^{n} \xi_j(x)\psi_i(x)g_{U_iV_j}(x)t_{V_j}(\gamma(x))\,dx \\ &= \int_{\tilde{K}} \sum_{j=1}^{m} t_{V_j}(\xi_j(x)\gamma(x))\,dx \\ &\quad + \int_{K\setminus\tilde{K}} \sum_{j=1}^{m} \sum_{i=1}^{n} \xi_j(x)\psi_i(x)g_{U_iV_j}(x)t_{V_j}(\gamma(x))\,dx.\end{aligned}$$

However, since $\xi_j(x)$, $\psi_i(x) \geqq 0$ for all i, j, and since each $g_{U_iV_j}$ is an isometry,

$$\left\|\sum_{j=1}^{m} \sum_{i=1}^{n} \xi_j(x)\psi_i(x)g_{U_iV_j}(x)t_{V_j}(\gamma(x))\right\| \leqq \sum_{j=1}^{m} \sum_{i=1}^{n} \xi_j(x)\psi_i(x)\|\gamma\| = \|\gamma\|,$$

where $\|\gamma\|$ is the norm of γ in $\Gamma_0(\mathscr{E})$ (see [3]): $\|\gamma\| = \sup_x |\gamma(x)|$, where $|\gamma(x)| = \|t_U(\gamma(x))\|_A$, $x \in U \in \tilde{\mathscr{U}}$. We now write, since $\operatorname{supp}\gamma = K$,

$$\begin{aligned}&\int_{\tilde{K}} \sum_{j=1}^{m} t_{V_j}(\xi_j(x)\gamma(x))\,dx + \int_{K\setminus\tilde{K}} \sum_{j=1}^{m} \sum_{i=1}^{n} \xi_j(x)\psi_i(x)g_{U_iV_j}(x)t_{V_j}(\gamma(x))\,dx \\ &\quad = \int_G \sum_{j=1}^{m} \xi_j(x)t_{V_j}(\gamma(x))\,dx - \int_{K\setminus\tilde{K}} \sum_{j=1}^{m} \xi_j(x)t_{V_j}(\gamma(x))\,dx \\ &\qquad + \int_{K\setminus\tilde{K}} \sum_{j=1}^{m} \sum_{i=1}^{n} \xi_j(x)\psi_i(x)g_{U_iV_j}(x)t_{V_j}(\gamma(x))\,dx.\end{aligned}$$

Because of the obvious inequality: $\left\|\sum_{j=1}^{m} \xi_j(x)t_{V_j}(\gamma(x))\right\| \leqq \|\gamma\|$, we find

$$\|I(\gamma; \{U_i\}_{i=1}^n, \{\psi_i\}_{i=1}^n) - I(\gamma; \{V_j\}_{j=1}^m, \{\xi_j\}_{j=1}^m)\|_A < 2\mu(K\setminus\tilde{K})\|\gamma\|.$$

Consider the set of all triples $\tau = (\mathscr{B}, \{U_i\}_{i=1}^n, \{\psi_i\}_{i=1}^n)$ where

(i) $\mathscr{B}$ is a coordinate bundle in the equivalence class of $\mathscr{E}$; $\mathscr{B}$ is engendered by a refinement $\mathscr{U}$ of $\tilde{\mathscr{U}}$; the coordinate maps of $\mathscr{B}$ are found by restricting those of $\tilde{\mathscr{B}}$.

(ii) $\{U_i\}_{i=1}^n \subset \mathscr{U}$ is a covering of K.

(iii) $\{\psi_i\}_{i=1}^n$ is a partition of unity subordinate to $\{U_i\}_{i=1}^n$.

We partially order $\{\tau\}$ by the inclusion order among the associated $\{\mathscr{U}\}$.

We now assume that the measure μ on G is regular and that for all N in some neighborhood base $\mathscr{N} = \{N\}$ for G, $\mu(\partial N) = 0$, where $\partial N = \bar{N} \cap (G\setminus N)$ = boundary of N. If G is a locally compact group, these conditions obtain (see [3], [6]). According to the main result in [3], for $\varepsilon > 0$ there is a refinement $\mathscr{U}_0$ of $\tilde{\mathscr{U}}$, a coordinate bundle $\mathscr{B}^0$ in the equivalence class of $\mathscr{E}$ such that $\mathscr{U}_0$ is the associated covering and such that the associated coordinate maps are derived from those of $\tilde{\mathscr{B}}$ by restriction and there is a compact $\tilde{K} \subset K$ such that

(a) $\mu(\tilde{K}) > \mu(K) - \varepsilon/2\|\gamma\|$,
(b) for $U_0', U_0'' \in \mathscr{U}_0$, $x \in U_0' \cap U_0'' \cap \tilde{K}$ there obtains $g_{U_0'U_0''}(x) = e$.

Furthermore, if $\mathscr{B}$ is any coordinate bundle whose associated covering $\mathscr{U}$ is a refinement of $\mathscr{U}_0$ and whose associated coordinate maps are derived from those of $\mathscr{B}_0$ by restriction, then (b) holds with "$U_0', U_0'' \in \mathscr{U}_0$," etc., replaced by "$U', U'' \in \mathscr{U}$," etc. In consequence, for any such $\mathscr{B}$, $\mathscr{U}$ we find

$$\|I(\gamma; \{U_i\}_{i=1}^n, \{\psi_i\}_{i=1}^n) - I(\gamma; \{V_j\}_{j=1}^m, \{\xi_j\}_{j=1}^m\|_A < \varepsilon.$$

We shall write $I_\tau(\gamma)$ for $I(\gamma; \{U_i\}_{i=1}^n\{\psi_i\}_{i=1}^n)$. These remarks are the basis for the following theorem:

THEOREM 1. *Let $\gamma \in \Gamma_0(\mathscr{E})$ and let $\mathscr{B}$ be a coordinate bundle (in the equivalence class of $\mathscr{E}$) with associated covering $\mathscr{U}$. Then for $\varepsilon > 0$ there is a $\tau(\varepsilon)$ such that if $\tau_1, \tau_2 \geqq \tau(\varepsilon)$ then*

$$\|I_{\tau_1}(\gamma) - I_{\tau_2}(\gamma)\|_A < \varepsilon.$$

PROOF. Let $\tau(\varepsilon) = \tau_0 \equiv (\mathscr{B}^0, \{U_{i0}\}_{i=1}^n, \{\psi_{i0}\}_{i=1}^n\}$ where $\mathscr{B}^0$ is chosen as in the argument above, $U_{i0} \in \mathscr{U}_0$, etc.

If $\tau_1, \tau_2 \geqq \tau(\varepsilon)$, let τ_3 arise from any common refinement $\mathscr{U}_3$ of the coverings $\mathscr{U}_1$ and $\mathscr{U}_2$ associated to τ_1 and τ_2. In other words, $\tau_3 = (\mathscr{B}_3, \{W_k\}_{k=1}^q, \{\eta_k\}_{k=1}^m)$, where, in particular, the open covering associated to $\mathscr{B}_3$ is some common refinement $\mathscr{U}_3$ of $\mathscr{U}_1$ and $\mathscr{U}_2$. We let $\tau_1 = (\mathscr{B}_1, \{U_i\}_{i=1}^n, \{\psi_i\}_{i=1}^n)$, $\tau_2 = (\mathscr{B}_2, \{V_j\}_{j=1}^m, \{\xi_j\}_{j=1}^m)$ and we calculate

$$\begin{aligned} I_{\tau_1}(\gamma) - I_{\tau_3}(\gamma) &= \int_G \sum_{i=1}^n \psi_i(x) t_{U_i}(\gamma(x))\, dx - \int_G \sum_{k=1}^q \eta_k(x) t_{W_k}(\gamma(x))\, dx \\ &= \int_G \sum_{i=1}^n \sum_{k=1}^q \psi_i(x)\eta_k(x) t_{U_i}(\gamma(x))\, dx - \int_G \sum_{k=1}^q \eta_k(x) t_{W_k}(\gamma(x))\, dx. \end{aligned}$$

We show now that $g_{U_iW_k}(x) = e$ on $U_i \cap W_k \cap \tilde{K}$. Indeed we recall that implicit in the definition of each τ is the fact that the associated coordinate maps are restrictions of those given for $\tilde{\mathscr{B}}$. This means that for $\tau_0, \tau_1, \tau_2, \tau_3$ there are maps $\sigma_r\colon \mathscr{U}_r \to \tilde{\mathscr{U}}$ such that $\sigma_r(U) \supset U$ and such that by definition $\varphi_U^{-1} = \varphi_{\sigma_r(U)}^{-1}$ on $p^{-1}(U)$, $r = 0, 1, 2, 3$. Let us consider t_{U_i} and t_{W_k}. By definition

$$t_{U_i} = t_{\sigma_1(U_i)}, \qquad t_{W_k} = t_{\sigma_3(W_k)}.$$

If $\sigma_1(U_i) = \sigma_3(W_k)$, then $\psi_i(x)\eta_k(x)t_{U_i}(\gamma(x)) = \psi_i(x)\eta_k(x)t_{W_k}(\gamma(x))$. If $\sigma_1(U_i) \cap \sigma_3(W_k) = \phi$ then $\psi_i(x)\eta_k(x) = 0$ and $\psi_i(x)\eta_k(x)t_{U_i}(\gamma(x)) = \psi_i(x)\eta_k(x)t_{W_k}(\gamma(x))$. If $x \in \sigma_1(U_i) \cap \sigma_3(W_k) \cap \tilde{K}$, where $\tilde{K} \subset K$ and $g_{U_{i_0}U_{j_0}}(x) = e$ on $U_{i_0} \cap U_{j_0} \cap \tilde{K}$, then $t_{U_i} = t_{\sigma_1(U_i)} = g_{\sigma_1(U_i)\sigma_3(W_k)}t_{\sigma_3(W_k)} = g_{\sigma_1(U_i)\sigma_3(W_k)}t_{W_k}$. However, $g_{\sigma_1(U_i)\sigma_3(W_k)}$ is generated by $\varphi^{-1}_{\sigma_1(U_i)}\varphi_{\sigma_3(W_k)}$ and, when restricted to $\tilde{K}$, the latter map is the identity. Thus on $\tilde{K}$, $t_{U_i}(\gamma(x)) = t_{W_k}(\gamma(x))$ and again $\psi_i(x)\eta_k(x)t_{U_i}(\gamma(x)) = \psi_i(x)\eta_k(x)t_{W_k}(\gamma(x))$.

In conclusion, we may write $I_{\tau_1}(\gamma) - I_{\tau_3}(\gamma)$ as

$$\int_G = \int_{\tilde{K}} + \int_{G\backslash\tilde{K}} = \int_{\tilde{K}} + \int_{K\backslash\tilde{K}}.$$

For $x \in \tilde{K}$,

$$\sum_{i=1}^{n}\sum_{k=1}^{q} \psi_i(x)\eta_k(x)t_{U_i}(\gamma(x)) = \sum_{i=1}^{n}\sum_{k=1}^{q} \psi_i(x)\eta_k(x)t_{W_k}(\gamma(x))$$
$$= \sum_{k=1}^{q} \eta_k(x)t_{W_k}(\gamma(x)).$$

Thus $\int_{\tilde{K}} = 0$. Since $\mu(K\backslash\tilde{K}) < \varepsilon/2\|\gamma\|$ we see that

$$\|I_{\tau_1}(\gamma) - I_{\tau_3}(\gamma)\|_A < \frac{\varepsilon}{2}.$$

Using a similar argument for I_{τ_2} and I_{τ_3} and then combining inequalities we conclude that

$$\|I_{\tau_1}(\gamma) - I_{\tau_2}(\gamma)\|_A < \varepsilon.$$

Since A is complete the Cauchy net $\{I_\tau(\gamma)\}$ has a limit, which we denote by $I(\gamma)$.

Before proceeding to discuss the properties of $I(\gamma)$ we note that the value of $I(\gamma)$ is, *a priori*, dependent upon $\tilde{\mathscr{B}}$. We shall show that despite this situation, the space $L^1(G, \mathscr{E})$ based on the "integral" I is isometrically isomorphic to $L^1(G, A)$. Furthermore, if G is a locally compact group then $L^1(G, \mathscr{E})$ as an algebra is isometrically isomorphic to $L^1(G, A)$. The informal conclusion we reach is that the bundle structure of $\mathscr{E}$ is irrelevant to the structure of $L^1(G, \mathscr{E})$.

We now prove that

(i) $I(\alpha_1\gamma_1 + \alpha_2\gamma_2) = \alpha_1 I(\gamma_1) + \alpha_2 I(\gamma_2)$, $\alpha_1, \alpha_2 \in \mathbf{C}$.
(ii) $\|I(\gamma)\| \leqq \|\gamma\|\mu(\operatorname{supp}\gamma)$.

(i) Let $K_i = (\operatorname{supp}\gamma_i)$; $i = 1, 2$, $\gamma_3 = \alpha_1\gamma_1 + \alpha_2\gamma_2$, then $\operatorname{supp}\gamma_3 \equiv K_3 \subset K_1 \cup K_2$. For $\tau = (\mathscr{B}, \{U_i\}_{i=1}^n, \{\psi_i\}_{i=1}^n)$ the expressions $I_\tau(\gamma_3)$, $I_\tau(\gamma_1)$, and $I_\tau(\gamma_2)$ are meaningful, if $\bigcup_{i=1}^{n} U_i \supset K_3$. For such τ it is clear that

$$I_\tau(\alpha_1\gamma_1 + \alpha_2\gamma_2) = \alpha_1 I_\tau(\gamma_1) + \alpha_2 I_\tau(\gamma_2).$$

On the other hand if τ_i are associated to γ_i, let the underlying coverings be $\mathscr{U}_i$, $i = 1, 2$. We can then construct in an obvious fashion a τ whose underlying covering is a refinement of $\mathscr{U}_1$ and $\mathscr{U}_2$ and whose $\{U_i\}_{i=1}^n$ covers $K_1 \cup K_2$. For such a τ the equation just given holds. The existence of $I(\gamma)$ for all $\gamma \in \Gamma_0(\mathscr{E})$ implies that ultimately $I_\tau(\gamma_i)$ and $I_{\tau_i}(\gamma_i)$ are close. Thus for τ associated to K we find

$$\lim_\tau I_\tau(\alpha_1\gamma_1 + \alpha_2\gamma_2) = I(\alpha_1\gamma_1 + \alpha_2\gamma_2).$$

For τ arising from τ_1 and τ_2 as described, we note that they are cofinal in the set of τ associated to K and that

$$I_\tau(\alpha_1\gamma_1 + \alpha_2\gamma_2) = \alpha_1 I_\tau(\gamma_1) + \alpha_2 I_\tau(\gamma_2) = \alpha_1 I_{\tau_1}(\gamma_1) + \alpha_2 I_{\tau_2}(\gamma_2).$$

Passage to the limit now yields the required linearity of I.

(ii) For all τ, $\|I_\tau(\gamma)\| \leqq \|\gamma\|\mu(\operatorname{supp}\gamma)$. Hence the conclusion follows.

Since $\mathscr{A}$ consists of isometries, for $\gamma \in \Gamma_0(\mathscr{E})$ and $x \in G$, $|\gamma(x)| \equiv \|t_U(\gamma(x))\|_A$ is independent of U. Furthermore $|\gamma(x)|$ is continuous on G, $\operatorname{supp}|\gamma(x)| = \operatorname{supp}\gamma(x)$.

We define

$$|\gamma| = \int_G |\gamma(x)|\,dx,$$

and thereby define a norm on $\Gamma_0(\mathscr{E})$. The completion of $\Gamma_0(\mathscr{E})$ in this norm will be denoted by $L^1(G, \mathscr{E})$. We observe that $L^1(G, \mathscr{E})$ is a Banach space insensitive to the bundle structure of $\mathscr{E}$.

For $\gamma \in \Gamma_0(\mathscr{E})$ we show

$$\|I(\gamma)\|_A \leqq |\gamma|.$$

Indeed $\|I_\tau(\gamma)\|_A \leqq |\gamma|$ by virtue of the definition of I_τ. The result follows since $I(\gamma) = \lim_\tau I_\tau(\gamma)$.

2. $L^1(G, \mathscr{E})$ amd $L^1(G, A)$

By $L^1(G, A)$ we mean the set of Bochner-integrable A-valued functions on G. We shall establish an isometric isomorphism between $L^1(G, \mathscr{E})$ and $L^1(G, A)$. For this purpose we denote by $C_0(A)$ the set of continuous compactly supported A-valued functions on G.

Let $\tilde{\mathscr{B}}$ be an arbitrary coordinate bundle in the equivalence class of $\mathscr{E}$. Let $\tilde{\mathscr{U}}$ be the underlying open covering of G, $\{\varphi_{\tilde{U}}\}$ the set of coordinate maps, and $\{g_{\tilde{U}'\tilde{U}''}\}$ the set of coordinate transformations (transition functions) defined by $\{\varphi_{\tilde{U}}\}$. For $\gamma \in \Gamma_0(\mathscr{E})$, let K be the support of γ and let

$\{\tilde{U}_i\}_{i=1}^n \subset \tilde{\mathscr{U}}$ be a covering of K with an associated subordinate partition of unity $\{\psi_i\}_{i=1}^n$. As we did earlier, we denote by τ a triple $(\mathscr{B}, \{U_i\}_{i=1}^n, \{\psi_i\}_{i=1}^n)$ consisting of a coordinate bundle $\mathscr{B}$ in the equivalence class of $\mathscr{E}$, an open covering $\{U_i\}_{i=1}^n$ of K_γ where $\{\tilde{U}_i\}_{i=1}^n \subset \tilde{\mathscr{U}}$, $\mathscr{U}$ is a refinement of $\tilde{\mathscr{U}}$ and $\{\varphi_U\}$ for $\mathscr{B}$ arise from $\{\varphi_{\tilde{U}}\}$ for $\tilde{\mathscr{B}}$ by restriction. For each τ let

$$T_\tau(\gamma)(x) = \sum_{i=1}^n \psi_i(x) t_{U_i}(\gamma(x)) \in C_0(A).$$

Clearly T_τ is linear.

For $\varepsilon > 0$ there is a $\tilde{K} \subset K$ such that $\mu(\tilde{K}\backslash K) < \varepsilon/2\|\gamma\|$ and there is a τ_0 such that for $\tau \geqq \tau_0$

$$T_\tau(\gamma)(x) = T_{\tau_0}(\gamma)(x_0), \qquad x \in \tilde{K}.$$

This follows from the argument used in the establishment of I. Thus, we see that for $\tau_1, \tau_2 \geqq \tau_0$

$$\int_G \|T_{\tau_1}(\gamma) - T_{\tau_2}(\gamma)\|_A \, dx = \int_{\tilde{K}} + \int_{K\backslash\tilde{K}} = \int_{K\backslash\tilde{K}} < \varepsilon.$$

Thus $\{T_\tau(\gamma)\}$ is a Cauchy net in $L^1(G, A)$ and we denote its limit by $T(\gamma)$. From the construction of $T(\gamma)$ we see that $T(\gamma)(x) = 0$ if $x \notin K$ and for each $\varepsilon > 0$ and corresponding $\tilde{K}$, τ_0, $T_\tau(\gamma)(x) = T(\gamma)(x)$, $x \in \tilde{K}$. If $\varepsilon_n \to 0$ we may choose corresponding $\tilde{K}_n$, τ_n so that $\tilde{K}_n \subset \tilde{K}_{n+1}$, and $\mu(K\backslash\tilde{K}_n) < \varepsilon_n$. Thus

$$T(\gamma)(x) = T_{\tau_n}(\gamma)(x) = T_{\tau_{n+k}}(\gamma)(x), \qquad k \in \mathbf{N},\ x \in \tilde{K}_n.$$

Hence $T(\gamma)(x)$ is continuous on $\bigcup_{n=1}^{\infty} \tilde{K}_n$ and thus a.e. on K. Since all T_τ are linear, so is T.

Next we note that

$$\int_G \|T(\gamma)(x)\|_A \, dx = \int_G |\gamma(x)| \, dx = |\gamma|.$$

Hence T is an isometry on $\Gamma_0(\mathscr{E})$ (in particular T is one-to-one). Furthermore, if $H(x) \in C_0(A)$, we may write

$$H(x) = \sum_{i=1}^n \psi_i(x) H(x)$$

and then find $\gamma_i(x)$ (see [2]) such that $t_{U_i}(\gamma_i(x)) = \psi_i(x)H(x)$. Thereupon, if $\gamma(x) = \sum_{i=1}^n \gamma_i(x)$ then

$$T_\tau(\gamma)(x) = \sum_{i=1}^n T_\tau(\gamma_i)(x) = \sum_{i=1}^n \left(\sum_{j=1}^n \psi_j t_{U_j}(\gamma_i(x))\right) = \sum_{i=1}^n \sum_{j=1}^n \psi_j g_{U_j}(x_{U_j}) t_{U_j}(\gamma_i(x)).$$

If $\operatorname{supp} H = K$, then $\operatorname{supp} \gamma \subset K$ and by arguments used earlier, for $\varepsilon > 0$, there is a compact $\tilde{K} \subset K$, $\mu(K \backslash \tilde{K}) < \varepsilon$ so that for all $\tau \geqq \tau_0$ and $x \in \tilde{K}$

$$T_\tau(\gamma)(x) = \sum_{i=1}^{n} \sum_{j=1}^{n} \psi_j t_{U_i}(\gamma_i(x)) = \sum_{i=1}^{n} t_{U_i}(\gamma_i(x)) = H(x).$$

Hence $T(\gamma)(x) = H(x)$ a.e. and thus, viewed as a subspace of $L^1(G, A)$, $C_0(A)$ is the complete image by T of $\Gamma_0(\mathscr{E})$, viewed as a subspace of $L^1(G, \mathscr{E})$. Thus we have

$$L^1(G, \mathscr{E}) \supset \Gamma_0(\mathscr{E}) \xrightarrow{T} C_0(A) \subset L^1(G, A),$$

where T is an isometric isomorphism between dense sets. Clearly T is extendable to $L^1(G, \mathscr{E})$ and thus this extension, again denoted by T, implements an isometric isomorphism between $L^1(G, \mathscr{E})$ and $L^1(G, A)$.

The map $I: \Gamma_0(\mathscr{E}) \to A$ may be extended uniquely to a map denoted again by $I: L^1(G, \mathscr{E}) \to A$. In terms of this map I we may define a multiplication in $L^1(G, \mathscr{E})$ when G is a locally compact group and μ is Haar measure. The natural definition extends classical convolution.

We begin by fixing a coordinate bundle $\tilde{\mathscr{B}}$ and all the associated apparatus used in the definitions of T_τ and I_τ, T and I. Since

$$T_\tau(\gamma)(x) = \sum_{i=1}^{n} \psi_i(x) t_{U_i}(\gamma(x)),$$

we see that

$$I_\tau(\gamma) = \int_G T_\tau(\gamma)(x)\, dx,$$

and, in consequence of earlier inequalities, we conclude that

$$I(\gamma) = \int_G T(\gamma)(x)\, dx.$$

We show that, if $\gamma, \delta \in \Gamma_0(\mathscr{E})$, then $T(\gamma\delta) = T(\gamma)T(\delta)$. Indeed, if $\tau = (\mathscr{B}, \{U_i\}_{i=1}^n, \{\psi_i\}_{i=1}^n)$ is given we observe that $\{\psi_i\psi_j\}_{i,j=1}^n$ is also a partition of unity subordinate to $\{U_i\}_{i=1}^n$. We let $\tau' = (\mathscr{B}, \{U_i\}_{i=1}^n, \{\psi_i\psi_j\}_{i,j=1}^n)$. The set of all τ' is cofinal in the set of all τ. Then

$$\begin{aligned} T_{\tau'}(\gamma\delta) &= \sum_{i=1}^{n} \sum_{j=1}^{n} \psi_i(x)\psi_j(x) t_{U_i}(\gamma(x)) t_{U_j}(\delta(x)) \\ &= \sum_{i=1}^{n} \sum_{j=1}^{n} \psi_i(x)\psi_j(x) g_{U_i U_j}(x) t_{U_j}(\gamma(x)) t_{U_j}(\delta(x)) \\ &= \sum_{i=1}^{n} \psi_i(x) g_{U_i U_j}(x) t_{U_j}(\gamma(x)) \sum_{j=1}^{n} \psi_j(x) t_{U_j}(\delta(x)). \end{aligned}$$

As in earlier arguments, for suitable $\tilde{K}$ in the union of the supports of γ and δ and for suitable τ we find

$$T_{\tau'}(\gamma\delta)(x) = T_\tau(\gamma)(x)T_\tau(\delta)(x) \quad \text{for } x \in \tilde{K}.$$

Using the cofinality of $\{\tau'\}$ in $\{\tau\}$ we conclude that

$$T(\gamma\delta)(x) = T(\gamma)(x)T(\delta)(x) \quad \text{a.e. in } x.$$

Thus we see that if for any $\eta \in \Gamma_0(\mathscr{E})$ we write $\eta_x(y) \equiv \eta(y^{-1}x)$, then

$$\begin{aligned} I(\gamma\delta_x) &= \int_G T(\gamma\delta_x)(y)\,dy \\ &= \int_G T(\gamma)(y)T(\delta_x)(y)\,dy \\ &= \int_G T(\gamma)(y)T(\delta)(y^{-1}x)\,dy \\ &= T(\gamma)*T(\delta)(x) \in C_0(A). \end{aligned}$$

Thus we *define*

$$(\gamma*\delta)(x) = T^{-1}[I(\gamma\delta_x)]$$

and conclude that

$$T(\gamma*\delta)(x) = T(\gamma)*T(\delta)(x).$$

It is now clear that T is an *algebraic* isometric isomorphism between $L^1(G, \mathscr{E})$ and $L^1(G, A)$.

UNIVERSITY OF CALIFORNIA
IRVINE

REFERENCES

1. GELBAUM, B. R., "Tensor products and related questions," *Trans. Amer. Math. Soc.*, *103* (1962), 525–548.
2. ———, "Banach algebra bundles," *Pacific Journal of Mathematics*, *28* (1969), 337–349.
3. ———, "Fibre bundles and measure," *Proceedings of the American Mathematical Society*, *21* (1969), 603–607.
4. ———, "*Q*-uniform Banach algebras" *Proc. Amer. Math. Soc.*, *24* (1970), 344–353.
5. GROTHENDIECK, A., "Produits tensoriels topologiques et espaces nucléaires," *Mem. Amer. Math. Soc.*, No. 16 (1955).
6. HERZ, C. S., "The spectral theory of bounded functions," *Trans. Amer. Math. Soc.*, *94* (1960), 181–232.
7. JOHNSON, G. P., "Spaces of functions with values in a Banach algebra," *Trans. Amer. Math. Soc.*, *92* (1959), 411–429.
8. STEENROD, N., *The Topology of Fibre Bundles*. Princeton, N.J.: Princeton University Press, 1960.

Quadratic Periods of Hyperelliptic Abelian Integrals

R. C. GUNNING[1]

1.

Consider a compact Riemann surface M of genus $g > 0$, represented in the familiar manner as the quotient space of its universal covering surface $\tilde{M}$ by the covering translation group Γ. For the present purposes the only thing one needs to know about the surface $\tilde{M}$ is that it is a simply connected noncompact Riemann surface. The group Γ is properly discontinuous and has no fixed points; and the quotient space $\tilde{M}/\Gamma$ is analytically equivalent to the Riemann surface M. The image of a point $z \in \tilde{M}$ under an automorphism $T \in \Gamma$ will be denoted by Tz. Some fixed but arbitrary point $z_0 \in \tilde{M}$ will be selected as the base point of the covering space $\tilde{M}$.

The Abelian differentials on the surface M can be viewed as the Γ-invariant complex analytic differential forms of type $(1, 0)$ on the surface $\tilde{M}$. Since $\tilde{M}$ is simply connected, any such differential form θ can be written as $\theta = dh$, for some complex analytic function h on the surface $\tilde{M}$; the function h is then uniquely determined by the normalization condition $h(z_0) = 0$. Since θ is Γ-invariant, it is apparent that $h(Tz) = h(z) - \theta(T)$ for some complex constant $\theta(T)$, for any element $T \in \Gamma$; and clearly $\theta(ST) = \theta(S) + \theta(T)$ for any two elements $S, T \in \Gamma$. The mapping $T \to \theta(T)$ is thus a homomorphism from the group Γ into the additive group **C** of complex numbers; this homomorphism will be called the period class of the Abelian differential θ. As is well known, the period classes of the Abelian differentials on the Riemann surface M form a g-dimensional subspace of the $2g$-dimensional complex vector space Hom $(\Gamma, \mathbf{C})$ consisting of all group homomorphisms from Γ into **C**.

Now suppose that $\theta_1 = dh_1$ and $\theta_2 = dh_2$ are two Abelian differentials on M, where the functions h_1, h_2 on $\tilde{M}$ are both normalized by requiring

[1] This work was partially supported by NSF Grant 3453 GP 6962.

that $h_1(z_0) = h_2(z_0) = 0$. The product $\sigma = h_1\theta_2 = h_1\, dh_2$ is an analytic differential form of type $(1, 0)$ on the surface $\tilde{M}$ and clearly satisfies the functional equation

$$\sigma(Tz) = \sigma(z) - \theta_1(T)\theta_2(z) \tag{1}$$

for any element $T \in \Gamma$. Again, since M is simply connected, this differential form can be written as $\sigma = ds$ for some complex analytic function on the surface $\tilde{M}$; the function s is uniquely determined by the normalization condition $s(z_0) = 0$. It follows from (1) that the function s satisfies the functional equation

$$s(Tz) = s(z) - \theta_1(T)h_2(z) - \sigma(T) \tag{2}$$

for some complex constant $\sigma(T)$, for any element $T \in \Gamma$. The mapping $T \to \sigma(T)$ will be called the *period class of the quadratic expression* $\sigma = h_1\, dh_2$ in the Abelian integrals on the surface M; the set of all such mappings will be called, for short, the *quadratic period classes* of the Abelian differentials on M. The aim of this paper is the explicit determination of these quadratic period classes for the special case of a hyperelliptic Riemann surface; in this case, the quadratic periods are quadratic expressions involving the ordinary period classes of the Abelian differentials on M. In general, the quadratic period classes may be transcendental functions of the Abelian period classes. Section 2 of this paper is devoted to a general discussion of the properties of these quadratic period classes, for an arbitrary Riemann surface; Section 3 contains the explicit calculations for a hyperelliptic surface; and Section 4 provides the motivation for studying these quadratic period classes, with a discussion of their role in the general theory of Riemann surfaces.

2.

A quadratic period class is not a homomorphism from the group Γ into the complex numbers; but it follows readily from (2) that it does satisfy the relation

$$\sigma(ST) = \sigma(S) + \sigma(T) - \theta_1(S)\theta_2(T) \tag{3}$$

for any two elements $S, T \in \Gamma$. This sort of relation is a familiar one in the cohomology theory of abstract groups (see [2]). In general, for any two group homomorphisms $\theta_1, \theta_2 \in \mathrm{Hom}(\Gamma, \mathbf{C})$, the function $\theta_1 \cup \theta_2$ defined on $\Gamma \times \Gamma$ by $(\theta_1 \cup \theta_2)(S, T) = \theta_1(S)\theta_2(T)$ satisfies the relation

$$(\theta_1 \cup \theta_2)(S, T) - (\theta_1 \cup \theta_2)(RS, T) + (\theta_1 \cup \theta_2)(R, ST) - (\theta_1 \cup \theta_2)(R, S) = 0$$

for any three elements $R, S, T \in \Gamma$; that is to say, in the cohomological language, $\theta_1 \cup \theta_2$ is a 2-cocycle of the group Γ with coefficients in the trivial Γ-module **C**. Equation (3) is simply the assertion that this cocycle is cohomologous to zero, indeed, that it is the coboundary of the 1-cochain σ.

Perhaps it should be noted in passing that requiring the cocycle $\theta_1 \cup \theta_2$ to be cohomologous to zero does impose nontrivial restrictions on the homomorphisms θ_1 and θ_2; indeed, these restrictions are precisely the Riemann bilinear equalities on the periods of the Abelian differentials θ_1 and θ_2. This can be verified by a straightforward calculation, choosing the canonical presentation for the group Γ and examining the restrictions imposed by (3) and the group relations in Γ. However, it is more instructive to note that the correspondence $\theta_1 \times \theta_2 \to \theta_1 \cup \theta_2$ can be viewed as a bilinear mapping

$$H^1(\Gamma, \mathbf{C}) \otimes H^1(\Gamma, \mathbf{C}) \to H^2(\Gamma, \mathbf{C}),$$

since Hom $(\Gamma, \mathbf{C}) \cong H^1(\Gamma, \mathbf{C})$. Since $\tilde{M}$ is contractible, there are further canonical isomorphisms $H^q(\Gamma, \mathbf{C}) \cong H^q(\tilde{M}/\Gamma, \mathbf{C}) \cong H^q(M, \mathbf{C})$, for $q = 1, 2$, and with these isomorphisms, the bilinear mapping just given can be identified essentially with the usual cup product operation in the cohomology of the space M. The Riemann bilinear equalities amount precisely to the conditions that the cup products of any two analytic cohomology classes in $H^1(M, \mathbf{C})$ are zero (see [1]), which yields the desired assertion.

The collection of all the 1-cochains $\sigma \in C^1(\Gamma, \mathbf{C})$ satisfying merely the coboundary relation (3) form a complex linear set of dimension $2g$; for this set is nonempty, and any two elements of the set differ by a 1-cocycle, that is, by an element of Hom $(\Gamma, \mathbf{C})$. The problem is to select the unique quadratic period class in this set; this appears to be a nontrivial problem in general. A much simpler problem that one might imagine tackling first is the problem of determining the analytic 1-cochains in this set, where a 1-cochain σ satisfying (3) is called analytic if there is some complex analytic function s on the Riemann surface $\tilde{M}$ for which the functional equation (2) holds. Note that the analytic 1-cochains form a complex linear set of dimension g; for it is clear from (2) that two analytic 1-cochains differ by an analytic class in Hom $(\Gamma, \mathbf{C})$, that is, by the period class of an Abelian differential on the surface M. It is actually quite easy to determine the analytic 1-cochains, using the Serre duality theorem and the familiar technique for calculating that duality (see [1]). The details will be given elsewhere.

In at least one special case it is very easy to calculate the quadratic periods for an arbitrary Riemann surface; for when $\theta_1 = \theta_2$ it is clear that $s = h_1^2/2$, and hence that $\sigma(T) = -\theta_1(T)^2/2$ for any element $T \in \Gamma$. This observation can be extended to yield a *symmetry principle* for the quadratic periods in general. Let $\theta_i = dh_i$, $i = 1, \ldots, g$, be a basis for the Abelian differentials on M; and let $T \to \sigma_{ij}(T)$ be the period class of the quadratic expression $\sigma_{ij} = h_i\theta_j$, $i, j = 1, \ldots, g$. Then since, clearly, $s_{ij} + s_{ji} = h_i h_j$, it follows immediately that

$$\sigma_{ij}(T) + \sigma_{ji}(T) = -\theta_i(T)\theta_j(T) \tag{4}$$

for any element $T \in \Gamma$.

Finally, it should be noted in the general discussion that the quadratic periods, unlike the ordinary Abelian periods, depend on the choice of the base point $z_0 \in \tilde{M}$. To examine this dependence, select another base point z_0^*; and indicate the various expressions just considered, renormalized to vanish at z_0^*, with an asterisk. Thus clearly $h_i^*(z) = h_i(z) - h_i(z_0^*)$ and $s_{ij}^*(z) = s_{ij}(z) - h_i(z_0^*)h_j(z) - s_{ij}(z_0^*) + h_i(z_0^*)h_j(z_0^*)$; and it follows immediately for the quadratic periods that

$$\sigma_{ij}^*(T) = \sigma_{ij}(T) + \theta_i(T)h_j(z_0^*) - \theta_j(T)h_i(z_0^*) \tag{5}$$

for any element $T \in \Gamma$. Note that the change is skew-symmetric in the indices i, j, as of course it must be in consequence of the symmetry principle (4).

3.

Suppose now that the Riemann surface M is hyperelliptic, so that it can be represented as a two-sheeted branched covering of the Riemann sphere $\mathbf{P}$, branched over points $t_0, t_1, \ldots, t_{2g+1}$ in $\mathbf{P}$; and for simplicity, suppose further that the base point z_0 of the universal covering surface $\tilde{M}$ is chosen to lie over the point $t_0 \in \mathbf{P}$. Select disjoint smooth oriented paths τ_k in $\mathbf{P}$ from the point t_0 to the various points t_k, $k = 1, \ldots, 2g + 1$. Each path τ_k in $\mathbf{P}$ lifts to two separate paths τ_k^1, τ_k^2 in M; these two paths are disjoint except for their common initial point, the single point $p_0 \in M$ lying over t_0, and their common terminal point, the single point $p_k \in M$ lying over t_k. In turn, each path τ_k^ν lifts to a unique path $\tilde{\tau}_k^\nu$ in $\tilde{M}$ with initial point the base point $z_0 \in \tilde{M}$ and terminal point denoted by z_k^ν. Since the points z_k^1 and z_k^2 lie over the same point of M, they are necessarily congruent under the covering translation group Γ; therefore, there is a uniquely determined element $T_k \in \Gamma$ such that $z_k^1 = T_k z_k^2$. Note that the paths $\tilde{\tau}_k^1$ and $T_k\tilde{\tau}_k^2$ have the same terminal point; hence the path $(\tilde{\tau}_k^1)\cdot(T_k\tilde{\tau}_k^2)^{-1}$, obtained by traversing first the path $\tilde{\tau}_k^1$ with its given orientation and then the path $T_k\tilde{\tau}_k^2$ with the opposite of its given orientation, is a connected piecewise-smooth oriented path from z_0 to $T_k z_0$ in $\tilde{M}$.

THEOREM 1. *For a hyperelliptic Riemann surface represented as above, the quadratic periods for a basis $\theta_i = dh_i$ of the Abelian differentials on the surface are given by*

$$\sigma_{ij}(T_k) = -\theta_i(T_k)\theta_j(T_k)/2 \tag{6}$$

for $i, j = 1, \ldots, g$ and $k = 1, \ldots, 2g + 1$.

PROOF. Introduce the hyperelliptic involution $P\colon M \to M$, the analytic automorphism of the Riemann surface M corresponding to the interchange of the sheets in the covering $M \to \mathbf{P}$; this automorphism has period 2, and the quotient space M/P under the cyclic group generated by P is analytically equivalent to the Riemann sphere. The key to the proof is the observation that $\theta_i(z) + \theta_i(Pz) = 0$ for any Abelian differential θ_i on the surface M. To see this, note that $\theta_i(z) + \theta_i(Pz)$ is an analytic differential form on M which is invariant under the automorphism P, hence corresponds to an Abelian differential on the quotient space $M/P \cong \mathbf{P}$, and thus must be identically zero. Now let $z_k^\nu(t) \in \tilde{\tau}_k^\nu$ be the point lying over $t \in \tau_k$, so that the functions z_k^ν parametrize the paths $\tilde{\tau}_k^\nu$ by the coordinates along the paths τ_k. Then $\theta_i(z_k^1(t)) + \theta_i(z_k^2(t)) = 0$ for all $t \in \tau_k$, as above, and consequently $h_i(z_k^1(t)) + h_i(z_k^2(t)) = \text{constant}$, for all $t \in \tau_k$; since $z_k^1(t_0) = z_k^2(t_0) = z_0$, it follows from the normalizations adopted that actually $h_i(z_k^1(t)) + h_i(z_k^2(t)) = 0$ for all $t \in \tau_k$. For the quadratic expressions $\sigma_{ij} = h_i\theta_j$ it then follows also that $\sigma_{ij}(z_k^1(t)) = \sigma_{ij}(z_k^2(t))$ for all $t \in \tau_k$. Applying these and the earlier observations, note that

$$\begin{aligned}
\sigma_{ij}(T_k) &= -\int_{z_0}^{T_k z_0} \sigma_{ij}(z) = \int_{T_k\tilde{\tau}_k^2 - \tilde{\tau}_k^1} \sigma_{ij}(z) \\
&= \int_{\tilde{\tau}_k^2} [\sigma_{ij}(z) - \theta_i(T_k)\theta_j(z)] - \int_{\tilde{\tau}_k^1} \sigma_{ij}(z) \\
&= \int_{\tau_k} [\sigma_{ij}(z_k^2(t)) - \sigma_{ij}(z_k^1(t))] - \theta_i(T_k)\int_{\tilde{\tau}_k^2} \theta_j(z) \\
&= -\theta_i(T_k)h_j(z_k^2).
\end{aligned}$$

Similarly note that

$$\begin{aligned}
\theta_j(T_k) &= -\int_{z_0}^{T_k z_0} \theta_j(z) = \int_{T_k\tilde{\tau}_k^2 - \tilde{\tau}_k^1(z)} \theta_j(z) \\
&= \int_{\tau_k} [\theta_j(z_k^2(t)) - \theta_j(z_k^1(t))] = 2\int_{\tilde{\tau}_k^2} \theta_j(z) \\
&= 2h_j(z_k^2).
\end{aligned}$$

The desired result follows immediately from these two formulas.

To see that the preceding theorem gives a complete description of the quadratic periods of the Abelian differentials on a hyperelliptic Riemann surface, it is necessary to show that the elements T_k generate the entire covering translation group Γ. It is quite easy not only to do this, but also to determine the relations between these generators; the results are well known, but since it appears hard to find a sufficiently explicit reference for them, the details will be included here. Matters are somewhat simplified if a bit more care is taken with the notation. In particular, suppose that the branch points $t_k \in \mathbf{P}$ are so numbered that the paths τ_k occur in counter-clockwise order around the point t_0. The domain $\mathbf{P} - \bigcup_{k=1}^{2g+1} \tau_k$ is simply connected; thus the portion of the Riemann surface lying over it consists of two disjoint sets each homeomorphic to that domain. These two sets will be labeled sheets 1 and 2 of the covering, in some fixed but arbitrary order. This having been done, let τ_k^ν be that path in M lying over τ_k which has sheet ν along its right-hand side, for $\nu = 1, 2$.

Having fixed a base point $z_0 \in \tilde{M}$ lying over the point $p_0 \in M$, there is a canonical isomorphism between the covering translation group Γ and the fundamental group $\pi_1(M, p_0)$. Specifically, for an element $T \in \Gamma$, select any path in $\tilde{M}$ from z_0 to Tz_0; this path covers a closed path at p_0 in M, representing the element in $\pi_1(M, p_0)$ associated to T under this isomorphism. It is clear that the element $T_k \in \Gamma$ is associated to the element of $\pi_1(M, p_0)$ represented by the path $(\tau_k^1)(\tau_k^2)^{-1}$ obtained by traversing first the path τ_k^1 in the direction of its orientation and then the path τ_k^2 in the opposite direction to its orientation; this element of $\pi_1(M, p_0)$ will also be denoted by T_k. The problem is now to show that these elements generate the entire fundamental group $\pi_1(M, p_0)$. Note that any closed path based at p_0 in M can be deformed continuously into a path contained in the subset $K = \bigcup_{k=1}^{2g+1} (\tau_k^1 \cup \tau_k^2)$; for clearly the set $K_0 = \bigcup_{k=1}^{2g+1} \tau_k$ is a deformation retraction of $\mathbf{P} - t$ for any point $t \in \mathbf{P} - K_0$, and the lifting of this retraction to M accomplishes the desired result. Since K is a wedge of circles, $\pi_1(K, p_0)$ is the free group generated by the paths tracing out these component circles; but these are precisely the paths representing the elements T_k, hence these elements T_k generate the group $\pi_1(M, p_0)$, as asserted. Moreover, since the T_k are free generators for the group $\pi_1(K, p_0)$, the relations between these elements in $\pi_1(M, p_0)$ all follow from the exact homotopy sequence of the pair (M, K), which clearly has the form

$$\pi_2(M, K, p_0) \xrightarrow{\partial} \pi_1(K, p_0) \to \pi_1(M, p_0) \to 0.$$

The pair (M, K) is evidently 1-connected; thus it follows from the Hurewicz isomorphism and excision that, with the notation of [4],

$$\pi_2'(M, K, p_0) \cong H_2(M, K) \cong H_2(M/K, K/K) \cong \mathbf{Z} \oplus \mathbf{Z},$$

noting that M/K is a wedge of 2-spheres. Thus the full group $\pi_2(M, K, p_0)$ is generated by two elements Δ_1, Δ_2, together with the natural action of the fundamental group; and since this action of the fundamental group commutes with the exact homotopy sequence, all relations among the generators of T_k of $\pi_1(M, p_0)$ are consequences of the relations derived from the elements $\partial\Delta_1, \partial\Delta_2$ of $\pi_1(K, p_0)$. It is clear geometrically that Δ_1, Δ_2 correspond to the obvious mappings from the closed unit disk to sheets 1 and 2 of the covering respectively, with their common boundary K; and that $\partial\Delta_1 = T_1T_2\cdots T_{2g+1}$ and $\partial\Delta_2 = T_1^{-1}T_2^{-1}\cdots T_{2g+1}^{-1}$. Therefore the relations in $\pi_1(M, p_0)$ are consequences of the following two relations:

$$T_1T_2\cdots T_{2g+1} = T_{2g+1}\cdots T_2T_1 = 1. \tag{7}$$

It must be pointed out that the quadratic periods are not quite so simple as they might first appear to be from the symmetry formula (4) and formula (6) of Theorem 1. These periods are essentially non-Abelian in nature, and for a general element of the group Γ they have a rather more complicated form than they have for the special generators T_k. For instance, one may wish to calculate the quadratic periods for a canonical set of generators of the fundamental group $\pi_1(M, p_0)$. Introducing the elements $A_k, B_k \in \pi_1(M, p_0)$ defined by

$$\begin{aligned} A_k &= T_{2k}T_{2k+1}\cdots T_{2g+1}T_{2k}^{-1}, \\ B_k &= T_{2k+1}^{-1}T_{2k}^{-1}, \quad \text{for } k = 1, \ldots, g, \end{aligned}$$

a simple calculation shows that these are also generators for $\pi_1(M, p_0)$, and that the relations between them are consequences of the single relation

$$A_1B_1A_1^{-1}B_1^{-1}A_2B_2A_2^{-1}B_2^{-1}\cdots A_gB_gA_g^{-1}B_g^{-1} = 1.$$

Thus these elements can be viewed as a canonical set of generators of the fundamental group of the surface M. It then follows by a straightforward calculation from formulas (3) and (6) that the quadratic periods for these generators have the form

$$\begin{aligned} -2\sigma_{ij}(A_k) &= \theta_i(A_k)\theta_j(A_k) + \theta_i(A_k)\theta_j(B_k\cdots B_g) - \theta_i(B_k\cdots B_g)\theta_j(A_k) \\ &\quad + \sum_{l=k}^{g} [\theta_i(A_l)\theta_j(B_l) - \theta_i(B_l)\theta_j(A_l)] \\ -2\sigma_{ij}(B_k) &= \theta_i(B_k)\theta_j(B_k) + \theta_i(B_k)\theta_j(A_kB_{k+1}\cdots B_g) \\ &\quad - \theta_i(A_kB_{k+1}\cdots B_g)\theta_j(B_k). \end{aligned}$$

In each case the first term, which is symmetric in i and j, can be determined directly from the symmetry formula (4); the interest actually lies in the remaining terms, which are skew-symmetric in i and j, and which reflect the non-Abelian nature of the quadratic periods.

4.

That all the quadratic Abelian periods with base point a Weierstrass point can be expressed so simply in terms of the usual period matrix of the Riemann surface reflects the particular simplicity of hyperelliptic surfaces. Even allowing an arbitrary base point, the expressions remain quite simple, in view of (5). For a general Riemann surface the situation is rather more involved. Even sidestepping the matter of the choice of base point, by considering only those linear combinations of the quadratic Abelian periods which are independent of the base point, there remains the question whether the resulting expressions are actually functions of the period matrix of the surface; that is to say, any expressions in the quadratic Abelian periods which are independent of the base point can be viewed as functions on the Teichmuller space, but need not be functions on the Torelli space (see [3]). The detailed investigation of this matter is of some interest, involving as it does the relatively lightly explored area of non-Abelian problems on Riemann surfaces, but will be reported elsewhere.

The mere fact that these quadratic periods are of a non-Abelian character is, of course, an insufficient justification for considering them extensively. However, on the one hand, the quadratic periods do arise naturally in studying some problems on compact Riemann surfaces. For example, any finitely sheeted topological covering space of a compact Riemann surface has a natural Riemann surface structure; fixing the topological type of the covering, the period matrix of the covering space is a well defined function of the period matrix of the base space, and it is of some interest to know this function more explicitly. Equivalently, given a compact Riemann surface M of genus g, the problem is to determine explicitly the period matrix of the unique compact Riemann surface M' of the same genus g such that a fixed subset $G \subset M \times M'$, the graph of the topological covering, is homeomorphic to an analytic subvariety of the manifold $M \times M'$. The Abelian part of the problem can be handled quite easily, for Lefschetz' theorem determines the period matrices of the surfaces M' such that the graph G is homologous to an analytic cycle on the manifold M'. There are numerous surfaces M' admitting correspondences onto M having the topological type of the covering mapping, however. A detailed examination of the analogue of Riemann's bilinear relations picks out the period matrix of the unique covering space among all such surfaces M' in terms of the quadratic periods. In order for this approach to be very useful, more detailed knowledge of the quadratic periods on arbitrary Riemann surfaces seems necessary.

On the other hand, the quadratic periods do have some relevance to the old problem of determining which Riemann matrices are the period matrices of Riemann surfaces. The Riemann bilinear equalities correspond to the vanishing of the cup product of any two analytic period classes, or equivalently, to the fact that the product $\theta_1 \wedge \theta_2$ of any two Abelian differentials is cohomologous to zero. Actually, of course, the product $\theta_1 \wedge \theta_2$ vanishes identically; this is equivalent to the fact that the differential form $\sigma = h_1\theta_2$ is closed, that is, that the quadratic periods are well defined. Note that the Abelian differentials $\theta = dh$ have simple critical points in the sense that h is a finite branched covering map at any zero of θ; and conversely, if $\theta = dh$ is any closed complex-valued differential form with simple critical points on a two-dimensional differentiable manifold, the manifold has a unique complex structure for which θ is an Abelian differential form. Now if $\theta_1, \ldots, \theta_g$ are any closed complex-valued differential forms with simple critical points on a two-dimensional differentiable manifold, and if $\theta_i \wedge \theta_j = 0$ for all i, j, it follows immediately that all the forms determine the same complex structure; that is, that these are precisely the Abelian differential forms on some Riemann surface. Thus one is led to suspect that the period conditions beyond the Riemann bilinear relations may involve the quadratic periods; but again, some more detailed knowledge of the quadratic periods on arbitrary Riemann surfaces seems necessary before pursuing this line of investigation further.

PRINCETON UNIVERSITY

REFERENCES

1. GUNNING, R. C., *Lectures on Riemann Surfaces*. Princeton, N.J.: Princeton University Press, 1966.
2. HALL, MARSHALL, *The Theory of Groups*. New York: The Macmillan Company, 1959.
3. RAUCH, H. E., "A transcendental view of the space of algebraic Riemann surfaces," *Bull. Amer. Math. Soc.*, *71* (1969), 1–39.
4. SPANIER, EDWIN H., *Algebraic Topology*. New York: McGraw-Hill Book Co., Inc., 1966.

The Existence of Complementary Series

A. W. KNAPP[1] AND E. M. STEIN[2]

1. Introduction

Let G be a semisimple Lie group. The principal series for G consists of unitary representations induced from finite-dimensional unitary representations of a certain subgroup of G. These representations are not all mutually inequivalent, and their study begins with a study of the operators that give the various equivalences—the so-called intertwining operators.

For $G = SL(2, R)$, these operators are classical transformations. The principal series can be viewed conveniently as representations on L^2 of the line or L^2 of the circle. In the first case, the operators are given formally by scalar multiples of

$$f(x) \to \int_{-\infty}^{\infty} f(x-y)|y|^{-1+it}\,dy \tag{1.1a}$$

and

$$f(x) \to \int_{-\infty}^{\infty} f(x-y)(\operatorname{sign} y)|y|^{-1+it}\,dy. \tag{1.1b}$$

The operator (1.1a) is fractional integration of the imaginary order it and is also known as a Riesz potential operator of imaginary order; for $t = 0$, the operator (1.1b) is the Hilbert transform. If the principal series instead is viewed on the circle, the operators are less familiar analogs of these, given formally in the case of (1.1a) by

$$f(\theta) \to \int_0^{2\pi} f(\theta - \varphi)(1 - \cos\varphi)^{-(1-it)/2}\,d\varphi. \tag{1.2}$$

In [3] the authors investigated the operators that generalize (1.1) to an arbitrary group G of real-rank one in order to determine which representations of the principal series are irreducible. The idea was roughly that

[1] Supported by National Science Foundation Grant GP 7952X.
[2] Supported by AFOSR Grant AF68-1467.

reducibility occurs exactly when the operator generalizing (1.1) can be interpreted as a bounded operator given as a principal-value integral.

Here we shall study these operators for the same groups G for a different purpose. We wish to determine what unitary representations of G can be obtained by inducing from *nonunitary* finite-dimensional representations of the special subgroup. In other words, we ask what the representations are of the complementary series of G.

We shall treat the problem of existence of complementary series by considering analytic continuations of the operators generalizing (1.1) and (1.2). The essential question will be to determine which of the continued operators are positive-definite in a suitable sense. [In (1.1) the operators (1.1a) are positive-definite when it is replaced by a *real* number between 0 and 1, and the operators (1.1b), with it replaced by a complex parameter, are never positive-definite.] The ideas used in answering this question will be given in Section 2, and a more precise exposition will follow in the later sections. Most of the arguments will involve operators $A(z)$ generalizing (1.2), rather than (1.1), but at one point indicated in Section 3 we shall pass to the operators generalizing (1.1). This passage back and forth between integration on a compact group and integration on a noncompact group appears to play an important role in our work.

Our results are special, in that we work only with semisimple groups of real rank 1. Among other results concerning existence of complementary series in special situations are those of Kostant [4] (for general G but only for "class 1" induced representations) and Kunze [5] (for complex semisimple groups G).

The sections of the paper are arranged as follows. The notation and motivation are in Section 2, the precise definition of complementary series and the main theorem (Theorem 3.3) are in Section 3, and a discussion of the applicability of the main theorem is in Section 4. Since it is our intention to present here only the main ideas, we defer most proofs until another time.

The authors wish to thank S. Rallis for stimulating conversations about this work.

2. Notation and heuristics

In what follows, G will denote a connected semisimple Lie group with finite center. Let $G = ANK$ be an Iwasawa decomposition of G, let θ be the Cartan involution of G corresponding to K, let M be the centralizer of A in K, let M' be the normalizer of A in K, let ρ be half the sum of the positive restricted roots, and let $\bar{N} = \theta N$. Then MAN is a closed subgroup whose finite-dimensional irreducible unitary representations are all of the form $man \rightarrow \lambda(a)\sigma(m)$, where λ is a unitary character of A and σ is an irreducible representation of M. The *principal series* of unitary representations of G is parametrized by (σ, λ) and is obtained by inducing these representations of MAN to G.

These representations may be viewed as operating on a space of functions on K by restriction. That is, let σ operate on the finite-dimensional space V_σ and let H^σ be the subspace of members f of $L^2(K) \otimes V_\sigma$ such that, for each m in M,

$$f(mk) = \sigma(m)f(k)$$

for almost all k in K. Define operators on H^σ by

$$(U^{\sigma,\lambda}(x)f)(k) = e^{\rho H(kx)}\lambda(\exp H(kx))f(\kappa(kx)) \qquad x \in G \tag{2.1}$$

where the notation on the right refers to the Iwasawa decomposition $kx = \exp H(kx) \cdot n \cdot \kappa(kx)$. The representation $U^{\sigma,\lambda}$ is unitarily equivalent with the member of the principal series corresponding to the pair (σ, λ).

The definition (2.1) of a representation in the Hilbert space H^σ also makes sense when λ is a nonunitary character of A. In this case, $U^{\sigma,\lambda}(x)$ is a bounded operator with norm $\leqq \sup_{k \in K} |\lambda(\exp H(kx))|$, but it is not unitary. We call these representations the *nonunitary principal series*. Somewhat imprecisely, the complementary series consists of those representations of the nonunitary principal series that can be made unitary by redefining the inner product. (A precise definition will be given in Section 3.)

Temporarily we shall proceed only formally and see what has to happen for a representation to be in the complementary series. Suppose $\langle \cdot, \cdot \rangle$ is an inner product for which $U^{\sigma,\lambda}$ is unitary. This inner product will be given by an operator, possibly unbounded, say

$$\langle f, g \rangle = (Lf, g).$$

Here $(\cdot, \cdot)$ is the usual inner product on $L^2(K)$ given by integration. The condition that $\langle U^{\sigma,\lambda}(x)f, U^{\sigma,\lambda}(x)g \rangle = \langle f, g \rangle$ means that

$$LU^{\sigma,\lambda}(x) = U^{\sigma,\lambda}(x^{-1})^*L,$$

where the adjoint is defined relative to $(\cdot, \cdot)$. On the other hand, we have the lemma below, which follows from a change of variables.

LEMMA 2.1. $U^{\sigma,\lambda}(x^{-1})^* = U^{\sigma,\bar{\lambda}^{-1}}(x)$.

We conclude that

$$LU^{\sigma,\lambda}(x) = U^{\sigma,\bar{\lambda}^{-1}}(x)L \tag{2.2}$$

with $L \not\equiv 0$. There is a theorem of F. Bruhat [1], in the case that λ is unitary, that most $U^{\sigma,\lambda}$ are irreducible and that $U^{\sigma,\lambda}$ and $U^{\tau,\mu}$ are equivalent if and only if there is some member m' of M' such that $\sigma^{m'}$, defined by $\sigma^{m'}(m) = \sigma(m'mm'^{-1})$, is equivalent with τ and such that $\lambda^{m'} = \mu$. If we assume (slightly inaccurately) that these facts persist for nonunitary λ, then we expect that, for most λ, if (2.2) holds, then L is unique up to a scalar and there exists some m' in M' with $\sigma^{m'}$ equivalent with σ and $\lambda^{m'}$ equal to $\bar{\lambda}^{-1}$.

From now on, assume that dim $A = 1$ (the *real-rank one* case). Then M'/M has order 2. Fix an m' in M' but not in M, and introduce a complex parameter z by the definition

$$\lambda(a) = e^{z\rho H(a)}.$$

If λ corresponds to z, then $\bar{\lambda}^{-1}$ corresponds to $-\bar{z}$ and $\lambda^{m'}$ corresponds to $-z$. From what we have just said, there are only two possibilities:

(i) $z = -\bar{z}$. That is, z is imaginary and λ is unitary; hence $U^{\sigma,\lambda}$ is in the principal series.

(ii) $\sigma^{m'}$ is equivalent with σ, and $-z = -\bar{z}$. That is, it is possible to define $\sigma(m')$, and z is real.

Thus we are looking for an operator L such that $LU^{\sigma,z} = U^{\sigma,-z}L$, and we expect it to be unique up to a scalar for most z. Such an operator was obtained by Kunze and Stein [6] for Re $z > 0$. It is

$$A(z)f(k_0) = \int_K e^{(1-z)\rho \log a(km')}\sigma(m')\sigma^{-1}(m(km'))f(kk_0)\,dk, \tag{2.3}$$

where the notation on the right refers to the decomposition of G into $MAN\bar{N}$, namely

$$km' = m(km')\cdot a(km')\cdot n\cdot \bar{n};$$

this decomposition exists uniquely for all k not in M.

In short, for (σ, z) to give a representation of the complementary series, we expect that σ must be equivalent with $\sigma^{m'}$ and that z must be real. In this case if $z > 0$, the inner product should be a multiple of $(A(z)f, g)$. That is, a multiple of $A(z)$ must be a positive Hermitian operator. $A(z)$ is always Hermitian for z real, and it is positive if and only if its kernel is a positive-definite function.

For $z > 1$, we can settle the question of positivity immediately. For such a value of z, the kernel vanishes at the identity and is continuous and bounded on K; since it is not identically 0, it is not positive-definite. Thus there should be no complementary series for $z > 1$.

Our approach to the question of positivity when $0 < z < 1$ involves complex methods. To begin with, $z \to A(z)f$ is analytic for $\operatorname{Re} z > 0$, and, if f is smooth, we show that this function of z extends to be meromorphic in the whole plane. Denoting the new operators, defined for $\operatorname{Re} z \leqq 0$, by $A(z)$ also, we shall see that $z \to A(-z)A(z)f$ is meromorphic in the whole plane. For z purely imaginary and not 0, $A(-z)A(z)$ is an intertwining operator for the unitary representation $U^{\sigma,z}$, which is irreducible by Bruhat's theorem. Thus $A(-z)A(z)f = c(z)f$ with $c(z)$ scalar for z imaginary. If we introduce a suitable normalization $B(z) = \gamma(z)^{-1}A(z)$, we shall obtain $B(-z)B(z)f = f$ for imaginary z, hence for all z by analytic continuation. Suppose $B(0)$ is the identity. Then for $B(z_0)$ to fail to be positive-definite for some positive z_0, the equality $B(-z)B(z)f = f$ says that either $B(z)$ or $B(-z)$ must have a singularity for some z with $0 < z < z_0$. Thus an investigation of the singularities of $B(z)$ will be the key to the whole problem of the existence of complementary series associated with σ.

3. The existence theorem

We continue to assume that G has real-rank one. The representations $U^{\sigma,z}(x)$ of the nonunitary principal series, which was defined in Section 2, are parametrized by the finite-dimensional irreducible unitary representations σ of M and by the complex number z, which corresponds to the character $a \to e^{z\rho H(a)}$ of A.

The space H^σ on which $U^{\sigma,z}$ operates is a subspace of $L^2(K) \otimes V_\sigma$ that depends on σ but not on z, and the action of K, by right translation, is independent of z. The space of C^∞ vectors for $U^{\sigma,z}$ is the subspace of C^∞ functions in H^σ; thus, it too is independent of z. We denote this subspace by $C^\infty(\sigma)$.

We shall say that $U^{\sigma,z}$ is a member of the *complementary series* if there exists a positive-definite continuous inner product $\langle \cdot,\cdot \rangle$ on $C^\infty(\sigma) \times C^\infty(\sigma)$ such that

$$\langle U^{\sigma,z}(x)f, U^{\sigma,z}(x)g \rangle = \langle f, g \rangle \tag{3.1}$$

for all x in G and all f and g in $C^\infty(\sigma)$. If there is a nontrivial positive-semidefinite continuous inner product on $C^\infty(\sigma) \times C^\infty(\sigma)$ such that (3.1) holds, we shall say that $U^{\sigma,z}$ is a member of the *quasi-complementary series.*

In either case, the continuity of the inner product, equation (3.1) for x in K, and the Schwartz Kernel Theorem together imply the existence of a continuous operator L mapping $C^\infty(\sigma)$ into itself such that

$$\langle f, g \rangle = (Lf, g)$$

for f and g in $C^\infty(\sigma)$. Here $(\cdot,\cdot)$ denotes the usual inner product on $L^2(K) \otimes V_\sigma$.

As in Section 2, equation (3.1) translates into the fact that L intertwines $U^{\sigma,z}$ and its contragredient. Applying Lemma 2.1, we see that

$$LU^{\sigma,z}(x) = U^{\sigma,-\bar{z}}(x)L \tag{3.2}$$

for all x in G.

Recall that m' is a fixed member of M' that is not in M. If σ is equivalent with $\sigma^{m'}$, then it is possible to extend σ to a representation of all of M' on V_σ; that is, we can regard $\sigma(m')$ as defined. In this case, we define the operator $A(z)$ for $\operatorname{Re} z > 0$ by equation (2.3).

The operator $A(z)$ (actually a slight variant of it) was considered in [6]. It was shown that the kernel

$$e^{(1-z)\rho \log a(km')}\sigma(m')\sigma^{-1}(m(km')) \tag{3.3}$$

is an integrable function of k for $\operatorname{Re} z > 0$ and hence that $A(z)$ is a bounded operator on H^σ. Moreover, $A(z)$ satisfies

$$A(z)U^{\sigma,z}(x) = U^{\sigma,-z}(x)A(z). \tag{3.4}$$

(For $\operatorname{Re} z < 0$, the expression (3.3) is not an integrable function, and we consequently shall not deal directly with this case. In any event, one expects that $U^{\sigma,z}$ is in the complementary series if and only if $U^{\sigma,-z}$ is and that, in this case, $U^{\sigma,z}$ and $U^{\sigma,-z}$ lead to the same unitary representation.)

LEMMA 3.1. *Fix σ and z, and suppose* $\operatorname{Re} z > 0$. *Unless σ is equivalent with $\sigma^{m'}$ and z is real, the only continuous linear operator L on $C^\infty(\sigma)$ satisfying* (3.2) *is* 0. *If σ is equivalent with $\sigma^{m'}$ and z is real, then the continuous operators on $C^\infty(\sigma)$ satisfying* (3.2) *are exactly the scalar multiples of* $A(z)$. $A(z)$ *is bounded and Hermitian.*

Before we pass to a study of the analyticity of $A(z)$, let us observe that the $A(z)$ have a common finite-dimensional resolution. Specifically, let H_D^σ be the subspace of H^σ of functions that transform under K according to the equivalence class D of irreducible representations of K. H_D^σ is finite-dimensional since $H^\sigma \subseteq L^2(K) \otimes V_\sigma$, and it is independent of z. Then each $A(z)$ maps each H_D^σ into itself, by equation (3.4) for x in K.

If f is in $C^\infty(\sigma)$, the mapping $z \to A(z)f$ is an analytic mapping of $\{\operatorname{Re} z > 0\}$ into $C^\infty(\sigma)$. We shall be concerned with extending this mapping to a meromorphic function defined in the whole complex plane. It will be enough to consider the simpler function $z \to A(z)f(1)$, where 1 is the identity of K, provided we prove joint continuity of this function in z and f. Since the singularities of the kernel (3.3) occur only for k in M, we can suppose that f is supported near M, particularly away from $M' - M$.

This turns out to mean that we can transform the whole problem to a problem about the simply connected nilpotent group $\bar{N}$. In fact, using the change-of-variables formula of [2, p. 287], we find that

$$A(z)f(1) = \int_{\bar{N}} e^{(1-z)\rho \log a(ym')}\sigma(m')\sigma^{-1}(m(ym'))\{e^{(1+z)\rho H(y)}f(\kappa(y))\}\, dy. \tag{3.5}$$

The notation here is the same as in formulas (2.1) and (2.3). The ingredients of this formula are technically much simpler than those of formula (2.3), and we consider them one at a time.

First we make some comments about $\bar{N}$. The restricted roots of the Lie algebra $\mathfrak{g}$ of G are either $2\alpha, \alpha, 0, -\alpha, -2\alpha$ or $\alpha, 0, -\alpha$. In this notation, $\bar{N} = \exp(\mathfrak{g}_{-\alpha} \oplus \mathfrak{g}_{-2\alpha})$. Let $p = \dim \mathfrak{g}_{-\alpha}$ and $q = \dim \mathfrak{g}_{-2\alpha}$. The group A acts on $\bar{N}$ by conjugation; geometrically this action looks like dilations, except that the $\mathfrak{g}_{-2\alpha}$ directions are dilated twice as fast as the $\mathfrak{g}_{-\alpha}$ directions.

Now consider the first factor in the integrand of (3.5). Although it is not necessary to do so for the present problem, one can compute this factor explicitly. If $y = \exp(X + Y)$ with $X \in \mathfrak{g}_{-\alpha}$ and $Y \in \mathfrak{g}_{-2\alpha}$, then[3]

$$e^{(1-z)\rho \log a(ym')} = (\tfrac{1}{4}c^2\|X\|^4 + 2c\|Y\|^2)^{-(p+2q)(1-z)/4}, \tag{3.6}$$

where the norm is that induced by the Killing form of $\mathfrak{g}$ and where $c = (2p + 8q)^{-1}$. Put $|y| = e^{-\rho \log a(ym')}$. Then the function (3.6) has an important property of homogeneity relative to A: if b is in A, then

$$|byb^{-1}| = e^{-2\rho H(b)}|y|.$$

Next we consider $\sigma(m')\sigma^{-1}(m(ym'))$, which we shall denote $\sigma(y)$. This is a matrix-valued function defined everywhere but at the identity and satisfying the homogeneity property $\sigma(byb^{-1}) = \sigma(y)$ for all b in A.

Finally we consider the factor $e^{(1+z)\rho H(y)}f(\kappa(y))$. This function is a smooth function of compact support in $\bar{N}$ because f is assumed to be supported away from $M' - M$. The function depends on the complex parameter z but is entire in the variable z since $\rho H(y)$ has no singularities.

To see that (3.5) extends to be meromorphic in the whole z-plane, we choose a continuous function $\varphi(r)$ of compact support on $[0, \infty)$ so that $\varphi(|y|)f(\kappa(y)) = f(\kappa(y))$, we expand $e^{(1+z)\rho H(y)}f(\kappa(y))$ about $y =$ identity in a finite Taylor series with remainder term, we collect the polynomial terms of the same homogeneity relative to A, we multiply both sides of the expansion by $\varphi(|y|)$, and we substitute into (3.5). The terms of the expansion can be computed well enough to conclude the following: each term but the remainder has a meromorphic extension with at most one pole, that one simple and occurring at an integral multiple of $z = -(p + 2q)^{-1}$, and the

[3] That this explicit formula holds might be guessed from an earlier formula that S. Helgason had derived for $\exp\{-2\rho H(y)\}$.

remainder term gives a contribution analytic in a large right-half plane. Collecting these results, we have the following theorem:

THEOREM 3.2. *Let f be in $C^\infty(\sigma)$. As a mapping into $C^\infty(\sigma)$, the function $z \to A(z)f$ has a meromorphic extension to the whole complex plane with singularities only at the non-negative integral multiples of $-(p + 2q)^{-1}$. The singularities at these points are at most simple poles. The poles can occur only at integral multiples of $-2(p + 2q)^{-1}$ if $\sigma(y) = \sigma(m')\sigma^{-1}(m(ym'))$ satisfies*

$$\sigma(\exp(-X + Y)) = \sigma(\exp(X + Y)) \tag{3.7}$$

for $X \in \mathfrak{g}_{-\alpha}$ and $Y \in \mathfrak{g}_{-2\alpha}$. Moreover, the mapping $(z, f) \to A(z)f$ for z in the regular set and f in $C^\infty(\sigma)$ is a continuous mapping to $C^\infty(\sigma)$.

REMARKS. Condition (3.7) holds for all σ for the Lorentz groups $SO_e(n, 1)$, the Hermitian Lorentz groups $SU(n, 1)$, and the symplectic Lorentz groups $Sp(n, 1)$, but it fails for the spin groups Spin$(n, 1)$. In any case, the parameter z is normalized so that $z = 1$ corresponds to ρ; therefore, $z = 2(p + 2q)^{-1}$ corresponds to the restricted root α. The result for $SO_e(n, 1)$ that the only poles of $A(z)f$ are simple and are at multiples of $-\alpha$ was obtained by Schiffmann [7].

Using Theorem 3.2, we can now define $A(z)$ for all z. To proceed further, however, we need more information about σ. It is possible to show, under the additional assumptions on G that G is simple and has a faithful matrix representation, that *some representation D of K, when restricted to M, contains σ exactly once.*[4] By the reciprocity theorem, this means that K acts irreducibly on some $H_D^\sigma \neq 0$. Fix such a $D = D_0$, and let $v(k)$ be a nonzero member of $H_{D_0}^\sigma$. Since $A(z)$ commutes with K, we obtain

$$A(z)v = \gamma(z)v \tag{3.8}$$

for a complex-valued meromorphic function $\gamma(z)$. Define

$$B(z) = \gamma(z)^{-1}A(z).$$

As we shall see in Section 4, there is no complementary series associated with σ near $z = 0$ unless the unitary representation $U^{\sigma,0}$ is irreducible. [And for $G = \text{Spin}(n, 1)$ or $SU(n, 1)$ there is no complementary series for any z unless this condition is satisfied.] We therefore assume now that $U^{\sigma,0}$ is irreducible. A necessary and sufficient condition for this irreducibility is given as Theorem 3 of [3]. The condition implies that $\gamma(z)$ does have a pole at $z = 0$, which implies that the operators $B(z)$ are uniformly bounded on compact subsets of $0 \leqq \text{Re}\, z \leqq c$ if c is sufficiently small. The irreducibility and equation (3.4) then imply that $B(0) = I$.

[4] Independently J. Lepowsky has obtained this result and a generalization in his thesis at the Massachusetts Institute of Technology.

The definition of $\gamma(z)$ is arranged so that $B(-z)B(z)f = f$ for all f in $C^\infty(\sigma)$ and for all z. In fact, Theorem 3.3 implies that $B(-z)B(z)f$ is meromorphic in the whole plane. But $B(-z)B(z)$ for purely imaginary $z \neq 0$ intertwines $U^{\sigma,z}$ with itself. By Bruhat's irreducibility theorem in [1], $B(-z)B(z) = c(z)I$ with $c(z)$ scalar for z imaginary. Applying both sides to v, we see that $c(z) = 1$ for z imaginary. That is, $B(-z)B(z)f = f$ for z imaginary. By analytic continuation, $B(-z)B(z)f = f$ for all z.

$B(z)$ preserves each H_D^g, and $B(0) = I$. Fix D, and suppose $B(z_0)|_D$ is not positive-definite for some $z_0 > 0$. Then either $B(z)$ has a pole nearer 0, or some $B(z)f$ has a 0, in which case $B(-z)$ has a pole, because $B(-z)B(z) = I$. The poles of $B(z)$ (and similarly for $B(-z)$) arise when $A(z)f$ has, for some f, a pole of higher order (possibly negative) than does $\gamma(z)$. These are the ideas behind the main theorem:

THEOREM 3.3. *Suppose that G is simple, that G has a faithful matrix representation, and that* $\dim A = 1$. *Let σ be an irreducible finite-dimensional unitary representation of M satisfying the necessary conditions above, namely, that*

(i) *σ is equivalent with $\sigma^{m'}$, where m' is a member of M that is not in M, and*

(ii) *the unitary representation $U^{\sigma,0}$ is irreducible.*

Define $A(z)$ by (2.3) *and $\gamma(z)$ by* (3.8). *Let z_0 be the least number* $\geqq 0$ *such that, for some $f \in C^\infty(\sigma)$, $z \to A(z)f$ has a pole at $-z_0$ and $\gamma(z)$ does not or such that γ has a zero at z_0 or at $-z_0$. Then $z_0 > 0$ and the parameters (σ, z) give rise to representations of the complementary series for* $0 < z < z_0$ *with inner product*

$$\langle f, g\rangle = \gamma(z)^{-1} \int_K (A(z)f, g)_{V_\sigma}\, dk$$

for f and g in $C^\infty(\sigma)$.

It is a simple matter to see also that the parameters (σ, z_0) give rise to a representation of the quasi-complementary series. It can happen that this representation is the trivial representation of G.

We should emphasize why the number z_0 in the theorem is strictly positive. The set whose least member is z_0 consists at most of the positive nonzero integral multiples of $(p + 2q)^{-1}$ and the non-negative values z such that one of the meromorphic functions $\gamma(z)$ and $\gamma(-z)$ vanishes. This set is discrete, and it does not contain $z = 0$ because $\gamma(z)$ has a pole at $z = 0$.

4. Further investigation of the singularities of $B(z)$

For the case that σ is the trivial representation of M, we can choose the eigenvector v of the $A(z)$ to be a constant function. The associated function $\gamma(z)$ is closely related to Harish–Chandra's **c** function (see [2]), and we can obtain very explicit results as a consequence. For G of general real rank, the **c** function can be defined as the analytic continuation in μ to the whole complexified dual of the Lie algebra of A of the function

$$\mathbf{c}(\mu) = \int_{\bar{N}} e^{(i\mu + \rho)H(x)}\, dx.$$

Let us return to the real-rank one case.

PROPOSITION 4.1. *Let σ be the trivial representation, and choose v to be a constant function. Then the function $\gamma(z)$ is given by $\gamma(z) = \mathbf{c}(-iz\rho)$. Therefore*

$$\gamma(z) = 2^{(p+2q)(1-z)/2} \frac{\Gamma(\frac{1}{2}(p+q+1))\Gamma(\frac{1}{2}(p+2q)z)}{\Gamma(\frac{1}{4}(p+2q)(1+z))\Gamma(\frac{1}{4}(p+2+(p+2q)z))},$$

where $p = \dim \mathfrak{g}_{-\alpha}$ and $q = \dim \mathfrak{g}_{-2\alpha}$.

PROOF.

$$\begin{aligned}\gamma(z) &= \int_K e^{(1-z)\rho \log a(km')}\, dk = \int_K e^{(1-z)\rho \log a(k)}\, dk \\ &= \int_{K/M} e^{(1-z)\rho \log a(k)}\, dk = \int_{\bar{N}} e^{(1-z)\rho \log a(\kappa(x))} e^{2\rho H(x)}\, dx,\end{aligned}$$

the last equality following from [2] (see p. 287). If $x = ank \in ANK$, then $k = a^{-1}(an^{-1}a^{-1})x \in MAN\bar{N}$. Hence $a(\kappa(x)) = \exp(-H(x))$, and we obtain $\gamma(z) = \int_{\bar{N}} e^{(1+z)\rho H(x)}\, dx$, as required. The formula for $\gamma(z)$ then follows from [2] (see p. 303).

In [4], B. Kostant obtained the existence of complementary series for σ trivial and G of any real rank. For a first application of Proposition 4.1, we shall compare his results in the rank-one case with what we can prove from Theorem 3.3 and Proposition 4.1 when σ is trivial.

Using the explicit value of $\gamma(z)$ in the proposition, we see that $\gamma(z)$ is nonvanishing for $-1 < z < 1$ and that the only poles of $\gamma(z)$ for $-1 < z \leqq 0$ occur at the non-negative integral multiples of $-2(p + 2q)^{-1}$. If $q = 0$, there is a pole at every multiple of $-2p^{-1}$ less than 1. If $q > 0$, then p is even and a pole occurs at every multiple satisfying $-(p + 2)(p + 2q)^{-1} < z \leqq 0$. By Theorem 3.2 the poles of $A(z)f$ occur only at multiples of $-2(p + 2q)^{-1}$. Thus by Theorem 3.3 there is a complementary series for

$$(4.1) \qquad 0 < z < z_0 = \begin{cases} 1 & \text{if } q = 0 \\ \dfrac{p+2}{p+2q} & \text{if } q > 0. \end{cases}$$

Kostant has this estimate (in [4], see Section 3.1 and Theorem 10). Kostant shows further that there is no (positive-definite) complementary series to the right of the point z_0 in this inequality. We can obtain this result by our method if $q = 0$, 1, or 3. But if $q = 7$, we obtain only the weaker result that there is no complementary series immediately to the right of z_0. This weaker result comes from comparing the signs near z_0 of $\gamma(z)$ and $(A(z)f, f)$ for an f such that $A(-z_0)f$ has a pole. (Such an f exists.)

We turn to other applications of Proposition 4.1. When $q = 0$ or 1, we have $z_0 = 1$ in (4.1). If $\exp\{(1 - z_1)\rho \log a(km')\}\sigma(m')\sigma^{-1}(m(km^{-1}))$ is a positive-definite function and if $0 < z < z_1 \leqq 1$, then the product with $\exp[\{1 - (1 - z_1 + z)\}\rho \log a(km')]$ is also positive-definite. We obtain the following corollary:

COROLLARY 4.2. *Let* $G = SO_e(n, 1)$, Spin$(n, 1)$, *or* $SU(n, 1)$. *Let* σ *be an irreducible unitary representation of* M *such that* σ *is equivalent with* $\sigma^{m'}$. *Then the positive* z *such that* $U^{\sigma,z}$ *is in the quasi-complementary series form an interval with* 0 *as left endpoint.*

COROLLARY 4.3. *Let* $G = SO_e(n, 1)$, Spin$(n, 1)$, *or* $SU(n, 1)$. *Let* σ *be an irreducible unitary representation of* M *such that* σ *is equivalent with* $\sigma^{m'}$ *and such that* $U^{\sigma,0}$ *is reducible. Then there is no positive* z *such that* $U^{\sigma,z}$ *is in the quasi-complementary series.*

For the proof of the second corollary, it is possible to use Theorems 1 and 3 of [3] to show from the reducibility of $U^{\sigma,0}$ that $B(0)$ is unitary and not scalar. But if U^{σ,z_0} is in the quasi-complementary series, $A(z)$ is semidefinite for $0 < z < z_0$, by Corollary 4.2; this fact implies that $B(0)$ is semidefinite, which is a contradiction.

THE INSTITUTE FOR ADVANCED STUDY
PRINCETON UNIVERSITY

REFERENCES

1. BRUHAT, F., "Sur les représentations induites des groups de Lie," *Bull. Soc. Math. France*, *84* (1956), 97–205.
2. HARISH-CHANDRA, "Spherical functions on a semisimple Lie group, I," *Amer. J. Math.*, *80* (1958), 241–310.
3. KNAPP, A. W., and E. M. STEIN, "Singular integrals and the principal series," *Proc. Nat. Acad. Sci.*, *63* (1969), 281–284.
4. KOSTANT, B., "On the existence and irreducibility of certain series of representations," *Bull Amer. Math. Soc.*, *75* (1969), 627–642.
5. KUNZE, R., *Analytic Continuation of Intertwining Operators: Construction of Complementary Series for Complex Semisimple Groups* (unpublished manuscript).
6. KUNZE, R., and E. M. STEIN, "Uniformly bounded representations, III," *Amer. J. Math.*, *89* (1967), 385–442.
7. SCHIFFMANN, G., "Intégrales d'entrelacement: cas des groupes de Lorentz," *C. R. Acad. Sci. Paris*, Serié A, *266* (1968), 859–861.

Some Recent Developments in the Theory of Singular Perturbations

P. A. LAGERSTROM[1]

1. Historical introduction

In the nineteenth century celestial mechanics played an essential role in developing perturbation methods and asymptotic theory. This work culminated in Poincaré's great treatise, *Les Méthodes Nouvelles de la Méchanique Céleste*. In the twentieth century fluid mechanics has played a somewhat similar role. The present paper attempts to draw mathematicians' attention to some important ideas in the theory of singular perturbations developed recently by workers in applied fields. The ideas will be illustrated by simple model equations; the references to fluid mechanics are merely historical and are not needed for the mathematical discussion which starts in Section 2.

We first give a brief historical sketch. The equations of motion for stationary viscous flow may be written (in suitable nondimensional variables)[2]

$$\alpha \sum_{j=1}^{n} v_j \frac{\partial v_k}{\partial x_j} = \sum_{j=1}^{n} \frac{\partial^2 v_k}{\partial x_j^2} - \frac{\partial p}{\partial x_k}, \tag{1.1a}$$

$$\sum_{j=1}^{n} \frac{\partial v_j}{\partial x_j} = 0, \tag{1.1b}$$

where the v_j are the velocity components, p is the pressure, and

$$\alpha = \text{Reynolds number} = \frac{\text{length} \times \text{velocity}}{\text{viscosity}}. \tag{1.1c}$$

In a typical problem one wants the solution outside a closed $(n-1)$-dimensional surface S. Typical boundary conditions are then

$$\text{on } S: \quad v_j = 0, \tag{1.2a}$$

$$\text{at infinity:} \quad v_1 = 1; \quad v_j = 0 \quad \text{for } j \neq 1. \tag{1.2b}$$

[1] Work on this paper was supported by NSF Grant No. GP 9335.

[2] For details concerning fluid mechanics, see [3].

In 1850 Stokes treated the case of small α by simply putting $\alpha = 0$ in (1.1a). While this worked for $n = 3$, it was found that for $n = 2$ any nonzero solution grows logarithmically at infinity, thus violating (1.2b). This is the famous Stokes paradox. It was later found that an attempt to find a second approximation for $n = 3$ also led to logarithmic divergence (Whitehead's paradox). The contribution to asymptotic theory was entirely of a negative nature.

It was the problem of large values of α which led to significant progress in asymptotic methods. In 1904 Prandtl introduced his famous boundary-layer theory to find approximate solutions for α large. His method of using different length-scales and different approximations in different regions of space has had a great although belated influence on what is now known as the theory of singular perturbations.

In the 1950's the range of application of Prandtl's technique was extended to many fields of applied mathematics; at the same time efforts were made to understand and enlarge the technique by fitting it into some orderly asymptotic scheme. The ideas advanced by Saul Kaplun[3] about 1955 were crucial. Kaplun made a very profound analysis of the ideas underlying the Prandtl technique (developed for large α) and showed that the same ideas may be used to get asymptotic solutions of the same equation with small α. This not only essentially solved the Stokes paradox (after more than a hundred years) and the Whitehead paradox but also greatly increased the understanding of singular perturbations and advanced the technique for handling them. Concepts and techniques such as "matching," "overlap," and "intermediate limit," which are now in standard use, are due to Kaplun. Two recent books on perturbation methods ([1] and [4]) give abundant evidence of the increase in techniques (Prandtl–Kaplun technique as well as many others) and the extension of the range of applications. But while mathematicians have given the older Prandtl technique some mathematical respectability by rigorously proving its correctness in special cases (see, for example, [5]), the more recent and much deeper Kaplun ideas are practically unknown among pure mathematicians.

The first model equation which follows is an illustration of Prandtl's technique (the parameter ϵ corresponds to $\alpha^{-1/2}$ in Prandtl's case). The second model example illustrates some of the important ideas introduced by Kaplun, in particular his method of clarifying the Stokes paradox.

2. Posing the problem: Prandtl's method

In mathematical physics one frequently encounters the problem of constructing an asymptotic expansion of a function, with the requirement that

[3] The late Saul Kaplun's work is collected in [2].

it be uniformly valid over a given region in space and time as certain parameters tend to limiting values. Thus (restricting ourselves to one coordinate and one parameter) we seek to construct functions $f_k(x, \epsilon)$ such that, for a given $f(x, \epsilon)$,

$$\lim \frac{f(x, \epsilon) - \sum_{k=0}^{n} f_k(x, \epsilon)}{\mu_n(\epsilon)} = 0, \tag{2.1}$$

as ϵ tends to zero; the limit is required to be uniform for all values of x which lie in a given closed set. The $\mu_n(\epsilon)$ are gauge functions which form an asymptotic sequence

$$\mu_{n+1}(\epsilon) = o(\mu_n(\epsilon)), \tag{2.2}$$

as ϵ tends to zero. The physical situation may be such that ϵ is always real and positive. In fact, in the type of problem to be discussed here $f(x, \epsilon)$ often has an essential singularity in ϵ at $\epsilon = 0$; hence the path of approach to zero must be specified. We shall always assume that "lim" means "limit as ϵ tends to zero through positive values." In a typical situation the function $f(x, \epsilon)$ is defined implicitly by a differential equation and boundary conditions. The f_k are then found by solving approximating equations. In order to find these equations one normally makes specific assumptions about how ϵ occurs in the f_k. (We shall see at the end of Section 4 that in a certain sense this procedure should be reversed.) The simplest assumption is, of course,

$$f_k(x, \epsilon) = \mu_k(\epsilon) g_k(x), \tag{2.3}$$

(in particular μ_k may be ϵ^k). A problem for which this assumption works is called a regular perturbation problem. A trivial example illustrates how a regular perturbation method may fail. Let f be defined by

$$\epsilon f'' + f' = a; \qquad f(0) = 0; \qquad f(1) = 1, \tag{2.4}$$

where a = constant $\neq 1$. Putting $f \sim \sum_{k=0} \epsilon^k g_k(x)$ we find that the equation for g_0 is

$$g_0' = a. \tag{2.5}$$

Only one boundary condition may be imposed. Standard methods show that (since $\epsilon > 0$) we should retain the condition at $x = 1$. Hence

$$g_0 = ax + (1 - a). \tag{2.6}$$

Since $a \neq 1$ this approximation is not uniformly valid in [0, 1]. (Such a solution is usually called an outer solution.) In *some* sense, to be discussed later, it is valid if we exclude the origin. A solution valid near the origin (called an inner solution) may be obtained by introducing a new variable (a "stretched coordinate" or "inner variable"), defined by

$$\epsilon \bar{x} = x. \tag{2.7}$$

Putting

$$\bar{f}(\bar{x}, \epsilon) = f(x, \epsilon), \tag{2.8}$$

we seek an expansion

$$\bar{f} \sim \sum_{k=0} \epsilon^k h_k(\bar{x}). \tag{2.9}$$

The equation for h_0 is then

$$\frac{d^2 h_0}{d\bar{x}^2} + \frac{dh_0}{d\bar{x}} = 0. \tag{2.10}$$

We retain the boundary condition at $x = 0$ which gives

$$h_0 = A(1 - e^{-\bar{x}}). \tag{2.11}$$

The constant A cannot be determined by the boundary condition at $x = 1$ since we expect h_0 to be valid only near the origin. Instead we determine A by *matching* the inner solution h_0 with the outer solution g_0, using the matching condition

$$h_0(\infty) = g_0(0), \tag{2.12}$$

which gives

$$A = 1 - a. \tag{2.13}$$

Condition (2.12) may seem surprising; for the time being we note only that a comparison with the exact solution shows that it is correct in the present case.

Although the example is trivial, it illustrates a situation typical for a singular perturbation problem: one constructs (at least!) two asymptotic expansions valid in different regions. They are not constructed independently; in order to determine the terms of both expansions one needs to compare them with the aid of matching principles. The main purpose of this article is to point out the underlying ideas which lead to a matching condition of the type (2.12). These ideas also show how to do the matching correctly in cases for which a simple matching condition of the type (2.12) does not hold. As a preliminary, the concepts of valid "near the origin" or "away from the origin" have to be defined. The basic reference for the following sections is [2].

In passing we note that by adding the approximations g_0 and h_0 and then subtracting their common part, one obtains an approximation f_0 which is uniformly valid in [0, 1]:

$$f_0(x, \epsilon) = (1 - a)(1 - e^{-x/\epsilon}) + ax. \tag{2.14}$$

Actually it is valid to all orders ϵ^n, the mistake being of order $e^{-1/\epsilon}$.

3. The domain of validity: overlap matching

ORDER SPACES. Comparison with the exact solution, or with (2.14), shows that g_0 is a uniformly valid approximation in an interval $[x_0, 1]$, for any $x_0 > 0$. This, however, is insufficient for matching. We need in a sense its maximal domain of uniform validity. As is easily seen, g_0 is uniformly valid in any interval $[\eta(\epsilon), 1]$, where $\eta(\epsilon)$ is >0 for $\epsilon > 0$ and $\epsilon = o(\eta(\epsilon))$. Thus we may characterize the domain of validity by a function class. The class of functions $\zeta(\epsilon)$, defined on some open interval $0 < \epsilon < \epsilon_0$, which are positive and continuous will be denoted by $\mathscr{F}$. Only the behavior of ζ as ϵ tends to zero will matter; hence we may group the elements of $\mathscr{F}$ into equivalence classes. The relation between ζ and η,

$$\frac{\zeta}{\eta} \text{ tends to a nonzero finite limit,} \tag{3.1}$$

is an equivalence relation; the corresponding class, called an o-class, is denoted by "ord η." We may make a broader identification, into O-classes, by the equivalence relation:

$$\begin{aligned}&\text{There exist constants } c_1, c_2 \text{ and a number } \epsilon_0 > 0\\ &\text{such that } 0 < c_1 < \zeta/\eta < c_2 \text{ for } 0 < \epsilon < \epsilon_0.\end{aligned} \tag{3.2}$$

The corresponding class is denoted by "Ord η" and the space of such classes will be denoted by $\mathscr{F}_0$. A partial ordering may be introduced. Let Z and Y be two classes. Then

$$Z < Y \quad \text{means} \quad \zeta = o(\eta) \quad \text{for } \zeta \text{ in } Z \text{ and } \eta \text{ in } Y. \tag{3.3}$$

(The same relation could be defined for o-classes.) In the obvious sense we may define intervals; $(Z, Y]$, $[Z, Y]$, etc. The space of O-classes (or of o-classes) has an interesting topology, with open intervals as basic neighborhoods, which does not seem to have been sufficiently explored.

DOMAIN OF VALIDITY. Let $\mathscr{V}$ be a set of O-classes and let $F(\mathscr{V})$ be the set of all elements of $\mathscr{F}$ whose O-classes are in $\mathscr{V}$. Denoting the difference

$|f(x, \epsilon) - g(x, \epsilon)|$ by $w(x, \epsilon)$ we say that g is an approximation to f uniformly valid to order unity on the domain $\mathscr{V}$ if

$$\text{(3.4)} \qquad \begin{array}{l} \zeta, \eta \text{ in } F(\mathscr{V}) \text{ implies that } w(x, \epsilon) \text{ tends to zero} \\ \text{uniformly on the interval}^4\ \zeta(\epsilon) \leqq x \leqq \eta(\epsilon); \end{array}$$

$\mathscr{V}$ is then called a domain of validity of the approximation g.

To give examples of domains of validity we return to the functions discussed in Section 2. By comparing g_0 with the exact solution [for which we may substitute f_0 as defined by (2.14)], we see that the half-open interval (Ord ϵ, Ord 1] is a set of validity of g_0. It is in fact the maximal set of validity if we assume that g_0 and f are defined only for $0 \leqq x \leqq 1$; similarly [Ord ϵ, Ord 1) is a set of validity of h_0.

OVERLAP MATCHING. We notice that the domains of validity of g_0 and h_0 have a common part, namely the open O-interval

$$\text{(3.5)} \qquad \mathscr{I} = (\text{Ord } \epsilon, \text{Ord } 1).$$

It is the existence of a *domain of overlap* which makes matching possible. If there is an O-set where both g_0 and h_0 are valid,[5] then on this set $|g_0 - h_0|$ tends to zero with ϵ. This may be conveniently formalized with the aid of a limit process. We define x_η by

$$\text{(3.6)} \qquad \eta(\epsilon)x_\eta = x,$$

and the η-limit of a function $a(x, \epsilon)$ as

$$\text{(3.7)} \qquad \lim_\eta a(x, \epsilon) = \lim a(x, \epsilon) \quad \text{as } \epsilon \downarrow 0,\ x_\eta \text{ fixed}.$$

In other words $\lim_\eta$ means that we approach $\epsilon = 0$ along the curves $x_\eta =$ constant in the (x, ϵ)-plane. It is clear that if $b(x, \epsilon)$ and $c(x, \epsilon)$ are two approximations to $d(x, \epsilon)$ which have an overlap domain $\mathscr{D}$, then

$$\text{(3.8)} \qquad \text{Ord } \eta \text{ in } \mathscr{D} \quad \text{implies} \quad \lim_\eta |b - c| = 0.$$

Such a limit is called an intermediate limit.

We shall consider (3.8) as the fundamental matching principle. For the example in Section 1 we find

$$g_0(x) - h_0(\bar{x}) = (a\eta x_\eta + 1 - a) - A(1 - e^{-\eta x_\eta/\epsilon}).$$

Applying $\lim_\eta$ for any η such that Ord η is in the overlap domain we find

$$(1 - a) - A = 0,$$

[4] When such an interval is empty we say that $w = 0$ for all x in this interval. Also, one should of course consider only the intersection of the interval and the domain of definition of $w(x, \epsilon)$.

[5] Throughout this paper we restrict ourselves to the lowest order and assume $\mu_0(\epsilon) = 1$.

which is nothing but the matching principle (2.12). However, the validity of (2.12) is rather restricted; it is a special case of the basic matching principle (3.8).

The clarification of the nature of matching has proved of great importance in asymptotic constructions. However, we are still faced with the problem how to determine the domain of validity a priori. This will now be discussed with the aid of a second model equation.

4. Maximal domains of validity. Extension theorem. Stokes paradox

STRUCTURE OF DOMAINS OF VALIDITY. We first discuss some properties of domains of validity which are independent of the special nature of the function $f(x, \epsilon)$.

Let $\mathscr{V}$ be a domain of validity. We shall see how it may be enlarged. It may be easily proved that if $\mathscr{V}_1$ and $\mathscr{V}_2$ are domains with an O-class in common, then the union of $\mathscr{V}_1$ and $\mathscr{V}_2$ is a domain of validity. Thus one may form maximal domains of validity. In practical cases there will generally be one such maximal domain. A maximal domain $\mathscr{V}$ has the obvious property that if X and Z are in $\mathscr{V}$ and $X < Y < Z$ then Y is in $\mathscr{V}$. Various other properties may be found although the intrinsic characterization of a maximal domain of validity remains an unsolved problem. (For comments and speculation, see [2], Chapter VI.) One property, however, will later be of essential use; it is expressed by Kaplun's extension theorem (see [2]):

(4.1) Let $\mathscr{V}$ be a maximal domain of validity containing the O-class Y. Then there exists a class X in $\mathscr{V}$ such that $X < Y$.

PROOF. Assume first that $Y = \text{Ord } 1$. Then, by assumption, given a constant $c > 0$ and a $\delta > 0$, one may find an $\epsilon_\delta > 0$ such that $w(x, \epsilon) < \delta$ for $c \leqq x \leqq 1$, $0 < \epsilon \leqq \epsilon_\delta$. The trick of the proof is to use this property for a sequence of constants c_n and a sequence δ_n, both tending to zero. One may, for instance, put $\delta_n = c_n = \dfrac{1}{n}\cdot$ Since all constants $\dfrac{1}{n}$ belong to Ord (1) we may for each n pick an ϵ_n such that $w < \dfrac{1}{n}$ for $\dfrac{1}{n} \leqq x \leqq 1; 0 < \epsilon \leqq \epsilon_n$. We now define a function $\eta(\epsilon)$ by the requirement $\eta(\epsilon_{n+1}) = \dfrac{1}{n}$ and assume that η is a linear function of ϵ between any two adjacent values ϵ_k and ϵ_{k+1}. We can now find an X such that $\text{Ord } \eta < X < Y$. It is easily seen that any such X is in $\mathscr{V}$. In the general case, when $Y = \text{Ord } \zeta$, we replace x by x/ζ in the preceding argument. Obviously the theorem is true if "$<$" is replaced by "$>$."

An important consequence of the theorem is that a maximal domain of validity cannot contain an infimum or a maximum;[6] it is an open set if the topology is based on intervals (Z, W), $Z < W$.

THE STOKES PARADOX: DETERMINATION OF DOMAINS OF VALIDITY. As will be seen, the extension theorem will play an important role in the matching. However, we obviously need some additional method for determining domains of validity. A fundamental heuristic principle will now be illustrated by an example. Let the function $h(\bar{x}, \epsilon)$ be defined by

$$\frac{d^2h}{d\bar{x}^2} + \frac{1}{\bar{x}}\frac{dh}{d\bar{x}} + \left(\frac{dh}{d\bar{x}}\right)^2 + \epsilon h \frac{dh}{d\bar{x}} = 0, \tag{4.2a}$$

$$h(1, \epsilon) = 0, \text{ and} \tag{4.2b}$$

$$h(\infty, \epsilon) = 1. \tag{4.2c}$$

This is a model for viscous flow at a low Reynolds number. In order to find a first approximation we put $\epsilon = 0$ and obtain an equation, corresponding to the Stokes equation[7]:

$$\frac{d^2h_0}{d\bar{x}^2} + \frac{1}{\bar{x}}\frac{dh_0}{d\bar{x}} + \left(\frac{dh_0}{d\bar{x}}\right)^2 = 0. \tag{4.3}$$

The solution of this which satisfies the condition $h_0 = 0$ at $\bar{x} = 1$ is

$$h_0 = \log(1 + A \log \bar{x}). \tag{4.4}$$

The remaining constant of integration, A, should be chosen so that $h(\infty) = 1$. The impossibility of doing so is the analogue of the Stokes paradox. In some sense, however, (4.4) should be an approximation, although it is not uniformly valid near $x = \infty$. An essential first step in resolving the problem is to transfer the nonuniformity to the origin. We define new variables by

$$x = \epsilon\bar{x}; \qquad f(x, \epsilon) = h(\bar{x}, \epsilon). \tag{4.5}$$

In the new variables the problem is

$$\frac{d^2f}{dx^2} + \frac{1}{x}\frac{df}{dx} + \left(\frac{df}{dx}\right)^2 + f\frac{df}{dx} = 0, \tag{4.6a}$$

and

$$f(\epsilon, \epsilon) = 0; \qquad f(\infty, \epsilon) = 1. \tag{4.6b}$$

[6] This statement, as well as the theorem itself, needs a proviso to cover exceptions arising from the fact that the domain of definition of w may be limited. We shall not belabor this trivial point. The requirement $x \leqq 1$ is purely arbitrary.

[7] The classical Stokes equation is linear. This helped to hide its true significance. It is accidentally linear but the asymptotic principle which leads to it is not, per se, a linearization. Actually, the Stokes equation for compressible flow is nonlinear. Therefore (4.2a) is chosen so that (4.3) retains a nonlinear term.

An easy physical analogy leads to the heuristic assumption that, for any fixed $x > 0$, $\lim f(x, \epsilon) = 1$. As a lowest-order approximation we then take

$$g_0(x) = 1,$$

and expect Ord 1 to be a domain of validity, i.e., $|f - g_0|$ tends to zero uniformly in the semi-infinite interval $x \geqq c$ where c is any constant >0. On the other hand, because of (4.6b), g_0 cannot be valid near $x = \epsilon$. We therefore introduce a stretched coordinate, using the scale ϵ. This leads us back to the original variable $\bar{x}$.[8] We now see that $g_0 = 1$ is the outer solution, analogous to (2.6). It is not uniformly valid near the origin and we need a correction h_0, i.e., a solution satisfying the condition at $x = \epsilon$ but which need not be valid at large x. The solution h_0 as given by (4.4) is an inner solution analogous to (2.11), and there is no a priori reason to expect it to satisfy the boundary condition at $x = \infty$. Thus the Stokes paradox is shown to be due to a faulty interpretation of the Stokes equation (4.3). Following a central idea of singular perturbation theory we instead expect to determine A in (4.4) by matching h_0 with g_0. The matching condition (2.12) obviously makes no sense in the present case and we have to use the more general formulation of Section 3. To do this we first have to determine the domain of validity of h_0.

We note that if we introduce the variable x_η, defined by $\eta(\epsilon)x_\eta = x$, into (4.6a) and then take the formal limit of the resulting equation, keeping x_η fixed, we obtain the "Stokes" equation (4.3) whenever $\eta = o(1)$. In an obvious sense the *formal* domain of validity of (4.3), relative to (4.6a), is the set of O-classes

$$\mathscr{V} = \{X | X < \text{Ord } 1\}. \tag{4.7}$$

We now invoke a heuristic principle (also due to Kaplun).

(4.8) There exists a solution $h_0(x, \epsilon)$ of (4.3) whose maximal domain of validity, as an approximation to $f(x, \epsilon)$, includes the formal domain of validity of (4.3).

This solution must be of the form (4.4) since validity on Ord ϵ necessitates that the boundary condition (4.2b) be satisfied. We now have to find A, possibly as a function of ϵ, by matching. Note that matching should be possible according to our heuristic principles. If Ord 1 is in the maxi-

[8] This circular argument serves to illustrate an important point. Historically, the Stokes equation was derived in the manner in which (4.3) was derived from (4.2). However, from the point of view of theory of singular perturbations, the formulation (4.6) is the principal formulation. It is the solution $g_0(x)$ which is being perturbed, not the solution $h_0(x)$; the latter one is a correction. This point of precedence is an important feature of the theory (see [2], p. 3).

mum domain of validity $\mathscr{U}$ of g_0, then by the extension theorem $\mathscr{U}$ intersects $\mathscr{V}$, which is a domain of validity of h_0 according to (4.8). To actually carry out the matching we use the technique of the intermediate limit. We note that $g_0 - h_0 = 1 - \log[1 + A(\log x_\eta + \log\eta - \log\epsilon)]$ and hence that $\lim_\eta |g_0 - h_0| = 0$ if

$$A = \frac{e-1}{-\log\epsilon}, \tag{4.9}$$

and if η is in the domain

$$\mathscr{D} = \{\text{Ord}\,\eta | \log\eta = o(\log\epsilon)\}. \tag{4.10}$$

We shall now show that $\mathscr{D}$ is actually the overlap domain. We first note that A is unique, within $o(A)$. We must require that: (1) as Ord η varies over the overlap domain, application of $\lim_\eta$ to $|g_0 - h_0|$ does not give contradictory results for A, and that: (2) the overlap domain contain order-classes arbitrarily close to Ord 1. By this is meant that it must have a nonzero intersection with any interval (Ord ζ, Ord 1) for any ζ which is $o(1)$. The requirement follows from the description of $\mathscr{U}$ and $\mathscr{V}$ above. It is seen that $\mathscr{D}$, as determined by (4.10), is the maximal domain satisfying both requirements. Note that, for instance, Ord $\epsilon^{1/2}$ could not be in the overlap domain since one may find functions η satisfying Ord $\epsilon^{1/2} <$ Ord $\eta <$ Ord 1 which give contradictory values of A.

The domain $\mathscr{D}$ contains Ord 1 and even classes greater than Ord 1. Thus the maximal domain of validity of h_0 is slightly larger than the formal domain of validity of (4.3).

In the present case the overlap domain is quite restricted. One finds, for instance, that Ord $\epsilon^\alpha < \mathscr{D} <$ Ord $\epsilon^{-\alpha}$ for any $\alpha > 0$. Still the fact that h_0 and g_0 have a nonempty overlap domain and satisfy the inner (4.6b) and outer (4.6c) boundary conditions, respectively, means that in principle we have obtained an approximation which is uniformly valid to order unity and also that the determination of A by matching is legitimate. Note that the manner in which ϵ occurred in A was not determined a priori; instead it followed from the form of the equation and the matching.

In the example discussed in Section 2 the functions g_0 and h_0 could be obtained by applying a suitable $\lim_\eta$ to the exact solution: g_0 is obtained by choosing $\eta = 1$ (outer limit) and h_0 is obtained by choosing $\eta = \epsilon$ (inner limit). The example discussed in the present section shows that this viewpoint is too narrow. If we apply $\lim_\eta$ with $\eta = \epsilon^\alpha$, $0 \leqq \alpha \leqq 1$, to the function f defined by (4.6), we obtain $\log[(1-\alpha)e + \alpha]$. (Note that in such a limit process f may be replaced by h_0.) Thus, the idea of applying limits to the equation and then finding solutions of the resulting approximate equations is more basic than looking for limits of the solution. In fact, the example in Section 2 could be studied from the point of view adopted in the present section. Also, the function $g_0 \equiv 1$ in this section was regarded as a limit and was determined by physical reasoning. However, if $\lim_\eta$ with $1 = o(\eta)$ is applied to (4.6a) one obtains

$$g_0 \frac{dg_0}{dx} = 0. \tag{4.10}$$

Solving this equation using the boundary condition at infinity yields $g_0 \equiv 1$. Applying $\lim_\eta$ with $\eta = 1$ leaves the equation (4.6a) invariant. Using the extension theorem and the heuristic principle (4.8) we see that we need a solution of (4.6a) which matches both with g_0 (valid at least for $\eta \gg 1$) and with h_0 (valid at least for $\eta \ll 1$). Such a solution is $f \equiv 1$.

COMMENTS. The applied mathematician is interested in actual asymptotic constructions rather than in rigorous theory. As just illustrated, this construction is usually based on physical or heuristic ideas. In the example just studied one may actually prove rigorously that the approximations g_0 and h_0 have the asymptotic validity which was claimed for them on heuristic grounds. However, a genuine mathematical investigation of the heuristic principles themselves seems desirable; in particular, (4.8) should be studied. The basis of practically all perturbation methods is a more or less vague notion that neighboring equations have neighboring solutions. This notion is expressed more concisely by (4.8), but the ideas for proving it (for a suitable class of equations) are still very incomplete (see [2, Chapter IV]). The study of the partially ordered spaces $\mathscr{F}_o$ and $\mathscr{F}_O$ not only has an intrinsic interest but also may be important in the discussion of (4.8).

CALIFORNIA INSTITUTE OF TECHNOLOGY

REFERENCES

1. COLE, JULIAN D., *Perturbation Methods in Applied Mathematics*. Waltham, Mass.: Blaisdell Publishing Company, 1968.
2. KAPLUN, SAUL, *Fluid Mechanics and Singular Perturbations*, A Collection of Papers, P. A. Lagerstrom, L. N. Howard, and C. S. Liu, eds. New York: Academic Press, Inc., 1967.
3. LAGERSTROM, P. A., "Laminar flow theory," in Vol. IV of *High Speed Aerodynamics and Jet Propulsion*, F. K. Moore, ed. Princeton, N.J.: Princeton University Press, 1964.
4. VAN DYKE, MILTON, *Perturbation Methods in Fluid Mechanics*. New York: Academic Press, Inc., 1964.
5. WASOW, WOLFGANG, *Asymptotic Expansions for Ordinary Differential Equations*. New York: John Wiley & Sons, Inc., 1965.

Sequential Convergence in Lattice Groups

SOLOMON LEADER[1]

1. Introduction and basic definitions

We introduce here the concept of *Riesz convergence*, any kind of sequential convergence in a lattice group subject to the four conditions given below. Riesz convergence includes as special cases order convergence and relative uniform convergence in lattice groups as well as sequential convergence in any locally *o*-convex topological vector lattice (see [4]). Each Riesz convergence $\mathscr{C}$ induces a number of related Riesz convergences which are either stronger or weaker than $\mathscr{C}$ and which in special cases coincide with $\mathscr{C}$. We shall be particularly concerned with Riesz convergences induced by seminorms.

All functions and scalars are assumed to be finite, real-valued except where otherwise indicated. For K, any subset of a lattice group, let K^+ consist of all f in K with $f \geqq 0$.

A *Riesz convergence* $\mathscr{C}$ in a lattice group L is a set of sequences in L satisfying:

(1.1) If $[f_n]$ and $[g_n] \in \mathscr{C}$ then $[f_n + g_n] \in \mathscr{C}$.
(1.2) If $f_n = 0$ ultimately then $[f_n] \in \mathscr{C}$.
(1.3) If $[f_n] \in \mathscr{C}$ and $[p(n)]$ is any sequence of positive integers tending to ∞ then $[f_{p(n)}] \in \mathscr{C}$.
(1.4) If $|f_n| \leqq |g_n|$ for all n and $[g_n] \in \mathscr{C}$ then $[f_n] \in \mathscr{C}$.

Define $f_n \xrightarrow{\mathscr{C}} f$ to mean $[f_n - f] \in \mathscr{C}$. In particular $f_n \xrightarrow{\mathscr{C}} 0$ means $[f_n] \in \mathscr{C}$.

PROPOSITION 1. *Let* $f_n \xrightarrow{\mathscr{C}} f$ *and* $g_n \xrightarrow{\mathscr{C}} g$. *Then*

(i) $f_n \wedge g_n \xrightarrow{\mathscr{C}} f \wedge g$,
(ii) $f_n \vee g_n \xrightarrow{\mathscr{C}} f \vee g$,
(iii) $|f_n| \xrightarrow{\mathscr{C}} |f|$.

[1] Work done under NSF-GP 7539.

PROOF. $|f_n \wedge g_n - f \wedge g| \leqq |f_n \wedge g_n - f_n \wedge g| + |f_n \wedge g - f \wedge g| \leqq |g_n - g| + |f_n - f|$. Hence (i) follows from (1.1) and (1.3), (ii) follows similarly with cap replaced by cup, and (iii) follows from (1.4) since $| |f_n| - |f| | \leqq |f_n - f|$.

PROPOSITION 2. *If* $[f_n] \in \mathscr{C}$ *and* $f_0 \in L$ *then* $[f_{n-1}] \in \mathscr{C}$. *Thus, adjoining finitely many terms to a sequence belonging to* $\mathscr{C}$ *always yields a sequence belonging to* $\mathscr{C}$.

PROOF. Let $p(1) = 1$ and $p(n) = n - 1$ for $n > 1$. Then $[f_{p(n)}] \in \mathscr{C}$ by (1.3). By (1.2) $[f_{n-1} - f_{p(n)}] \in \mathscr{C}$. Hence $[f_{n-1}] \in \mathscr{C}$ by (1.1).

PROPOSITION 3. *If* $0 \leqq u_{n+1} \leqq u_n$ *and some subsequence* $[u_{n_k}] \in \mathscr{C}$ *then* $[u_n] \in \mathscr{C}$.

PROOF. By Proposition 2 we may assume that $n_1 = 1$. Let $v_k = u_{n_k}$. Define $p(k)$ to be the last i for which $n_i \leqq k$. $p(k)$ is well defined since $n_1 = 1$ and $n_i < n_{i+1}$. Moreover $p(k) \to \infty$. Hence by (1.3) $[v_{p(k)}] \in \mathscr{C}$. Now $0 \leqq u_k \leqq v_{p(k)}$. Hence $[u_k] \in \mathscr{C}$ by (1.4).

The largest Riesz convergence consists of all sequences in L. The smallest, the discrete convergence $\mathscr{D}$, consists of all sequences which are ultimately zero.

A Riesz convergence $\mathscr{C}$ is *separated* if the only constant sequence belonging to $\mathscr{C}$ is the zero sequence.

PROPOSITION 4. *Let* $\mathscr{C}$ *be a separated Riesz convergence in L. Then*

(i) *If* $w_n \overset{\mathscr{C}}{\to} w$ *and* $w_n \geqq 0$ *then* $w \geqq 0$.

(ii) *If* $0 \leqq u_n \leqq u_{n+1}$ *and* $u_n \overset{\mathscr{C}}{\to} u$ *then* $u_n \leqq u$.

(iii) *If* $f_n \leqq g_n$, $f_n \overset{\mathscr{C}}{\to} f$, *and* $g_n \overset{\mathscr{C}}{\to} g$ *then* $f \leqq g$.

(iv) *If* $g_n \overset{\mathscr{C}}{\to} 0$ *and* $0 \leqq u \leqq |g_n|$ *then* $u = 0$.

PROOF. Given the hypothesis of (i), $w_n \wedge 0 \overset{\mathscr{C}}{\to} w \wedge 0$ by (1.2) and Proposition 1. Moreover $w_n \wedge 0 = 0$. Therefore by (1.2) the constant sequence $[w \wedge 0] \in \mathscr{C}$. Since $\mathscr{C}$ is separated $w \wedge 0 = 0$. Hence (i).

To prove (ii) consider any fixed positive integer r. By (1.3) $u_{r+n} \overset{\mathscr{C}}{\to} u$ since $u_n \overset{\mathscr{C}}{\to} u$. Apply (i) with $w_n = u_{r+n} - u_r$ and $w = u - u_r$.

To prove (iii) apply (i) with $w_n = g_n - f_n$ and $w = g - f$.

To prove (iv) apply (1.4) with $f_n = u$ and use the fact that $\mathscr{C}$ is separated.

A subset $\mathscr{B}$ of $\mathscr{C}$ is called a *base* for $\mathscr{C}$ if $[f_n] \in \mathscr{C}$ implies some $[g_n] \in \mathscr{B}$ with $|f_n| \leqq |g_n|$ for all n.

With each Riesz convergence $\mathscr{C}$ we associate the Riesz convergences $\mathscr{C}_s$ (*$\mathscr{C}$-super*), $\mathscr{C}_m$ (*$\mathscr{C}$-monotone based*), $\mathscr{C}_t$ (*$\mathscr{C}$-truncated*), $\mathscr{C}_d$ (*$\mathscr{C}$-drastic*), and $\mathscr{C}_*$ (*$\mathscr{C}$-star*) defined, respectively, as follows:

(1.5) $[f_n] \in \mathscr{C}_s$ if $[|f_n| \vee |f_{n+1}| \vee \cdots \vee |f_{n+p(n)}|] \in \mathscr{C}$ for every sequence $[p(n)]$ of positive integers.
(1.6) $[f_n] \in \mathscr{C}_m$ if there exists $[u_n] \in \mathscr{C}$ such that $|f_n| \leqq u_n \leqq u_{n-1}$.
(1.7) $[f_n] \in \mathscr{C}_t$ if $[|f_n| \wedge u] \in \mathscr{C}$ for all u in L^+.
(1.8) $[f_n] \in \mathscr{C}_d$ if there exists a sequence of positive integers $p_n \to \infty$ such that $[p_n f_n] \in \mathscr{C}$.
(1.9) $[f_n] \in \mathscr{C}_*$ if every subsequence has a subsequence belonging to $\mathscr{C}$.

Verification that these are Riesz convergences whenever $\mathscr{C}$ is a Riesz convergence is routine and is therefore omitted. Note that each of the operations $s, m, t, d, *$ is idempotent (e.g., $\mathscr{C}_{ss} = \mathscr{C}_s$) and isotone (e.g., $\mathscr{C} \subseteq \mathscr{B}$ implies $\mathscr{C}_s \subseteq \mathscr{B}_s$). $\mathscr{C}_s$, $\mathscr{C}_m$, and $\mathscr{C}_d$ are all contained in $\mathscr{C}$. $\mathscr{C}$ is contained in $\mathscr{C}_t$ and $\mathscr{C}_*$. Also $\mathscr{C}_m \subseteq \mathscr{C}_s$. We call $\mathscr{C}$ an *x-convergence* if $\mathscr{C} = \mathscr{C}_x$, where x can be any of the properties $s, m, d, t, *$. $\mathscr{C}_x$ is the largest x-convergence contained in $\mathscr{C}$ for x either s, m, or d. $\mathscr{C}_x$ is the smallest x convergence containing $\mathscr{C}$ for x either t or $*$. $\mathscr{C}_{st} = \mathscr{C}_{ts}$ since a lattice group is a distributive lattice (see [1]). Since $\mathscr{C}_{tm} = \mathscr{C}_m$, $\mathscr{C}_{tm} = \mathscr{C}_{mt}$ if and only if $\mathscr{C}_m = \mathscr{C}_{mt}$. From Proposition 3 we have $\mathscr{C}_{*m} = \mathscr{C}_m$. But we may have $\mathscr{C}_{m*} \nsubseteq \mathscr{C}_m$ as occurs when $\mathscr{C}$ is norm convergence in $L^1[0, 1]$.

By definition order convergence $\mathscr{O}$ is monotone-based. However if L is a lattice subgroup of a lattice group K then the sequences in L which order-converge to 0 in K form a super convergence in L which need not have a monotone base in L (see [6]).

Relative uniform convergence $\mathscr{U}$ in a lattice group L consists of all $[f_n]$ for which there exist u in L^+ and a sequence of positive integers $q_n \to \infty$ such that $q_n|f_n| \leqq u$ for all n. In vector lattices this agrees with the usual definition (see [1]). $\mathscr{U}$ is a drastic super convergence. $\mathscr{U}$ is separated if and only if L is Archimedean. $\mathscr{U}$ has a monotone base if L is a vector lattice.

Order drastic convergence easily implies relative uniform convergence. The converse holds in Archimedean vector lattices. In particular, for the Archimedean vector lattice of all functions on a set X, pointwise convergence is order convergence. Hence for such spaces pointwise drastic convergence is relative uniform convergence. Pointwise drastic convergence will be studied in Section 2.

In normed vector spaces weak drastic convergence is norm convergence since weakly convergent sequences are norm-bounded and norm convergence is drastic.

For the supremum norm on function spaces norm convergence is a super convergence. We shall characterize seminorms with this property in Theorem 4 of Section 4.

For the supremum-normed space $K(X)$ of compactly supported continuous functions on a locally compact Hausdorff space X, norm-truncated convergence is uniform convergence on each compact subset of X.

For $1 \leqq p < \infty$ norm-super convergence in the Lebesgue space L^p is dominated convergence almost everywhere. [See Corollary 5(b) in Section 4.] Thus the Dominated Convergence Theorem of Lebesgue is included in the triviality $\mathscr{C}_s \subseteq \mathscr{C}$. Norm-truncated super convergence in L^p is convergence almost everywhere.

2. Pointwise convergence in function spaces

In this section let L be a group of functions on a set X under pointwise addition. We assume also that L is a lattice under pointwise ordering. Let $\mathscr{P}$ be pointwise convergence. Then $\mathscr{P}$ is a separated, super, truncated, star convergence. However $\mathscr{P}$ need not be monotone-based.

In the function space $K(X)$ mentioned at the end of Section 1 $\mathscr{P}_m$ consists of all compactly supported sequences which converge uniformly to 0. If X is compact, then $\mathscr{P}_m$ is uniform convergence.

$\mathscr{P}$ need not be drastic. For example, consider the space $\mathbf{c}_0$ of all real sequences converging to 0. Define $f_n(x) = x_n$ for each $x = [x_n]$ belonging to $\mathbf{c}_0$. Then f_n converges pointwise to 0 on $\mathbf{c}_0$. But given any sequence of positive integers $p_n \to \infty$, the sequence $z = [1/p_n]$ belongs to $\mathbf{c}_0$ and $p_n f_n(z) = 1$ for all n. Hence $p_n f_n$ cannot converge pointwise to 0. (Specifically we have shown that w^*-convergence in the dual $\mathbf{l}^1$ of $\mathbf{c}_0$ is not drastic.)

In order to characterize $\mathscr{P}_d$ we need the following lemma:

LEMMA 1. *Every countable subset of* $\mathbf{c}_0$ *is contained in some principal* $\mathbf{l}$-*ideal. That is, given*

(2.1) $a_n(k) \to 0$ *as* $n \to \infty$ *for every positive integer* k, *there exists* $\epsilon_n \to 0$ *and* $M_k > 0$ *such that*

(2.2) $|a_n(k)| \leqq M_k \epsilon_n$ *for all* n, k.

PROOF. Define

$$b_n(k) = \sup_{m \geqq n} |a_m(k)| \tag{2.3}$$

and

$$c_n(k) = \max_{i \leqq k} b_n(i). \tag{2.4}$$

Note that (2.3) is finite by (2.1). Also $b_{n+1}(k) \leqq b_n(k)$ by (2.3). Hence (2.4) implies $c_{n+1}(k) \leqq c_n(k)$. (2.3) and (2.4) imply

$$|a_n(k)| \leqq c_n(k) \leqq c_n(k + 1). \tag{2.5}$$

Moreover (2.1), (2.3), and (2.4) imply

$$c_n(k) \to 0 \quad \text{as } n \to \infty \text{ for all } k. \tag{2.6}$$

Choose d_k such that $0 < d_k < 1$ and $d_k c_1(k) \to 0$ as $k \to \infty$. Then define

$$\epsilon_n = \sup_k d_k c_n(k) \tag{2.7}$$

which is finite since $d_k c_n(k) \leqq d_k c_1(k)$ with the right side tending to 0 as $k \to \infty$. Given $\epsilon > 0$, choose N such that $d_k c_1(k) < \epsilon$ for all $k > N$. Using (2.6) choose M such that $c_n(N) < \epsilon$ for all $n > M$. Then for all $n > M$ we have $d_k c_n(k) \leqq d_k c_n(N) < \epsilon$ for all $k \leqq N$ by (2.5) and $d_k c_n(k) \leqq d_k c_1(k) < \epsilon$ for $k > N$. In any case $n > M$ implies $d_k c_n(k) < \epsilon$ for all k. Hence (2.7) yields $\epsilon_n \leqq \epsilon$ for all $n > M$. That is, $\epsilon_n \to 0$. Set $M_k = 1/d_k$. Then (2.2) follows from (2.5) and (2.7).

THEOREM 1. *Given any sequence of functions f_n on a set X, the following are equivalent:*

(i) (*Pointwise-drastic convergence*) *There exists $a_n \to \infty$ such that $a_n f_n(x) \to 0$ for all x in X.*

(ii) (*Sigma-uniform convergence*) *X is a union of countably many subsets X_k such that $f_n(x) \to 0$ uniformly on each X_k.*

PROOF. Given (i) let X_k consist of all x in X such that $|a_n f_n(x)| < k$ for all n. Clearly each x in X belongs to X_k for k sufficiently large. So the X_k's cover X. Now for all x in X_k we have $|f_n(x)| < k/a_n$ for all n. Hence since $a_n \to \infty$ we have f_n converging uniformly to 0 on X_k.

Conversely let (ii) hold. Since f_n converges pointwise to 0 we may assume each f_n is bounded on each X_k because $X = \bigcup_{j=1}^{\infty} h^{-1}[0, j]$ for $h(x) = \max_n |f_n(x)|$. Then let $a_n(k) = \sup_{x \in X_k} |f_n(x)|$. Thus $0 \leqq a_n(k) < \infty$ and by (ii) we have (2.1) of Lemma 1; hence we have (2.2). Let $a_n = \epsilon_n^{-1/2}$. Given x in X choose k so that x is in X_k. Then $a_n |f_n(x)| \leqq a_n a_n(k) \leqq M_k \epsilon_n^{1/2}$ by (2.2). Since $\epsilon_n \to 0$, $a_n f_n(x) \to 0$.

3. Seminorms

A *seminorm* is an extended real-valued function β on a vector space V such that for f, g in V and c any scalar

$$0 \leqq \beta(f) \leqq \infty \tag{3.1}$$

$$\beta(0) = 0 \tag{3.2}$$

$$\beta(f + g) \leqq \beta(f) + \beta(g) \tag{3.3}$$

$$\beta(cf) = |c|\beta(f). \tag{3.4}$$

A seminorm β is *finite* if $\beta(f) < \infty$ for all f in V. β is *separating* if $\beta(f) > 0$ for all $f \neq 0$. A *norm* is a finite, separating seminorm.

A *Riesz seminorm* is an extended real-valued function β on a lattice group L satisfying (3.1), (3.2), (3.3), and (3.4) for every integer c, and

$$\text{(3.5)} \qquad \text{If} \quad |f| \leqq |g| \quad \text{then} \quad \beta(f) \leqq \beta(g).$$

It is a simple exercise to show that if L is a vector lattice then every Riesz seminorm β on L satisfies (3.4) for every scalar c.

PROPOSITION 5. *Let β on a lattice group L satisfy* (3.1), (3.2), (3.3), *and* (3.5). *Then β is Lipschitz-equivalent to some Riesz seminorm α if and only if there exists a constant $c \geqq 1$ such that*

$$\text{(3.6)} \qquad n\beta(f) \leqq c\beta(nf)$$

for every positive integer n and every f in L.

PROOF. The direct implication is trivial. To prove the converse define $\alpha(f) = \limsup_{n\to\infty} \beta(nf)/n$. Then $\alpha \leqq \beta \leqq c\alpha$ by (3.3) and (3.6). That α is a Riesz seminorm is easily verified.

Note that for a seminorm β on a vector space V β-convergence ($\beta(f_n) \to 0$) satisfies (1.1), (1.2), and (1.3). We are therefore led to investigate conditions under which (1.4) holds when V is a vector lattice.

THEOREM 2. *Let β be a seminorm on a vector lattice V. Then the following are equivalent:*

(i) *β-convergence is a Riesz convergence.* (*If* $|f_n| \leqq |g_n|$ *and* $\beta(g_n) \to 0$ *then* $\beta(f_n) \to 0$.)

(ii) *There exists a constant $c > 0$ such that $\beta(f) \leqq c\beta(g)$ whenever $|f| \leqq |g|$.*

(iii) *β is equivalent to a Riesz seminorm.*

PROOF. That (iii) implies (i) is trivial. Given (i) let S be the solid hull of the unit β-ball. That is, $f \in S$ whenever $|f| \leqq |g|$ for some g with $\beta(g) \leqq 1$. Let $c = \sup_{f\in S} \beta(f)$. Clearly $c \geqq 1$ since S contains the unit β-ball.

We contend $c < \infty$. Otherwise we could choose f_n in S with $\beta(f_n) > n$. Then there would exist g_n with $\beta(g_n) \leqq 1$ and $|f_n| \leqq |g_n|$. Then since $\beta\left(\frac{1}{n} g_n\right) \to 0$, (i) implies $\beta\left(\frac{1}{n} f_n\right) \to 0$. But $\beta\left(\frac{1}{n} f_n\right) = \frac{1}{n}\beta(f_n) > 1$, a contradiction.

Now given any f, g in V with $|f| \leqq |g|$, $\beta(g) = 0$ implies $\beta(f) = 0$ by (i) applied to constant sequences. So $\beta(f) \leqq c\beta(g)$ if $\beta(g) = 0$. For $\beta(g) = \infty$ the inequality is trivial. For $0 < \beta(g) < \infty$ let $h = \dfrac{f}{\beta(g)}$. Then $h \in S$. Hence $\dfrac{\beta(f)}{\beta(g)} = \beta(h) \leqq c$ which completes the proof of (ii).

Given (ii) define $\alpha(g) = \sup_{|w| \leqq |g|} \beta(w)$. Setting $w = g$ one sees that $\beta \leqq \alpha$. By (ii) we have $\alpha \leqq c\beta$. We need only show that α is a Riesz seminorm. (3.1), (3.2), (3.4), and (3.5) are trivial for α. Thus we have only to prove the triangle inequality (3.3) for α. We may assume $f, g \in V^+$ since $\alpha(f + g) \leqq \alpha(|f| + |g|)$ and $\alpha(|f|) = \alpha(f)$ by (3.5). Given w with $|w| \leqq f + g$ the Decomposition Lemma [8] yields u and v in V such that $w^+ = u^+ + v^+$, $w^- = u^- + v^-$, $|u| \leqq f$ and $|v| \leqq g$. Then $w = u + v$. Thus $\beta(w) \leqq \beta(u) + \beta(v) \leqq \alpha(f) + \alpha(g)$. Hence (3.3) holds for α.

Our next result shows that continuity of Riesz seminorms is determined by positive monotone sequences.

THEOREM 3. *Let α and β be Riesz seminorms on a lattice group L. Then the following two conditions are equivalent:*

(i) *Every positive increasing α-Cauchy sequence is β-Cauchy.*

(ii) *β is continuous with respect to α.*

Moreover if L is complete and separated under α then (i) *and* (ii) *are each equivalent to*

(iii) *If $0 \leqq u_{n+1} \leqq u_n$ and $\alpha(u_n) \to 0$ then $\beta(u_n) \to 0$.*

PROOF. That (ii) implies (i) is trivial. Conversely, if (ii) is false there exist $u_n > 0$ in L and $\epsilon > 0$ such that $\alpha(u_n) < 2^{-n}$, but $\beta(u_n) > \epsilon$. Let $s_n = u_1 + \cdots + u_n$. Then $[s_n]$ is α-Cauchy since $\sum_1^\infty \alpha(u_k) < \infty$. But $[s_n]$ is not β-Cauchy since $\epsilon < \beta(u_n) = \beta(s_n - s_{n-1})$. So (i) is false.

(ii) implies (iii) *a fortiori.*

Given (iii) with L complete and separated under α we contend (i) holds. Consider any α-Cauchy sequence $[w_n]$ with $0 \leqq w_n \leqq w_{n+1}$. By completeness w_n α-converges to some w in L. Moreover $w_n \leqq w$ for all n by (ii) of Proposition 4. Applying (iii) with $u_n = w - w_n$ we conclude that $[w_n]$ is β-convergent to w, hence is β-Cauchy. So (iii) implies (i).

4. Seminorm super convergence

Throughout this section β is a Riesz seminorm on a lattice group L except where otherwise indicated.

PROPOSITION 6. *A sequence $[f_n]$ in L is β-super convergent to* 0 *if and only if*

$$\lim_{n\to\infty} \lim_{k\to\infty} a(n, k) = 0, \tag{4.1}$$

where

$$a(n, k) = \beta(|f_n| \vee |f_{n+1}| \vee \cdots \vee |f_{n+k}|). \tag{4.2}$$

PROOF. The left side of (4.1) equals $\inf_n \sup_k a(n, k)$ since $a(n, k) \leqq a(n, k + 1)$ and $a(n + 1, k) \leqq a(n, k + 1)$. For any sequence $[k_n]$ of positive integers $a(n, k_n) \leqq \sup_k a(n, k)$ and therefore $\limsup_{n\to\infty} a(n, k_n) \leqq \limsup_{n\to\infty} \sup_k a(n, k)$. Thus (4.1) implies $\lim_{n\to\infty} a(n, k_n) = 0$ for all $[k_n]$. Conversely, given the latter, we must have $\sup_k a(n, k) < \infty$. Otherwise we could choose j_n so that $a(n, j_n) > n$, a contradiction. Now choose k_n large enough so that $\sup_k a(n, k) + \frac{1}{n} < a(n, k_n)$. Letting n go to ∞ we obtain (4.1).

PROPOSITION 7 (Borel–Cantelli). *If $[f_n]$ is a sequence in L such that $\sum_1^\infty \beta(f_n) < \infty$ then $[f_n]$ is β-super convergent to* 0.

PROOF. For f, g in L^+ we have $0 \leqq f \vee g \leqq f + g$. Thus $\beta(f \vee g) \leqq \beta(f) + \beta(g)$. Hence, in the notation (4.2), $a(n, k) \leqq \sum_n^\infty \beta(f_i)$ which gives (4.1).

PROPOSITION 8. *Every sequence which β-converges to* 0 *has a subsequence which β-super converges to* 0. *Thus β-convergence is equivalent to β-super star convergence.*

PROOF. Let $\mathscr{C}$ be β-convergence. Our first statement is obtained by choosing a β-summable subsequence and applying Proposition 7. Thus $\mathscr{C} \subseteq \mathscr{C}_{s*}$. Conversely $\mathscr{C}_s \subseteq \mathscr{C}$ so that $\mathscr{C}_{s*} \subseteq \mathscr{C}_* = \mathscr{C}$.

PROPOSITION 9. *If $[f_n]$ β-super converges to* 0, *then for every r the sequence $[|f_r| \vee |f_{r+1}| \vee \cdots \vee |f_{r+n}|]$ is β-Cauchy.*

PROOF. We may assume $f_n \geqq 0$. Then for $p \leqq q$ we have $0 \leqq f_r \vee f_{r+1} \vee \cdots \vee f_{r+q} - f_r \vee f_{r+1} \vee \cdots \vee f_{r+p} \leqq f_{r+p} \vee \cdots \vee f_{r+q}$. So $\beta(f_r \vee \cdots \vee f_{r+q} - f_r \vee \cdots \vee f_{r+p}) \leqq \beta(f_{r+p} \vee \cdots \vee f_{r+q})$. Thus Proposition 9 follows from Proposition 6.

A Riesz seminorm β on a lattice group L is called an *M-seminorm* (see [3], [8]) whenever

(4.3) $$\beta(f \vee g) = \beta(f) \vee \beta(g) \quad \text{for all } f, g \text{ in } L^+.$$

M-seminorm convergence is a super convergence. We now consider the converse.

THEOREM 4. *For β any seminorm on a vector lattice L the following are equivalent:*

(i) *β-convergence is a Riesz super convergence.*

(ii) *There exists a constant* $c \geqq 0$ *such that*

$$\beta(f) \leqq c \operatorname*{Max}_{i=1,\ldots,n} \beta(g_i) \quad \textit{whenever } |f| \leqq |g_1| \vee \cdots \vee |g_n|. \tag{4.4}$$

(iii) β *is equivalent to some M-seminorm.*

PROOF. Given (i) let S consist of all f in L for which there exists a finite subset $\{g_1, \ldots, g_n\}$ of the unit β-ball such that $|f| \leqq |g_1| \vee \cdots \vee |g_n|$. Define $c = \sup_{f \in S} \beta(f)$. Then $c \geqq 0$ since $0 \in S$.

We contend that $c < \infty$. Otherwise we could choose f_k in S with $\beta(f_k) > k$ and g_n for all n in a finite set I_k of consecutive integers $n_k, n_k + 1, \ldots, n_{k+1} - 1$ with $\beta(g_n) \leqq 1$ and $|f_k| \leqq \bigvee_{n \in I_k} |g_n|$. Define $h_n = \frac{1}{k} g_n$ for n in I_k. Then $\beta(h_n) = \frac{1}{k}\beta(g_n) \leqq \frac{1}{k}$ for $n \geqq n_k$. Hence $\beta(h_n) \to 0$. Moreover $\left|\frac{1}{k} f_k\right| \leqq \bigvee_{n \in I_k} |h_n|$. Therefore (i) implies $\beta\left(\frac{1}{k} f_k\right) \to 0$. But $\beta\left(\frac{1}{k} f_k\right) > 1$, a contradiction. Thus $c < \infty$.

If $\beta(f) = 0$, then (4.4) is trivial. For $\beta(f) > 0$ consider any $g_1, \ldots, g_n$ such that $|f| \leqq |g_1| \vee \cdots \vee |g_n|$. Let $b = \operatorname*{Max}_{k=1,\ldots,n} \beta(g_k)$. Then $b > 0$. Otherwise $\beta(g_1) = \cdots = \beta(g_n) = 0$ and the sequence $[f_i]$ defined by $f_i = g_k$ for $i \equiv k \pmod{n}$ β-converges to 0. But then (i) would imply $\beta(f) = 0$, contradicting our assumption. Thus $\frac{1}{b} f \in S$. Hence $\frac{1}{b}\beta(f) = \beta\left(\frac{1}{b} f\right) \leqq c$, which gives (4.4). Thus (i) implies (ii).

Let (ii) hold. Now (ii) with $n = 1$ in (4.4) is simply (ii) of Theorem 2. Hence β is equivalent to a Riesz seminorm. Thus we may assume without loss of generality that β is a Riesz seminorm. Given f in L denote by G_f any finite subset $\{g_1, \ldots, g_n\}$ of L^+ such that $|f| \leqq g_1 \vee \cdots \vee g_n$. Define

$$\alpha(f) = \inf_{G_f} \operatorname*{Max}_{g_k \in G_f} \beta(g_k). \tag{4.5}$$

From (4.5) and (4.4) we conclude that $\beta \leqq c\alpha$. For G_f consisting of the singleton f (4.5) implies $\alpha \leqq \beta$. We contend that α is an M-seminorm.

(3.1), (3.2), (3.4), and (4.3) clearly hold for α. To obtain (3.5), simply note that $g_1 \vee \cdots \vee g_n \geqq |g|$ implies $g_1 \vee \cdots \vee g_n \geqq |f|$ whenever $|f| \leqq |g|$. To obtain (3.3), consider any $G_f = \{f_i\}_{i=1,\ldots,m}$ and $G_g = \{g_j\}_{j=1,\ldots,n}$. Then $|f + g| \leqq |f| + |g| \leqq \bigvee_{i,j} (f_i + g_j)$. Thus by (4.5) $\alpha(f + g) \leqq \operatorname*{Max}_{i,j} \beta(f_i + g_j) \leqq \operatorname*{Max}_{i,j} \beta(f_i) + \beta(g_j) = \operatorname*{Max}_{i} \beta(f_i) + \operatorname*{Max}_{j} \beta(g_j)$. Taking infimums on the right over all G_f and G_g, we obtain (3.3) for α from (4.5).

That (iii) implies (i) is clear.

In connection with the proof of Theorem 4 we remark that for any seminorm β on a lattice group L there exists a largest M-seminorm such that $\beta_M \leqq \beta$, namely

$$\beta_M(f) = \operatorname*{Inf}_{G} \operatorname*{Max}_{g \in G} \beta(g) \tag{4.6}$$

where G is any finite subset $\{g_1, \ldots, g_n\}$ of L such that $|g_1| \vee \cdots \vee |g_n| \geqq |f|$. Note also that if β is a Riesz seminorm then (4.4) in Theorem 4 can be replaced by $\beta(u_1 \vee \cdots \vee u_n) \leqq c \operatorname*{Max}_{i=1,\ldots,n} \beta(u_i)$ for all finite subsets $\{u_1, \ldots, u_n\}$ of L^+.

THEOREM 5. *In any Banach lattice norm super convergence has a monotone base.*

PROOF. Let $[f_n]$ be norm super convergent to 0 in the Banach lattice L under the norm β. By Proposition 9 and the completeness of L under β there exists for each n some u_n which is the β-limit of $|f_n| \vee |f_{n+1}| \vee \cdots \vee |f_{n+k}|$ as $k \to \infty$. Then $|f_n| \vee \cdots \vee |f_{n+k}| \leqq u_n$ by (ii) of Proposition 4. Therefore $u_n = |f_n| \vee u_{n+1}$ by Proposition 2 and (ii) of Proposition 1. Thus $u_{n+1} \leqq u_n$. Now in the notation of (4.2), $\beta(u_n) = \lim_{k \to \infty} a(n, k)$ by continuity of norm. Hence $\beta(u_n) \to 0$ by (4.1) of Proposition 6.

COROLLARY 5(a). *In any Banach lattice norm super convergence implies order convergence.*

PROOF. Apply Theorem 5 and (iv) of Proposition 4.

COROLLARY 5(b). *In any Banach lattice the following are equivalent:*

(i) *Norm super convergence equals order convergence.*

(ii) *The norm is monotone-continuous with respect to order convergence: If $u_n \downarrow^{\mathcal{O}} 0$ then $\beta(u_n) \downarrow 0$.*

PROOF. Apply Theorem 5 and Corollary 5(a).

COROLLARY 5(c). *The norm in a Banach lattice is equivalent to an M-norm if and only if norm convergence has a monotone base.*

PROOF. Apply Theorems 4 and 5.

COROLLARY 5(d). *In any Banach lattice the following are equivalent:*

(i) *Norm convergence equals order convergence.*

(ii) *The norm is equivalent to a monotone-continuous M-norm.*

PROOF. Apply Corollaries 5(b) and 5(c).

Given a Banach lattice L let $\mathcal{N}$ be norm convergence, $\mathcal{O}$ order convergence, and $\mathcal{U}$ relative uniform convergence. Clearly

$$\mathcal{O}_{*d} \subseteq \mathcal{U}_{*}. \tag{4.7}$$

Now $\mathcal{N}_s \subseteq \mathcal{O}$ by Corollary 5(a). Starring and applying Proposition 8 we obtain $\mathcal{N} = \mathcal{N}_{s*} \subseteq \mathcal{O}_*$. Operating on this with d we obtain $\mathcal{N} = \mathcal{N}_d \subseteq \mathcal{O}_{*d}$ which, together with (4.7), yields $\mathcal{N} \subseteq \mathcal{U}_*$. On the other hand $\mathcal{U} \subseteq \mathcal{N}$. Thus $\mathcal{U}_* \subseteq \mathcal{N}_* = \mathcal{N}$. Thus we obtain the basic result $\mathcal{N} = \mathcal{U}_*$ of G. Birkhoff [1] for Banach lattices.

5. *L*-seminorms

A Riesz seminorm β on a lattice group L is an *L-seminorm* if (see [2], [8])

$$\beta(f+g) = \beta(f) + \beta(g) \quad \text{for all } f, g \text{ in } L^+. \tag{5.1}$$

THEOREM 6. *For β a seminorm on a vector lattice L the following are equivalent:*

(i) *Given sequences $[f_n]$ and $[g_k]$ in L and a strictly increasing sequence $[n_k]$ of positive integers such that $\beta(g_k) \to 0$ and $\sum_{n_k \leqq n \leqq n_{k+1}} |f_n| \leqq |g_k|$, then* $\lim_{k\to\infty} \sum_{n_k \leqq n < n_{k+1}} \beta(f_n) = 0$.

(ii) *There exists a constant c such that*

$$\sum_{n=1}^{m} \beta(f_n) \leqq c\beta(g) \tag{5.2}$$

for every g in L and

(5.3) *every finite sequence $[f_1, \ldots, f_m]$ in L satisfying* $\sum_{n=1}^{m} |f_n| \leqq |g|$.

(iii) *β is equivalent to an L-seminorm.*

PROOF. Given (i) define $c = \sup \sum_{n=1}^{m} \beta(f_n)$ where the supremum is taken over (5.3) for all g in L with $\beta(g) \leqq 1$.

We contend that $c < \infty$. Otherwise we could obtain $[g_k]$ and $[f_n]$ in L with a partition of the positive integers into intervals $I_k = [n_k, n_{k+1})$ such that $\sum_{n \in I_k} |f_n| \leqq |g_k|$, $\beta(g_k) \leqq 1$, and $\sum_{n\in I_k} \beta(f_n) \geqq k$. Multiplying these inequalities through by $1/k$ we obtain a contradiction to (i).

For g with $\beta(g) = 0$ (5.2) holds for (5.3) because β is equivalent to a Riesz seminorm by (i) and Theorem 2. For $\beta(g) > 0$ let $b = 1/\beta(g)$. Then for (5.3) we have $\sum_{n=1}^{m} |bf_n| \leqq |bg|$ with $\beta(bg) = 1$. Hence $b \sum_{n=1}^{m} \beta(f_n) = \sum_{n=1}^{m} \beta(bf_n) \leqq c$ which implies (5.2). Thus (i) gives (ii).

Given (ii) define

$$\sigma(g) = \sup_{(5.3)} \sum_{n=1}^{m} \beta(f_n). \tag{5.4}$$

Then $\sigma \leqq c\beta$ by (ii). Moreover, for $m = 1$ and $f_1 = g$ in (5.4) we have $\beta \leqq \sigma$. It is clear that σ satisfies (3.1), (3.2), and (3.4). To verify (3.5) let $\sum_1^m |f_n| \leqq |f| \leqq |g|$. Then $|g| - |f| + \sum_1^m |f_n| \leqq |g|$. Thus by (5.4) $\beta(|g| - |f|) + \sum_1^m \beta(f_n) \leqq \sigma(g)$. Hence $\beta(|g| - |f|) + \sigma(f) \leqq \sigma(g)$ by (5.4). Therefore $\sigma(f) \leqq \sigma(g)$. Thus (3.5) holds for σ.

To verify the triangle inequality (3.3) for σ consider f, g in L. Given $h_1, \ldots, h_n$ in L such that $\sum_1^n |h_i| \leqq |f + g|$ there exist f_i, g_i by the Decomposition Lemma (see [8]) such that $f_i + g_i = h_i$, $\sum_1^n |f_i| \leqq |f|$, and $\sum_1^n |g_i| \leqq |g|$. Then $\sum_1^n \beta(h_i) \leqq \sum_1^n \beta(f_i) + \sum_1^n \beta(g_i) \leqq \sigma(f) + \sigma(g)$. Hence $\sigma(f + g) \leqq \sigma(f) + \sigma(g)$ by (5.4) and (5.3).

Thus σ is a Riesz seminorm equivalent to β.

To show that σ is an L-seminorm consider f, g in L^+, $\sum_1^m |f_i| \leqq f$, and $\sum_{m+1}^n |f_i| \leqq g$. Then $\sum_1^n |f_i| \leqq f + g$. Hence by (5.3) and (5.4) $\sigma(f) + \sigma(g) \leqq \sigma(f + g)$. Thus (5.1) holds for σ. Thus (ii) implies (iii).

That (iii) implies (i) is trivial.

COROLLARY 6(a). *For a Riesz seminorm β on a vector lattice L the following are equivalent:*

(i) *If $u_n \in L^+$ and $\lim_{m\to\infty} \beta\left(\sum_1^m u_n\right) < \infty$, then $\sum_1^\infty \beta(u_n) < \infty$.*

(ii) *There exists a constant c such that $\sum_1^m \beta(u_n) \leqq c\beta\left(\sum_1^m u_n\right)$ for every finite sequence $[u_1, \ldots, u_m]$ in L^+.*

(iii) *β is equivalent to an L-seminorm.*

Note that (i) of Corollary 6(a) is equivalent to

(5.5) *If $0 \leqq w_n \leqq w_{n+1}$ and $\lim_{n\to\infty} \beta(w_n) < \infty$, then $\sum_1^\infty \beta(w_{n+1} - w_n) < \infty$.*

Now (5.5) implies

(5.6) *If $0 \leqq w_n \leqq w_{n+1}$ and $\lim_{n\to\infty} \beta(w_n) < \infty$, then* $[w_n]$ is *β-Cauchy.*

Property (5.6) holds for L^p norms with $1 \leqq p < \infty$.

6. *L*-norms and *M*-norms

PROPOSITION 10. *If a vector lattice V admits a norm σ which is both an L-norm and an M-norm, then V is at most one-dimensional.*

PROOF. If $f \wedge g = 0$ then $f + g = f \vee g$. Thus $\sigma(f) + \sigma(g) = \sigma(f + g) = \sigma(f \vee g) = \sigma(f) \vee \sigma(g)$. Hence either $f = 0$ or $g = 0$. Now for arbitrary f, g in V we have $(f - f \wedge g) \wedge (g - f \wedge g) = 0$. Thus the preceding argument implies either $f = f \wedge g$ or $g = f \wedge g$. That is, either $f \leqq g$ or $g \leqq f$. Thus V is simply ordered. Moreover V is Archimedean since σ is a Riesz norm. As is well known (see [1], Theorem 15, p. 226) every simply ordered Archimedean lattice group is isomorphic to a sub-lattice group of the reals.

THEOREM 7. *If a norm on a vector lattice V is equivalent to both an L-norm α and an M-norm β then V is finite dimensional.*

PROOF. We prove Theorem 7 by three lemmas. Since α and β are equivalent there exists a positive constant c such that $\alpha \leqq c\beta$ and $\beta \leqq c\alpha$. Given disjoint elements $u_1, \ldots, u_n$ of V^+ with $\alpha(u_i) = 1$ let

$$u = u_1 + \cdots + u_n = u_1 \vee \cdots \vee u_n. \tag{6.1}$$

Then $\alpha(u) = \alpha(u_1) + \cdots + \alpha(u_n) = n$ and $\beta(u) = \beta(u_1) \vee \cdots \vee \beta(u_n) \leqq c$. Hence $n = \alpha(u) \leqq c\beta(u) \leqq c^2$. We thus have

LEMMA 7(a). *There exists a finite subset $F = \{u_1, \ldots, u_n\}$ of V^+ of maximum cardinality n such that the u_i's are disjoint α-unit vectors. Then u, defined by (6.1), is a weak unit (see [1]) in V.*

Let V_i be the subspace of V for which u_i is a weak unit. That is,

$$f \in V_i \quad \text{if } x \wedge u_i = 0 \text{ implies } x \wedge |f| = 0. \tag{6.2}$$

Of course, V_i is a subvector lattice of V. Given f, g in V_i consider $(f - g)^+$ and $(f - g)^-$ in V_i^+. If neither of these were 0 we could multiply by positive scalars b, d to convert them into α-unit vectors and replace u_i in F by $b(f - g)^+$ and $d(f - g)^-$ to obtain $n + 1$ disjoint α-unit vectors, contradicting Lemma 7(a). Therefore either $(f - g)^+ = 0$ or $(f - g)^- = 0$. That is, either $f \leqq g$ or $g \leqq f$. We thus have

LEMMA 7(b). *V_i as defined by (6.2) is simply ordered and hence one-dimensional. Thus $f = \alpha(f)u_i$ for every f in V_i^+.*

Now consider any f in V^+. Since u is a weak unit and α is an L-norm, $\alpha(f - f \wedge ku) \downarrow 0$ as $k \uparrow \infty$. Applying Lemma 7(b) and summing over i we obtain

$$\sum_{i=1}^{n} \alpha(f \wedge ku_i)u_i = \sum_{i=1}^{n} f \wedge ku_i = f \wedge ku. \tag{6.3}$$

For fixed i, $\alpha(f \wedge ku_i)$ is nondecreasing in k and is bounded by $\alpha(f)$. Hence we can define

$$(6.4) \qquad a_i = \lim_{k\to\infty} \alpha(f \wedge ku_i).$$

Letting $k \to \infty$ in (6.3) we obtain

LEMMA 7(c). *If f is in V^+ then $f = \sum_{i=1}^{n} \alpha_i u_i$ with a_i defined by* (6.4).

7. Monotonely continuous seminorms

THEOREM 8. *Let $\mathscr{C}$ be a Riesz convergence in a lattice group L and α be a Riesz seminorm on L. Then the following are equivalent:*

(i) (α *is monotonely $\mathscr{C}$-continuous.*) *If $u_n \downarrow^{\mathscr{C}} 0$ then $\alpha(u_n) \downarrow 0$.*
(ii) *If $[f_n] \in \mathscr{C}_s$ and for each r the sequence $[|f_r| \vee |f_{r+1}| \vee \cdots \vee |f_{r+n}|]$ is α-Cauchy, then f_n α-super converges to* 0.

PROOF. That (ii) implies (i) is trivial. Given (i) and the hypothesis of (ii) consider any sequence $[k(n)]$ of positive integers. Let $g_n = |f_n| \vee |f_{n+1}| \vee \cdots \vee |f_{n+k(n)}|$. To prove (ii) we must show that $\alpha(g_n) \to 0$. Given $\epsilon > 0$ use the Cauchy condition in (ii) to choose $m(i)$ for each positive integer i such that

$$(7.1) \quad \alpha(|f_i| \vee \cdots \vee |f_{i+p}| - |f_i| \vee \cdots \vee |f_{i+q}|) < \frac{\epsilon}{2^i} \quad \text{for all } p, q \geqq m(i).$$

Let $r(i)$ be the larger of $k(i)$ and $m(i)$. Then let $h_n = |f_n| \vee |f_{n+1}| \vee \cdots \vee |f_{n+r(n)}|$. Since $[f_n] \in \mathscr{C}_s$, $[h_n] \in \mathscr{C}$. Therefore we can apply (i) to $u_n = h_1 \wedge \cdots \wedge h_n$ to conclude $\alpha(u_n) \downarrow 0$. Now $h_n - \sum_{i=1}^{n} (h_n - h_i)^+ \leqq h_n - (h_n - h_i)^+ \leqq h_i$ for $i = 1, \ldots, n$. Hence $h_n - \sum_{i=1}^{n} (h_n - h_i)^+ \leqq u_n$. Thus $0 \leqq g_n \leqq h_n \leqq u_n + \sum_{i=1}^{n} (h_n - h_i)^+$. Therefore

$$(7.2) \qquad \alpha(g_n) \leqq \alpha(u_n) + \sum_{i=1}^{n} \alpha(h_n - h_i)^+.$$

Now $(h_n - h_i)^+ = h_n \vee h_i - h_i \leqq h_i \vee h_{i+1} \vee \cdots \vee h_n - h_i$. Hence by (7.1) and the definition of h_i we conclude that

$$(7.3) \qquad \alpha(h_n - h_i)^+ < \frac{\epsilon}{2^i}.$$

Combining (7.2) and (7.3) we obtain $\alpha(g_n) \leqq \alpha(u_n) + \sum_{i=1}^{n} \frac{\epsilon}{2^i} < \alpha(u_n) + \epsilon$. Hence $\limsup_{n\to\infty} \alpha(g_n) \leqq \epsilon$ since $\alpha(u_n) \downarrow 0$. Since ϵ is arbitrary we have (ii).

Note that if we delete "super" in the conclusion of (ii) the modified (ii) is still equivalent to (i).

COROLLARY 8(a) (Dominated Convergence Theorem). *Let $\mathscr{C}$ be a Riesz convergence in a lattice group L and α a Riesz seminorm on L such that $0 \leqq u_n \leqq u_{n+1} \leqq u$ implies $[u_n]$ is α-Cauchy. Then the following are equivalent:*

(i) *α is monotonely $\mathscr{C}$-continuous.*

(ii) *If $[f_n] \in \mathscr{C}$ and is order-bounded then $\alpha(f_n) \to 0$.*

The next theorem was proved for order convergence by Zaanen and Luxemburg [6, Theorem 63.2 in Note XVIa]. With minor modifications their proof is valid for any Riesz convergence. However, we give a slightly different proof here.

THEOREM 9. *Let $\mathscr{C}$ be a Riesz convergence in a lattice group L and α a Riesz seminorm on L. Then the following are equivalent:*

(i) *(α is weakly monotonely $\mathscr{C}$-continuous.) If $u_n \downarrow^{\mathscr{C}} 0$ and $[u_n]$ is α-Cauchy then $\alpha(u_n) \downarrow 0$.*

(ii) *If $[f_n] \in \mathscr{C}$ and is α-Cauchy then $\alpha(f_n) \to 0$.*

PROOF. That (ii) implies (i) is trivial. Given (i) and the hypothesis of (ii), $[\alpha(f_n)]$ is a Cauchy sequence of real numbers since $|\alpha(f_n) - \alpha(f_m)| \leqq \alpha(f_n - f_m)$. Thus $\alpha(f_n) \to c$ for some real $c \geqq 0$. To show that $c = 0$ we shall find for any given $\epsilon > 0$ a subsequence $[f_{n(k)}]$ such that for $g_k = |f_{n(k)}|$,

$$c = \lim_{k \to \infty} \alpha(g_k) \leqq \epsilon. \tag{7.4}$$

We choose our subsequence so that

$$\alpha(g_k - g_{k+p}) < \frac{\epsilon}{2^k} \quad \text{for all } p. \tag{7.5}$$

This is done by choosing $n(k) > n(k-1)$ so that $\alpha(f_n - f_m) < \frac{\epsilon}{2^k}$ for all $m, n \geqq n(k)$. Since $g_k - \sum_{i=1}^{k} |g_k - g_i| \leqq g_k - |g_k - g_i| \leqq g_i$ for $i = 1, \ldots, k$ we conclude that the left side is a lower bound of $u_k = g_1 \wedge \cdots \wedge g_k$. Thus $0 \leqq g_k \leqq u_k + \sum_{i=1}^{k} |g_k - g_i|$. Hence by (7.5) we have

$$\left\{\begin{aligned} 0 \leqq \alpha(g_k) &\leqq \alpha(u_k) + \sum_{i=1}^{k} \alpha(g_k - g_i) < \alpha(u_k) \\ &+ \sum_{i=1}^{k} \frac{\epsilon}{2^i} < \alpha(u_k) + \epsilon. \end{aligned}\right. \tag{7.6}$$

Now $0 \leqq u_k - u_{k+p} = \sum_{i=k}^{k+p-1} (u_i - u_{i+1}) \leqq \sum_{i=k}^{k+p-1} (g_i - g_{i+1})$. Hence by (7.5), $\alpha(u_k - u_{k+p}) \leqq \sum_{i=k}^{k+p-1} \alpha(g_i - g_{i+1}) < \sum_{i=k}^{\infty} \frac{\epsilon}{2^i} \leqq \frac{\epsilon}{2^{k-1}}$ for all p. Thus $[u_k]$ is α-Cauchy. Moreover, since $0 \leqq u_{k+1} \leqq u_k \leqq g_k = |f_{n(k)}|$, $u_k \downarrow^{\mathscr{C}} 0$ by (1.3) and (1.4). Thus (i) and (7.6) give (7.4).

COROLLARY 9(a). *If an α-Cauchy sequence $[f_n]$ is α-super truncatedly convergent to* 0 *then it is α-convergent to* 0.

PROOF. This is (ii) of Theorem 9 with $\mathscr{C}$ taken to be α-super truncated convergence. Hence we need only verify (i) of Theorem 9 for such $\mathscr{C}$. But this is trivial since $\mathscr{C}_{tm} = \mathscr{C}_m$.

8. Limit units

Let $\mathscr{C}$ be a Riesz convergence in a lattice group L. An element u of L^+ *$\mathscr{C}$-absorbs* an element f of L if $[(|f| - nu)^+] \in \mathscr{C}$. That is,

$$|f| \wedge nu \xrightarrow{\mathscr{C}} |f|. \tag{8.1}$$

u is a *$\mathscr{C}$-limit unit* for L if u $\mathscr{C}$-absorbs every f in L. This concept of limit unit was introduced in [5] for topological Riesz convergences.

PROPOSITION 11. *If α is a Riesz seminorm on L, u an α-limit unit in L. and $[f_n]$ an α-Cauchy sequence in L, then u uniformly α-absorbs f_n.*

PROOF. Given $\epsilon > 0$ choose k large enough so that

$$\alpha(f_i - f_j) < \frac{\epsilon}{3} \quad \text{for all } i, j \geqq k. \tag{8.2}$$

Then choose M large enough so that

$$\alpha(f_j - f_j \wedge mu) < \frac{\epsilon}{3} \quad \text{for all } m \geqq M \text{ and } j \leqq k. \tag{8.3}$$

Now $0 \leqq (f_n - mu)^+ = f_n - f_n \wedge mu \leqq |f_n - f_k| + |f_k - f_k \wedge mu| + |f_k \wedge mu - f_n \wedge mu| \leqq 2|f_n - f_k| + (f_k - mu)^+$. Hence

$$\alpha(f_n - mu)^+ \leqq 2\alpha(f_n - f_k) + \alpha(f_k - mu)^+. \tag{8.4}$$

From (8.2), (8.3), and (8.4) we conclude that

$$\alpha(f_n - mu)^+ < \epsilon \quad \text{for all } m \geqq M \text{ and all } n. \tag{8.5}$$

PROPOSITION 12. *If f_n converges α-truncatedly to* 0 *and is uniformly α-absorbed by some u, then $\alpha(f_n) \to 0$.*

PROOF. From (8.5) and the triangle inequality $\alpha(f_n) \leqq \alpha(f_n - Mu)^+ + \alpha(f_n \wedge Mu)$ we have $\alpha(f_n) \leqq \epsilon + \alpha(f_n \wedge Mu)$ for all n. As $n \to \infty$ the last term goes to 0. Thus $\limsup_{n\to\infty} \alpha(f_n) \leqq \epsilon$ for all $\epsilon > 0$.

PROPOSITION 13. *If L has an α-limit unit then every α-Cauchy sequence which converges α-truncatedly to* 0 *must α-converge to* 0.

PROOF. Apply Propositions 11 and 12.

PROPOSITION 14. *A Banach lattice L has a norm-limit unit if and only if there exists a countable subset E of L whose solid hull H is norm-dense in L.*

PROOF. Let α be the norm in L. If u is an α-limit unit let E consist of all integral multiples of u. That E has the stated properties is then trivial.

Conversely, given E with the stated properties we may assume that E is contained in L^+. Enumerate the members of E to obtain a sequence $[u_n]$. Then choose coefficients $a_n > 0$ so that $\sum_{n=1}^{\infty} a_n\alpha(u_n) < \infty$. Define $u = \sum_{n=1}^{\infty} a_n u_n$ which exists because L is complete. Consider any f in L and any h in H. Then $|h| \leqq u_n$ for some n. Hence $|h| \leqq mu$ for $m \geqq 1/a_n$. Thus by (8.4) we have $\alpha(f - f \wedge mu) \leqq 2\alpha(f - h) + \alpha(h - h \wedge mu) = 2\alpha(f - h)$ for m sufficiently large. Therefore

$$\lim_{m\to\infty} \alpha(f - f \wedge mu) \leqq 2\alpha(f - h) \quad \text{for all } h \text{ in } H. \tag{8.6}$$

Since H is α-dense in L the left side of (8.6) must be 0. Thus u is an α-limit unit for L.

PROPOSITION 15. *Every separable Banach lattice has a norm-limit unit.*

PROOF. Apply Proposition 14.

PROPOSITION 16. *In a Banach lattice L with norm α every α-convergent sequence is uniformly α-absorbed by some u in L^+.*

PROOF. Apply Proposition 14 to the Banach lattice generated in L by the convergent sequence. Then apply Proposition 11.

RUTGERS UNIVERSITY
THE STATE UNIVERSITY OF NEW JERSEY

REFERENCES

1. BIRKHOFF, G., *Lattice Theory*. New York: A.M.S. Coll. Pub. 25, 1948.
2. KAKUTANI, S., "Concrete representations of abstract L-spaces and the mean ergodic theorem," *Ann. of Math., 42* (1941), 523–537.

3. Kakutani, S. "Concrete representations of abstract *M*-spaces," *Ann. of Math., 42* (1941), 994–1024.

4. Kist, J., "Locally *o*-convex spaces," *Duke Math. J., 25* (1958), 569–582.

5. Leader, S., "Separation and approximation in topological vector lattices," *Can. J. Math., 11* (1959), 286–296.

6. Luxemburg, W. A. J., and A. C. Zaanen, "Notes on Banach function spaces," *Proc. Acad. Sci. Amsterdam,*

A66 (1963)		A67 (1964)		A68 (1965)	
I	135–147	VIII	104–119	XIVa	229–239
II	148–153	IX	360–376	XIVb	240–248
III	239–250	X	493–506	XVa	415–429
IV	251–263	XI	507–518	XVb	430–446
V	496–504	XII	519–529	XVIa	646–657
VI	655–668	XIII	530–543	XVIb	658–667
VII	669–681				

7. Namioka, I., *Partially Ordered Linear Topological Spaces.* Providence: A.M.S. Memoir 24, 1957.

8. Peressini, A. L., *Ordered Topological Vector Spaces.* New York: Harper and Row, (1967).

A Group-theoretic Lattice-point Problem

BURTON RANDOL

1.

Let $G = SL(2, R)$, $\Gamma = SL(2, Z)$, and let P be the set of primitive integral lattice-points in R^2. For $g \in G$ and $r > 0$, let $N_r(g)$ be the number of points in the set $g(P)$ which intersect the disk $|x| < r$. Then, inasmuch as $\gamma(P) = P$ for any $\gamma \in \Gamma$, it is evident that $N_r(g)$ can be regarded as a function on the quotient space G/Γ, which has a normalized Haar measure dg. Moreover, as a result of Siegel, the mean, or integral, of $N_r(g)$ over G/Γ is simply $(\zeta(2))^{-1}\pi r^2$ (see [4], Formula 25).

Consider now the following question. What can we say about the variance of $N_r(g)$ over G/Γ? That is, what is the asymptotic behavior, as $r \to \infty$, of

$$v(r) = \int_{G/\Gamma} (N_r(g) - (\zeta(2))^{-1}\pi r^2)^2\, dg?$$

It will be shown here that, for any integer k, $v(r) = (\zeta(2))^{-1}\pi r^2 + 0(r^2 \log^{-k} r)$. Before passing to the proof, I would like to thank Robert Langlands for several informative conversations concerning $L^2(G/\Gamma)$.

2.

Suppose $f(x)$ is C^∞ with compact support in $R^2 - \{0\}$. Suppose, for the sake of simplicity, and since this is the only case with which we will be concerned, that $f(x)$ is radial, i.e., $f(Tx) = f(x)$ for $T \in SO(2)$. Define $\Theta(g) = \sum_{p \in P} f(g(p))$. Then

$$\int_{G/\Gamma} |\Theta(g)|^2\, dg = (2i\zeta(2))^{-1} \int_{\operatorname{Re} s = c} L_{\hat f}(2 - 2s) L_{\hat f}(2s)\, ds \tag{1}$$
$$+ (2i\zeta(2))^{-1} \int_{\operatorname{Re} s = c} \frac{\zeta(2 - 2s)}{\zeta(2s)} L_f(2 - 2s) L_f(2s)\, ds,$$

where $\hat f(x) = \int_{R^2} f(y) e^{-2\pi i(x,y)}\, dy$, c is any real number greater than 1, and $L_g(z) = 2 \int_0^\infty g(\rho)\rho^{z-1}\, d\rho$ for a radial function $g(\rho)$.

This result comes from combining Formulas 9 and 22 in [1]. Godement makes the additional hypothesis that $\int_{R^2} f(x)\,dx = 0$, but only to avoid the presence of a pole of $L_f(2-2s)$ at $s = 1$, and the correctness of (1) is not affected by removing this hypothesis (see also [3, 10a]).

The constant which precedes the integrals in (1) differs from the constant which precedes the integral in Formula 9 of [1]. The reason for this is two-fold. The latter constant should be $(2\pi i)^{-1}$ instead of π^{-2}, assuming that the measure of the circle-group is normalized, and we have, in addition, divided by $\pi^{-1}\zeta(2)$ in order to normalize the measure of G/Γ (see [2], p. 145).

Suppose now that $\{\phi_n(x)\}$ is a sequence of C^∞ radial functions having compact support in $R^2 - \{0\}$, and converging, except at the origin, to the indicator function of the unit disk, in the manner indicated cross-sectionally in Fig. 1.

Define $\phi_{n,r}(x) = \phi_n\left(\dfrac{x}{r}\right)$, and let $\phi_r(x)$ be the indicator function of the disk $|x| < r$. Then $\phi_{n,r}(x) \to \phi_r(x)$ monotonically, except at the origin. Define $N_{n,r}(g) = \sum_{p \in P} \phi_{n,r}\{g(p)\}$. Then it follows from Beppo Levi's theorem that

$$\int_{G/\Gamma} |N_{n,r}(g)|^2\,dg \to \int_{G/\Gamma} |N_r(g)|^2\,dg.$$

But by (1),

$$\int_{G/\Gamma} |N_{n,r}(g)|^2\,dg = (2i\zeta(2))^{-1} \int_{\operatorname{Re} s = c} L_{\phi_{n,r}}(2-2s)L_{\phi_{n,r}}(2s)\,ds$$

$$+ (2i\zeta(2))^{-1} \int_{\operatorname{Re} s = c} \frac{\zeta(2-2s)}{\zeta(2s)} L_{\hat{\phi}_{n,r}}(2-2s)L_{\phi_{n,r}}(2s)\,ds. \tag{2}$$

Now $L_{\phi_{n,r}}(z)$ is an entire function of z, inasmuch as $\phi_{n,r}(x)$ is C^∞ with compact support in $R^2 - \{0\}$. Moreover, repeated integration by parts

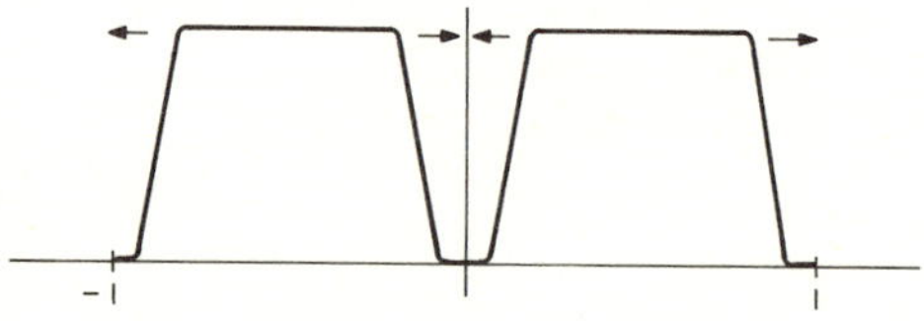

Fig. 1

shows that $L_{\phi_{n,r}}(z)$ is of rapid decrease in any vertical strip. The Fourier transform of $\phi_{n,r}(x)$ is again a radial function, C^{∞}, of rapid decrease as $|x| \to \infty$, and satisfying $\hat{\phi}_{n,r}(0) \to \pi r^2$ as $n \to \infty$. From this it follows by repeated integration by parts that $L_{\hat{\phi}_{n,r}}(2 - 2s)$ is of rapid decrease near infinity in any vertical strip and is meromorphic with a simple pole at $s = 1$, having residue c_n, with $c_n \to -2\pi r^2$, and possible additional poles at $s = 2, 3, 4, \ldots$. (There are no poles at $s = 3/2, 5/2, 7/2, \ldots$ since $\hat{\phi}_{n,r}(x)$ is radial, and hence its odd derivatives in the ρ-coordinate vanish at $\rho = 0$.) Now $\zeta(2 - 2s)$ has simple zeros at $s = 2, 3, 4, \ldots$, and $(\zeta(2s))^{-1}$ is holomorphic in a neighborhood of $\operatorname{Re} s \geqq \frac{1}{2}$ and vanishes only at $s = \frac{1}{2}$; thus we conclude that the expression $\dfrac{\zeta(2 - 2s)}{\zeta(2s)} L_{\hat{\phi}_{n,r}}(2 - 2s)L_{\phi_{n,r}}(2s)$ is meromorphic in a neighborhood of $\operatorname{Re} s \geqq \frac{1}{2}$, with a single simple pole at $s = 1$. Furthermore, since $L_{\hat{\phi}_{n,r}}(2 - 2s)$ is rapidly decreasing in any strip, and since $\dfrac{\zeta(2 - 2s)}{\zeta(2s)}$ is of no worse than polynomial growth in a neighborhood of any strip $\frac{1}{2} \leqq \operatorname{Re} s \leqq a$ (see [5], 3.6.5 and 5.1.1), it follows that $\dfrac{\zeta(2 - 2s)}{\zeta(2s)} L_{\phi_{n,r}}(2 - 2s)L_{\phi_{n,r}}(2s)$ is rapidly decreasing in a neighborhood of any strip $\frac{1}{2} \leqq \operatorname{Re} s \leqq a$. Moreover, since $L_{\hat{\phi}_{n,r}}(2 - 2s) \sim -\dfrac{\pi r^2}{s - 1}$ as $s \to 1$, $L_{\phi_{n,r}}(2s) \to \dfrac{r^{2s}}{s}$, and $\zeta(0) = -\frac{1}{2}$, the residue of

$$\frac{\zeta(2 - 2s)}{\zeta(2s)} L_{\hat{\phi}_{n,r}}(2 - 2s)L_{\phi_{n,r}}(2s)$$

at $s = 1$ tends to $\dfrac{\pi r^4}{2\zeta(2)}$. Thus, by moving the line of integration to $\operatorname{Re} s = \frac{1}{2}$, it follows from (2) that

$$\int_{G/\Gamma} |N_{n,r}(g)|^2 \, dg = C_n + (2i\zeta(2))^{-1} \int_{\operatorname{Re} s = 1/2} L_{\phi_{n,r}}(2 - 2s)L_{\phi_{n,r}}(2s) \, ds$$
$$+ (2i\zeta(2))^{-1} \int_{\operatorname{Re} s = 1/2} \frac{\zeta(2 - 2s)}{\zeta(2s)} L_{\hat{\phi}_{n,r}}(2 - 2s)L_{\phi_{n,r}}(2s) \, ds,$$

where $C_n \to (\zeta(2))^{-2}\pi^2 r^4$.

Now by an exponential change of coordinates, the Plancherel theorem for R^1, and the fact that $2s = \overline{2 - 2s}$ on $\operatorname{Re} s = \frac{1}{2}$, we find that in general

$$\begin{aligned} \int_{\operatorname{Re} s = 1/2} |L_f(2s)|^2 |ds| &= \int_{\operatorname{Re} s = 1/2} |L_f(2 - 2s)|^2 |ds| \\ &= 4\pi \int_0^{\infty} |f(\rho)|^2 \rho \, d\rho \\ &= 2 \int_{R^2} |f(x)|^2 \, dx. \end{aligned} \tag{3}$$

From this it is evident that in the sense of L^2 convergence on the line $\operatorname{Re} s = \frac{1}{2}$, $L_{\phi_{n,r}}(2s) \to L_{\phi_r}(2s) = \frac{r^{2s}}{s}$, and $L_{\phi_{n,r}}(2-2s) \to L_{\phi_r}(2-2s) = \frac{r^{2-2s}}{1-s}$. Moreover, since $\hat{\phi}_{n,r}(x) \to \hat{\phi}_r(x)$ in $L^2(R^2)$, it follows in a similar way that $L_{\hat{\phi}_r}(2-2s)$ and $L_{\hat{\phi}_{n,r}}(2-2s)$ are L^2 on $\operatorname{Re} s = \frac{1}{2}$, and that $L_{\hat{\phi}_{n,r}}(2-2s) \to L_{\hat{\phi}_r}(2-2s)$ in the sense of L^2 convergence on $\operatorname{Re} s = \frac{1}{2}$. This implies that

$$\int_{G/\Gamma} |N_r(g)|^2\, dg = (\zeta(2))^{-2}\pi^2 r^4 + (2i\zeta(2))^{-1} \int_{\operatorname{Re} s = 1/2} \frac{r^2}{s(1-s)}\, ds$$
$$+ (2i\zeta(2))^{-1} \int_{\operatorname{Re} s = 1/2} \frac{\zeta(2-2s)}{\zeta(2s)} L_{\hat{\phi}_r}(2-2s) \frac{r^{2s}}{s}\, ds.$$

Now

$$(2i\zeta(2))^{-1} \int_{\operatorname{Re} s = 1/2} \frac{r^2}{s(1-s)}\, ds = (\zeta(2))^{-1}\pi r^2,$$

so

$$\int_{G/\Gamma} |N_r(g)|^2\, dg = (\zeta(2))^{-2}\pi^2 r^4 + (\zeta(2))^{-1}\pi r^2$$
$$+ (4\zeta(2))^{-1} r \int_{-\infty}^{\infty} \frac{\zeta(1-it)}{\zeta(1+it)} L_{\hat{\phi}_r}(1-it) \frac{r^{it}}{1+it}\, dt.$$

But $L_{\hat{\phi}_r}(z) = r^{2-z} L_{\hat{\phi}_1}(z)$, and the mean of $N_r(g)$ is $(\zeta(2))^{-1}\pi r^2$, so we conclude that

$$\int_{G/\Gamma} (N_r(g) - \zeta(2)^{-1}\pi r^2)^2\, dg$$
$$= (\zeta(2))^{-1}\pi r^2 + (4\zeta(2))^{-1} r^2 \int_{-\infty}^{\infty} \frac{\zeta(1-it)}{\zeta(1+it)} L_{\hat{\phi}_r}(1-it) \frac{r^{2it}}{1+it}\, dt.$$

Now for any positive integer N,

$$\zeta(1 \pm it) = \sum_{n=1}^{N} \frac{1}{n^{1\pm it}} + (1 \pm it) \int_N^{\infty} \frac{[x] - x + \frac{1}{2}}{x^{2\pm it}}\, dx + \frac{N^{\pm it}}{\pm it} - \frac{N^{\pm it}}{2N}$$

(see [5], 3.5.3), and by differentiating this formula k times, and then setting $N = [t]$ in the result, we find that $\zeta^{(k)}(1 \pm it) = 0(\log^{k+1} t)$. This, combined with the fact that $\{\zeta(1+it)\}^{-1} = 0(\log^7 t)$ (see [5], 3.6.5), shows that any fixed derivative of $\frac{\zeta(1-it)}{\zeta(1+it)} \frac{1}{1+it}$ is in $L^2(-\infty, \infty)$.

On the other hand, $\hat{\phi}_1(x) = |x|^{-1} J_1(2\pi|x|) = 0(|x|^{-3/2})$, and since $\frac{d^k}{dt^k}\{L_{\hat{\phi}_1}(1-it)\} = (-i)^k L_g(1-it)$, where $g(x) = \hat{\phi}_1(x) \log^k |x|$, it is evident that $g(x)$ is in $L^2(R^2)$, and hence, by (3), any derivative of $L_{\hat{\phi}_1}(1 - it)$ is in $L^2(-\infty, \infty)$. It follows that we can integrate the expression

$$(4\zeta(2))^{-1}r^2 \int_{-\infty}^{\infty} \frac{\zeta(1 - it)}{\zeta(1 + it)} L_{\hat{\phi}_1}(1 - it) \frac{r^{2it}}{1 + it}\, dt$$

by parts, writing $r^{2it} = e^{2it \log r}$, and this shows that for any integer k, $v(r) = (\zeta(2))^{-1}\pi r^2 + 0(r^2 \log^{-k} r)$.

YALE UNIVERSITY AND
THE GRADUATE CENTER
CITY UNIVERSITY OF NEW YORK

REFERENCES

1. GODEMENT, R., "The decomposition of $L^2(G/\Gamma)$ for $\Gamma = SL(2, Z)$," *Proceedings of the Symposium on Algebraic Groups and Discontinuous Subgroups.* Providence, R.I.: American Mathematical Society, 1966, pp. 211–224.
2. LANGLANDS, R., "The volume of the fundamental domain for some arithmetical subgroups of Chevalley groups," *Proceedings of the Symposium on Algebraic Groups and Discontinuous Subgroups.* Providence, R.I.: American Mathematical Society, 1966, pp. 143–148.
3. ———, "Eisenstein series," *Proceedings of the Symposium on Algebraic Groups and Discontinuous Subgroups.* Providence, R.I.: American Mathematical Society, 1966, pp. 235–252.
4. SIEGEL, C. L., "A mean value theorem in geometry of numbers," *Ann. of Math.* (1945), 340–347.
5. TITCHMARSH, E. C., *The Theory of the Riemann Zeta-Function.* London: Oxford University Press, 1951.

The Riemann Surface of Klein with 168 Automorphisms

HARRY E. RAUCH[1] AND J. LEWITTES

1. Introduction

In the following we consider Klein's compact Riemann surface of genus three admitting a simple group of 168 automorphisms (conformal self-homeomorphisms) and adduce certain of its properties which to our knowledge and, indeed, surprise do not seem to be in the literature. Our main results are the explicit exhibition (i) of a canonical integral homology basis in dimension one (Section 3), (ii) of the action on this basis of Klein's group of automorphisms in the form of integral matrices representing a set of generators of the group in $Sp(3, \mathbf{Z})$, although actually we give more (Section 4), and (iii) of the ("right half" of the) period matrix of the normal Abelian integrals with respect to the homology basis in (i) [Section 5, Formula (10)].

In Section 2 we describe Klein's surface in the way needed for the remaining sections.

The paper is self-contained; however, for the benefit of readers who may wish to read more about a fascinating topic in which interest is currently reviving, we give some references. The primary references are to Klein's original papers [10] and [9], which are eminently readable. Additional material is in the books [8], [1], [21], and, particularly valuable here, [3]. Follow-up of Klein's projective-geometric treatment is in [2] and [4]. For the important liaison with algebraic equations of degree seven, culminating in Gordan's work, see [5].

From the time of Hurwitz's fundamental researches (see [6]), on Riemann surfaces with nontrivial automorphisms Klein's surface has served as the most pregnant illustration of the phenomenon; thus, for example, it was until recently the only known surface for which Hurwitz's bound of

[1] Research partially sponsored by the Air Force Office of Scientific Research, Office of Aerospace Research, United States Air Force, under AFOSR Grant No. AF-69-1641.

$84(g-1)$, $g \geqq 2$ is the genus, for the order of the automorphism group is attained (*vide infra*). Renewed interest attaches to it in light of the role played by surfaces with automorphisms in the structure of spaces of Riemann surfaces (see, for instance, [17], the references therein to both authors' work, and [18]).

The motivation for this paper is the desire to determine the action of Klein's group on the theta functions and constants attached to his surface. We plan a subsequent publication on this and other aspects of theta functions on Klein's surface.

It would be of interest to contemplate our process for a surface of higher genus. Macbeath, in [14] and [15] (see also [13]), has recently exhibited new examples of surfaces with automorphisms realizing Hurwitz's upper bound, in particular, one of genus seven with a simple group of 504 automorphisms.[2] The reader will see clearly that our methods can be used in principle to treat Macbeath's example, but he will also see that our geometrical method in that case would be quite tedious in application. After seeing our results Macbeath called our attention to Leech's work, in particular to papers [11] and [12], with the remark that answers to (i) and (ii) for Klein's and Macbeath's surfaces could be deduced from them. The first named author has since carried out the calculations with the aid of H. Lee Michelson. Our method here is more powerful in that we also obtain the homological action of the extended groups obtained by adjoining reflections. Leech's method, which is purely group-theoretical, on the other hand, utilizes a computer to do calculations that would otherwise be impractical.

We call attention to an incorrect answer to (iii) by Hurwitz ([7], p. 159, criticized in [1]) and an abortive attempt on (i), (ii), and (iii) by Poincaré in [16], p. 130, all noticed after the completion of our work.[2]

2. Klein's surface

Klein originally obtained his surface S in the form of the upper half-plane identified under the principal congruence subgroup of level seven, $\Gamma(7)$, of the modular group Γ. In this form it is necessary to compactify the fundamental domain at its cusps. Klein's group then appears as $\Gamma/\Gamma(7)$, which is simple and of order 168.

We need, however, another representation given by Klein, one which we recognize today as the unit circle uniformization of S. In the unit circle draw the vertical diameter L_1 and another diameter L_3 making an angle of $\pi/7$ with L_1 and going down to the right. In the lower semicircle draw

[2] See note added in proof at end of paper.

the arc L_2 of the circle which is orthogonal to the unit circle and to L_1 and which meets L_3 at the angle $\pi/3$. Let t be the non-Euclidean triangle enclosed by L_1, L_2, L_3, and let R_1, R_2, R_3 be the non-Euclidean reflections in L_1, L_2, L_3, respectively. R_1, R_2, R_3 generate a non-Euclidean crystallographic group, which we denote by $(2, 3, 7)'$, with t as a fundamental domain. The images of t under $(2, 3, 7)'$ are a set of non-Euclidean triangles each of which is congruent or symmetric to t according as the group element which maps t on it has an even or odd number of letters as a word in R_1, R_2, R_3. These triangles form a non-Euclidean plastering or tesselation of the interior of the unit circle. The union of t and its image under R_2 is a fundamental domain (with suitable conventions about edges) for the triangle group $(2, 3, 7)$, which is the group generated by $T = R_3R_1$, $V = R_3R_2$, $U = R_2R_1$ with the relations $T^7 = V^3 = U^2 = T^{-1}VU = I$. T, V, U are, respectively, non-Euclidean rotations of angles $2\pi/7$, $2\pi/3$, π counterclockwise about the vertices of t belonging to the angles $\pi/7$, $\pi/3$, $\pi/2$. Now the uniformizing Fuchsian group of S is a normal subgroup $N \subset (2, 3, 7)$ of index 168. The quotient group $\Gamma_{168} = (2, 3, 7)/N$ is Klein's simple group of automorphisms of S and is obtained from $(2, 3, 7)$ by imposing the relation $(UT^{-4})^4 = I$. A convenient fundamental domain for N is the circular arc (non-Euclidean) 14-gon Δ shown in Fig. 1. It will be noticed that t appears as the unshaded triangle immediately below and to the right of P^0. There are 168 unshaded triangles, which are the images of t under Γ_{168}, in Δ and 168 shaded triangles, which are the images of t under anticonformal elements of $(2, 3, 7)'$. We let Γ'_{168}, the *extended* Klein group, be the group of 336 proper and improper (sense reversing) non-Euclidean motions obtained by adjoining, say, R_1 to Γ_{168}. Then Δ can be viewed either as the union of 336 shaded and unshaded triangles consisting of t and its transforms by Γ'_{168} or as the union of 168 double triangles (one unshaded, one shaded) consisting of the union of t and its image under R_1 [thus a fundamental domain for $(2, 3, 7)$] and the transforms of this by Γ_{168}. If the sides of Δ are labeled as in Fig. 1, then, Klein has given the side identifications induced by N in Table 1. Δ with those identifications is Klein's model of his surface S.

TABLE 1

$1 \leftrightarrow 6$
$3 \leftrightarrow 8$
$5 \leftrightarrow 10$
$7 \leftrightarrow 12$
$9 \leftrightarrow 14$
$11 \leftrightarrow 2$
$13 \leftrightarrow 4$

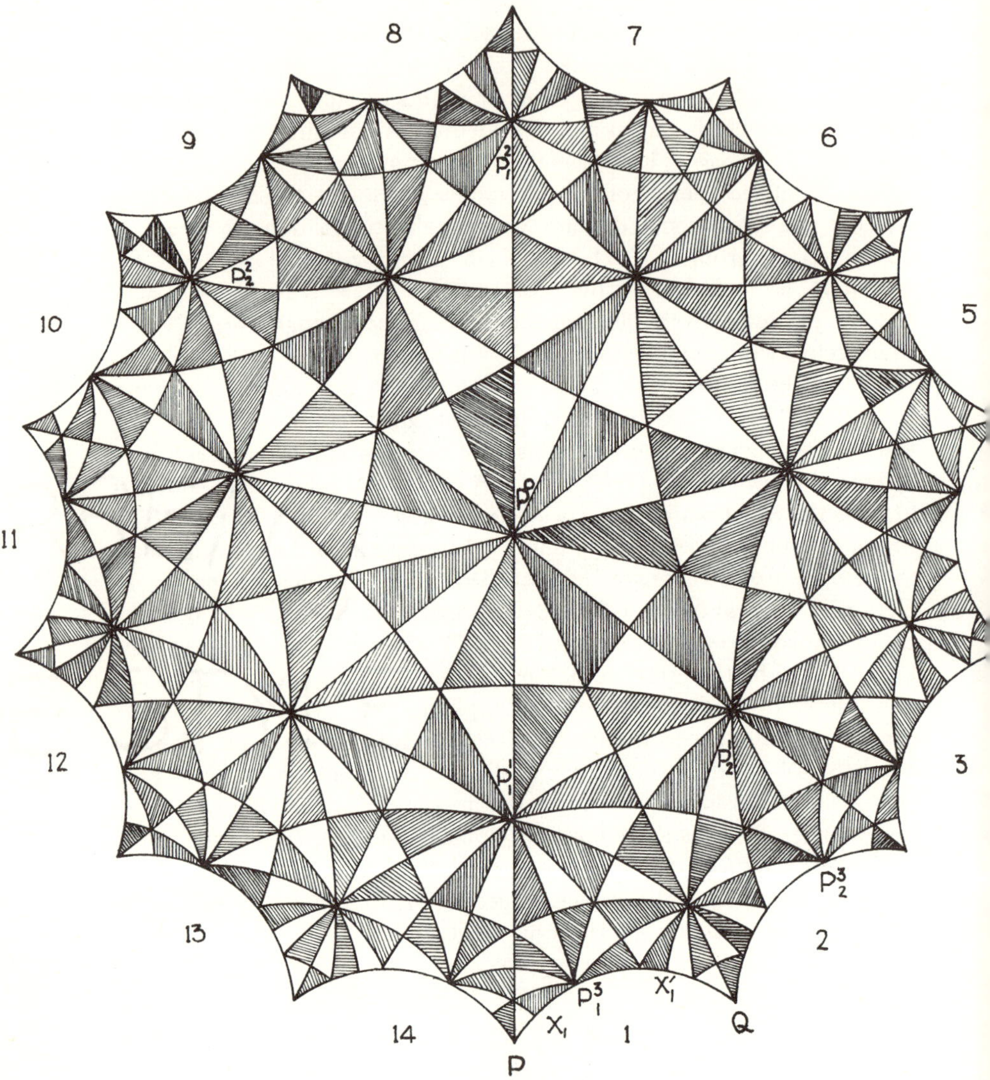

Fig. 1

One sees that each odd-numbered side is identified with the side 5 units around counterclockwise.

We need to label certain points in Δ for the sequel. The 14 external vertices, those in which pairs of sides meet, are identified under N in two sets of 7, which alternate. We have exhibited the labels of two successive vertices in Fig. 1 as P and Q. The remainder we think of as labeled P, Q, P, etc., in counterclockwise order. P and Q are 7-vertices (after identification), i.e., vertices where 7 double triangles meet. There are 22 other 7-vertices of which 7 have two representatives each on Δ. We label them as follows. P^0 is the center of Δ (the origin). Two units (triangles) out from P^0 is a layer of 7 which we label P_1^1, P_2^1, etc., counterclockwise, where the first two are actually shown in Fig. 1. Two units farther out is another layer of 7, P_1^2, P_2^2, etc., counterclockwise, where again only the first two are shown. Finally the 14 vertices (representing 7 when identified) P_1^3, P_2^3, etc., counterclockwise, are on the sides of Δ, the subscript denoting the side on which the vertex lies, and again only the first two are exhibited. In addition, it is necessary to label certain 2-vertices (4-vertices under identification) on the sides of Δ. There are two on each side; we call them X_1, X_1', etc., the subscript denoting the side and the unprimed point lying closer to P. The first two are shown in Fig. 1.

We call attention to the fact that L_1, L_3, L_2 appear in Fig. 1 as the two diameters and the circular arc, parts of which form the boundary of t.

3. The homology basis on S

We now use a standard technique in combinatorial surface topology (see, for example, [20]) to compute a canonical homology basis for S.

We make each side of Δ into an oriented path by assigning to it the counterclockwise direction as the positive orientation.

We introduce the surface symbol $12345\,1^{-1}\,73^{-1}\,95^{-1}\,2^{-1}\,7^{-1}\,4^{-1}\,9^{-1}$ deduced from Fig. 1 and Table 1.

We observe that 1 and 2 are linked. Write $B = 2345$ or $2^{-1} = 345B^{-1}$. One obtains $1\,B\,1^{-1}\,73^{-1}\,9\,5^{-1}\,345B^{-1}\,7^{-1}\,4^{-1}\,9^{-1}$. From B to B^{-1} one has $A^{-1} = 1^{-1}\,73^{-1}\,95^{-1}\,345$ or $1 = 73^{-1}\,95^{-1}\,345A$. One obtains $73^{-1}\,95^{-1}\,345\,[A, B]\,7^{-1}\,4^{-1}\,9^{-1}$, where $[A, B] = ABA^{-1}\,B^{-1}$. Observing that 7 and 9 are linked we write $C^{-1} = 3^{-1}\,95^{-1}\,345\,[A, B]$ or $9 = 3C^{-1}\,[B, A]\,5^{-1}\,4^{-1}\,3^{-1}\,5$ to obtain $C^{-1}\,7^{-1}\,4^{-1}\,5^{-1}\,345\,[A, B]\,C\,3^{-1}\,7$. Set $D = 3^{-1}\,7$ or $7 = D^{-1}\,3^{-1}$ to obtain $3^{-1}\,4^{-1}\,5^{-1}\,345\,[A, B]\,[C, D]$. 3 and 5 are linked, and so we set $E^{-1} = 45\,[A, B]\,[C, D]$ or $4^{-1} = 5\,[A, B]\,[C, D]\,E$ to obtain $3^{-1}\,5\,[A, B]\,[C, D]\,E\,5^{-1}\,3\,E^{-1}$. Thus, finally, setting $F = 5^{-1}\,3$, the surface symbol reduces to $[A, B]\,[C, D]\,[E, F]$, which is the desired canonical form.

Summarizing, we have

$$\begin{aligned} A &= 5^{-1}\,4^{-1}\,3^{-1}\,59^{-1}\,37^{-1}\,1 \\ B &= 2345 \\ C &= [B, A]\,5^{-1}\,4^{-1}\,3^{-1}\,59^{-1}\,3 \\ D &= 3^{-1}\,7 \\ E &= [D, C]\,[B, A]\,5^{-1}\,4^{-1} \\ F &= 5^{-1}\,3 \end{aligned}$$

If we look at Fig. 1, we find that all six paths are closed and begin and end at P. Hence, when we Abelianize paths to obtain chains, the six paths become the cycles of a canonical homology basis.

For any oriented path M on Δ let $\alpha(M)$ denote the chain obtained by Abelianizing. Then we have

TABLE 2

$$\begin{aligned} \gamma_1 &= \alpha(A) = 1 - 4 - 7 - 9 \\ \gamma_2 &= \alpha(C) = -4 - 9 \\ \gamma_3 &= \alpha(E) = -4 - 5 \end{aligned}$$

$$\begin{aligned} \delta_1 &= \alpha(B) = 2 + 3 + 4 + 5 \\ \delta_2 &= \alpha(D) = -3 + 7 \\ \delta_3 &= \alpha(F) = 3 - 5, \end{aligned}$$

where $KI(\gamma_i, \gamma_j) = KI(\delta_i, \delta_j) = 0$, $KI(\gamma_i, \delta_j) = \delta_{ij}$, $i, j = 1, 2, 3$, $KI(\gamma, \delta)$ being the intersection number of γ and δ.

4. Action of Γ'_{168} and Γ_{168} on homology basis

We restrict ourselves to the action of R_1, R_2, R_3, the generators of $(2, 3, 7)'$ and *a fortiori* Γ'_{168}. One sees immediately the symmetry of Δ under R_1 and R_3, and their action is analyzed in a few strokes.

Reference to Fig. 1 gives the following table of the actions of R_1 and R_3 on the sides of Δ.

TABLE 3

R_1:	R_3:
$1 \to -14 = 9$	$1 \to -2$
$2 \to -13 = 4$	$2 \to -1$
$3 \to -12 = 7$	$3 \to -14 = 9$
$4 \to -11 = 2$	$4 \to -13 = 4$
$5 \to -10 = 5$	$5 \to -12 = 7$
$7 \to -8 = 3$	$7 \to -10 = 5$
$9 \to -6 = 1$	$9 \to -8 = 3$

From Tables 2 and 3 one can read off the actions of R_1 and R_3 on the homology basis of Section 3.

TABLE 4

$$
\begin{aligned}
R_1\colon\ & \gamma_1 \to -3 - 2 - 1 + 9 = -\gamma_1 - \delta_1 - \delta_2 - \delta_3 \\
& \gamma_2 \to -1 - 2 = -\gamma_1 + \gamma_2 - \gamma_3 - \delta_1 - \delta_2 \\
& \gamma_3 \to -5 - 2 = -\gamma_3 - \delta_1 + \delta_3 \\
& \delta_1 \to 4 + 7 + 2 + 5 = \delta_1 + \delta_2 \\
& \delta_2 \to -7 + 3 = -\delta_2 \\
& \delta_3 \to -5 + 7 = \delta_2 + \delta_3 \\
R_3\colon\ & \gamma_1 \to -5 - 4 - 3 - 2 = -\delta_1 \\
& \gamma_2 \to -3 - 4 = \gamma_3 - \delta_3 \\
& \gamma_3 \to -7 - 4 = \gamma_3 - \delta_2 - \delta_3 \\
& \delta_1 \to -1 + 9 + 4 + 7 = -\gamma_1 \\
& \delta_2 \to -9 + 5 = \gamma_2 - \gamma_3 \\
& \delta_3 \to -7 + 9 = -\gamma_2 + \gamma_3 - \delta_2 - \delta_3
\end{aligned}
$$

Now Δ is not symmetric under R_2 although S is. The difficult part of this paper is, therefore, to compute the action of R_2. The difficulty is due to a certain complexity and tediousness in tracing the images of certain chains under R_2 and then reducing them homologically to chains on the boundary of Δ.

We now write a subdivided surface symbol for Δ: $a\,b\,c\,d\,e\,f\,g\,h\,i\,j\,a^{-1}\,k\,l\,m\,f^{-1}\,e^{-1}\,d^{-1}\,nj^{-1}\,i^{-1}\,c^{-1}\,b^{-1}\,m^{-1}\,l^{-1}\,k^{-1}\,j^{-1}\,g^{-1}\,n^{-1}$, where the subsymbols are explained by

TABLE 5

$a = P\,Q = 1$, $b = QX_2$, $c = X_2P$, $bc = 2$, $d = PP_3^3$, $e = P_3^3X_3'$,
$f = X_3'Q$, $def = 3$, $g = QX_4$, $h = X_4P$, $gh = 4$, $i = PP_5^3$,
$j = P_5^3Q$, $ij = 5$, $a^{-1} = QP = 6$, $k = PP_7^3$, $l = P_7^3X_7'$, $m = X_7'Q$,
$klm = 7$, $f^{-1} = QX_8'$, $e^{-1} = X_8'P_8^3$, $d^{-1} = P_8^3P$,
$f^{-1}e^{-1}d^{-1} = 8 = 3^{-1}$, $n = PQ = 9$, $j^{-1} = QP_{10}^3$,
$i^{-1} = P_{10}^3P$, $j^{-1}i^{-1} = 10 = 5^{-1}$, $k^{-1} = P_{12}^3P$,
$m^{-1}l^{-1}k^{-1} = 12 = 7^{-1}$, $h^{-1} = PX_{13}$, $g^{-1} = X_{13}Q$,
$h^{-1}g^{-1} = 13 = 4^{-1}$, $n^{-1} = QP = 14 = 9^{-1}$

In Table 5 each ordered pair of points denotes the oriented circular arc joining them.

Table 6 is constructed in the following way: the image of the arc in question is found by marking its end points and appropriately chosen (not unique) circular arcs λ joining them to L_2, if possible, and observing how many units, i.e., edges of triangles, the points are from L_2, and then marking the arcs corresponding to λ under R_2 and going out on the other side of L_2 the same number of units. In the course of doing so it often becomes necessary to use the identification of sides of Δ in Table 1 to re-enter Δ because of its lack of symmetry with respect to R_2. The interested reader will be able with perseverance and patience to verify the entries.

TABLE 6

$$
\begin{aligned}
R_2: \qquad & 1 \to P_1^2 P_5^3 \\
& 2 \to P_5^3 X_3' + X_8' P_1^2 \\
& 3 \to P_1^2 P_8^3 + P_3^3 X_2 + X_{11} P_{10}^3 \\
-13 = {} & 4 \to P_{10}^3 X_{12}' + X_7' P_1^2 \\
& 5 \to P_1^2 Q_7 + Q_5 P_5^3 \\
-12 = {} & 7 \to P_1^2 P_7^3 + P_{12}^3 X_{13} + X_4 P_5^3 \\
-14 = {} & 9 \to P_1^2 P_{10}^3,
\end{aligned}
$$

where Q_5, Q_7 denote the representatives of Q on sides 5 and 7.

From Tables 2, 5, and 6 and by the use of Fig. 1 to replace certain chains by homologous chains we have constructed Table 7 giving the action of R_2 on the homology basis. To avoid excessive length of the paper we give all the details for the first entry only, leaving the verification of the remainder again to the interested reader's perseverance and patience.

TABLE 7

$$
\begin{aligned}
R_2: \quad \gamma_1 &\to P_5^3 X_4 + X_{13} P_{12}^3 + P_7^3 P_1^2 + P_1^2 X_7' + P_{10}^3 P_1^2 + P_1^2 P_5^3 \\
&\sim -i - h + h + k + l + m + 2 + i + j - 9 + 3 - 7 + 1 - j \\
&= 2 - 9 + 3 + 1 = \gamma_1 + \delta_1 + \delta_2 + \delta_3 \\
\gamma_2 &\to P_{10}^3 P_1^2 + P_1^2 X_7' + X_{12}' P_{10}^3 \sim j - 9 - 8 - m + m - 11 + i \\
&= 5 - 9 + 3 + 2 = \gamma_2 + \delta_1 \\
\gamma_3 &\to Q_7 P_1^2 + P_1^2 X_7' + X_{12}' P_{10}^3 + P_5^3 Q_5 \sim -m + i - 11 + m + j \\
&= 5 + 2 = \gamma_3 + \delta_1 - \delta_3 \\
\delta_1 &\to P_5^3 X_3' + X_8' P_1^2 + P_1^2 P_8^3 + P_3^3 X_2 + X_{11} P_{10}^3 + P_{10}^3 X_{12}' \\
&\qquad + X_7' P_1^2 + P_1^2 Q_7 + Q_5 P_5^3 \\
&\sim -i - 4 - f - e - c - d + c + i - i + 11 - m + m - j \\
&= -2 - 3 - 4 - 5 = -\delta_1 \\
\delta_2 &\to P_{10}^3 X_{11} + X_2 P_3^3 + P_8^3 P_1^2 + P_1^2 P_7^3 + P_{12}^3 X_{13} + X_4 P_5^3 \\
&\sim -i - c + c + d - l - m + f + e - k - h + h + i \\
&= 3 - 7 = -\delta_2 \\
\delta_3 &\to Q_7 P_1^2 + P_1^2 P_8^3 + P_3^3 X_2 + X_{11} P_{10}^3 + P_5^3 Q_5 \\
&\sim -f - e - d - c + c + i + j = -3 + 5 = -\delta_3
\end{aligned}
$$

From Tables 2 and 6 we see that R_2 takes $\gamma_1 = 1 - 4 - 7 - 9$ into $P_5^3 X_4 + X_{13} P_{12}^3 + P_7^3 P_1^2 + P_1^2 X_7' + X_{12}' P_{10}^3 + P_{10}^3 P_1^2 + P_1^2 P_5^3$, where we have reversed the order of some pairs to have all positive signs and rearranged terms so that, with identifications, the terms fall in natural order, end-to-end. We next observe from Fig. 1 and Table 6 that $P_5^3X_4 \sim P_5^3P + PX_4 = -i - h$, $X_{13}P_{12}^3 \sim X_{13}P_{12}^3 \sim X_{13}P + PP_{12}^3 = h + k$, $P_7^3P_1^2 + P_1^2X_7' \sim P_7^3X_7' = l$, $X_{12}'P_{10}^3 \sim X_{12}'Q - 11 + PP_{10}^3 = m + 2 + i$, $P_{10}^3P_1^2 + P_1^2P_5^3 \sim P_{10}^3Q - 9 - 8 - 7 - 6 - P_5^3Q = j - 9 + 3 - 7 + 1 - j$. Hence $R_2\gamma_1 \sim -i - h + h + k + l + m + 2 + i + j - 9 + 3 - 7 + 1 - j = k + l + m + 2 - 9 + 3 - 7 + 1 = 7 + 2 - 9 + 3 - 7 + 1 = 2 - 9 + 3 + 1 = \gamma_1 + \delta_1 + \delta_2 + \delta_3$. In Table 7 $\sim$ means "homologous to."

Using Tables 4 and 7 to construct matrix representations $\mathscr{M}_1$, $\mathscr{M}_2$, $\mathscr{M}_3$ of the actions of R_1, R_2, R_3 on the homology basis of Table 2 we now write down by composition the matrix representations $\mathscr{M}_T^{-1}$, $\mathscr{M}_V$, $\mathscr{M}_U$ of the generators $T^{-1} = R_1R_3$, $V = R_3R_2$, $U = R_2R_1$ of Γ_{168} (we prefer for technical reasons to use T^{-1} instead of T as a generator). Observe that, consistent with the practice of [17], Lemma 8 and Proposition 9 (note the corrections in the reference in the bibliography), R_1R_3, for example, will be represented by $\mathscr{M}_3\mathscr{M}_1$, i.e., the reverse order of the geometric transformations.

$$\text{(1)} \qquad \mathscr{M}_T^{-1} = \left(\begin{array}{rrr|rrr} 0 & 0 & 0 & -1 & -1 & 0 \\ 0 & 0 & -1 & -1 & -1 & 0 \\ 0 & 0 & -1 & -1 & 0 & 0 \\ \hline 1 & 0 & 0 & 1 & 1 & 1 \\ -1 & 1 & 0 & 0 & -1 & -1 \\ 1 & -1 & 0 & 0 & 1 & 0 \end{array}\right)$$

$$\text{(2)} \qquad \mathscr{M}_V = \left(\begin{array}{rrr|rrr} -1 & 0 & 0 & -1 & -1 & -1 \\ -1 & 0 & 1 & 0 & 0 & -1 \\ -1 & 1 & 0 & 0 & 0 & 0 \\ \hline 1 & 0 & 0 & 0 & 0 & 0 \\ 0 & -1 & 1 & 0 & 0 & 0 \\ 0 & 1 & -1 & 0 & 1 & 1 \end{array}\right)$$

$$\text{(3)} \qquad \mathscr{M}_U = \left(\begin{array}{rrr|rrr} -1 & 0 & 0 & 0 & 0 & 0 \\ -1 & 1 & -1 & 0 & 0 & 0 \\ 0 & 0 & -1 & 0 & 0 & 0 \\ \hline 0 & 0 & 0 & -1 & -1 & 0 \\ 0 & 0 & 0 & 0 & 1 & 0 \\ 0 & 0 & 0 & 0 & -1 & -1 \end{array}\right)$$

5. Computation of the normal period matrix

We can now use the results of Section 4 to compute the period matrix π over $\delta_1, \delta_2, \delta_3$ of the normal Abelian integrals of the first kind on S with respect to the canonical homology basis $\gamma_1, \gamma_2, \gamma_3; \delta_1, \delta_2, \delta_3$ of Section 3 (see [17], p. 12 ff.). The technique is to observe that π is fixed under the transformations of the inhomogeneous Siegel modular group $\mathcal{M}^3$ corresponding to the matrices (1), (2), (3) in $Sp(3, \mathbf{Z})$ of Section 4 (see [19] and [17], Section 2, C, p. 16).

Thus we have

$$\pi = (A\pi + B)(C\pi + D)^{-1}$$

or

$$\pi(C\pi + D) = A\pi + B, \tag{4}$$

where

$$\mathcal{M} = \left(\begin{array}{c|c} D & C \\ \hline B & A \end{array}\right)$$

is any of the matrices in (1), (2), and (3). If we set

$$\pi = \begin{pmatrix} a & b & c \\ b & d & e \\ c & e & f \end{pmatrix}$$

(π is symmetric), and $\mathcal{M} = \mathcal{M}_U$ (here $B = C = 0$) we find from (4)

$$\begin{aligned} b &= -b - d \\ -b - c &= -c - e \\ -d - e &= e \end{aligned}$$

or

$$\begin{aligned} 2b &= -d \\ b &= e \\ -d &= 2e. \end{aligned} \tag{5}$$

Similarly, choosing $\mathcal{M} = \mathcal{M}_V$, one finds

$$\begin{aligned} &\text{(i) } a^2 + ab + ac + a + bc + b + c + 1 = 0 \\ &\text{(ii) } ab + ad + ae + be - c = 0 \\ &\text{(iii) } ab + b^2 + bc + b + cd + d + e = 0 \\ &\text{(iv) } b^2 + bd + be + de - e - 1 = 0 \\ &\text{(v) } f = eb + cd + ce + e^2 + d + e + 1. \end{aligned} \tag{6}$$

Setting (5) in (6, iv) one obtains

$$2e^2 + e + 1 = 0 \tag{7}$$

or

$$e = \frac{-1 \pm \sqrt{7}\, i}{4}. \tag{8}$$

We shall decide the choice of sign later.

Substituting (5) in (6, ii), (6, iii), and (6, v), we obtain

$$\begin{aligned} &c = e^2 \\ &ae - e^3 + e^2 = 0 \quad \text{or} \quad a = e^2 - e \qquad (e \neq 0) \\ &f = e^2 - e + 1. \end{aligned} \tag{9}$$

But from (7) one finds $e^2 = -\frac{1}{2} - e/2$ and hence $e^2 - e = -\frac{1}{2} - 3e/2$ and $e^2 - e + 1 = \frac{1}{2} - 3e/2$, so that

$$\pi = \begin{pmatrix} -\frac{1}{2} - \frac{3e}{2} & e & -\frac{1}{2} - \frac{e}{2} \\ e & -2e & e \\ -\frac{1}{2} - \frac{e}{2} & e & \frac{1}{2} - \frac{3e}{2} \end{pmatrix}$$

Now π as a normal period matrix must have a positive-definite imaginary part. Reference to (8) shows that we must pick the minus sign in e so that we have finally

$$\pi = \begin{pmatrix} -\frac{1}{8} + \frac{3\sqrt{7}\, i}{8} & -\frac{1}{4} - \frac{\sqrt{7}\, i}{4} & -\frac{3}{8} + \frac{\sqrt{7}\, i}{8} \\ -\frac{1}{4} - \frac{\sqrt{7}\, i}{4} & \frac{1}{2} + \frac{\sqrt{7}\, i}{2} & -\frac{1}{4} - \frac{\sqrt{7}\, i}{4} \\ -\frac{3}{8} + \frac{\sqrt{7}\, i}{8} & -\frac{1}{4} - \frac{\sqrt{7}\, i}{4} & \frac{7}{8} + \frac{3\sqrt{7}\, i}{8} \end{pmatrix} \tag{10}$$

We observe that the entries of π all lie in the field $k(\sqrt{-7})$, which is the field generated over the rational field k by the character of the representation (irreducible) induced on the differentials of first kind of S by Γ_{168} (see [18]).

NOTE ADDED IN PROOF (May 1970): Macbeath's surface of genus 7 in [14] was anticipated by Fricke in [22]. In pp. 265–270 of [1] Baker discusses Klein's surface without, as far as we can see, arriving at anything like our results. As he says, he does not construct any explicit homology basis, and the period matrix he obtains is not proved to correspond, and probably does not correspond, to a *canonical* homology basis.

THE CITY COLLEGE OF THE CITY UNIVERSITY OF NEW YORK
THE HERBERT H. LEHMAN COLLEGE
THE GRADUATE CENTER OF THE CITY UNIVERSITY OF NEW YORK

REFERENCES

1. BAKER, H. F., *Multiply Periodic Functions*. Cambridge, 1907.
2. ———, "Note introductory to the study of Klein's group of order 168," *Proc. Cambridge Philos. Soc., 31* (1935), 468–481.
3. BURNSIDE, W., *Theory of Groups of Finite Order*. Cambridge, 1911.
4. EDGE, W. L., "The Klein group in three dimensions," *Acta Math., 79* (1947), 153–223.
5. GORDAN, P., "Über Gleichungen siebenten Grades mit einer Gruppe von 168 Substitutionen," *Math. Ann., 20* (1882), 515–530; II, *ibid., 25* (1885), 459–521.
6. HURWITZ, A., "Über algebraische Gebilde mit eindeutigen Transformationen in sich," *Mathematische Werke*, Bd. I. Basel: Birkhäuser, 1932, pp. 392–430.
7. ———, "Über einige besondere homogene lineare Differentialgleichungen," *ibid.*, pp. 153–162.
8. KLEIN, F., and R. FRICKE, *Vorlesungen über die Theorie der elliptischen Modulfunctionen*, Bd. I. Leipzig, 1890.
9. KLEIN, F., "Über die Auflösung gewisser Gleichungen vom siebenten und achten Grade," *Math. Ann., 15* (1879), 251–282.
10. ———, "Über die Transformation siebenter Ordnung der elliptischen Funktionen," *Math. Ann., 14* (1879), 428–471.
11. LEECH, J., "Generators for certain normal subgroups of (2, 3, 7)," *Proc. Cambridge Philos. Soc., 61* (1965), 321–332.
12. LEECH, J., and J. MENNICKE, "Note on a conjecture of Coxeter," *Proc. Glasgow Math. Assoc.* (1961), 25–29.
13. MACBEATH, A. M., *Fuchsian Groups*, mimeographed notes, Queen's College, Dundee, University of St. Andrews.
14. ———, "On a curve of genus 7," *Proc. London Math. Soc., 15* (1965), 527–542.
15. ———, "On a theorem of Hurwitz," *Proc. Glasgow Math. Assoc., 5* (1961), 90–96.
16. POINCARÉ, H., *Sur l'intégration algébrique des équations linéaires et les périodes des intégrales abéliennes*, *Oeuvres*, t. III. Paris: Gauthier-Villars, 1934, pp. 106–166.
17. RAUCH, H(ARRY) E., "A transcendental view of the space of algebraic Riemann surfaces," *Bull. Amer. Math. Soc., 71* (1965), 1–39, Errata, *ibid., 74* (1968), 767.
18. ———, "The local ring of the genus three modulus space at Klein's 168 surface," *Bull. Amer. Math. Soc., 73* (1967), 343–346.
19. ———, "Variational methods in the problem of the moduli of Riemann surfaces," in *Contributions to Function Theory*. Bombay: Tata Institute of Fundamental Research, 1960, pp. 17–40.
20. SPRINGER, G., *Introduction to Riemann Surfaces*. Reading, Mass.: Addison-Wesley, 1957.
21. WEBER, H., *Lehrbuch der Algebra*, Bd. II. Braunschweig, 1908.
22. Fricke, R. "Ueber eine einfache Gruppe von 504 Operationen," *Math. Ann., 52* (1899), 319–339.

Envelopes of Holomorphy of Domains in Complex Lie Groups

O. S. ROTHAUS[1]

There are two well known results in the literature concerning envelopes of holomorphy that we want to single out here, namely the ones describing the completion of Reinhardt domains and tube domains. The statement for the latter type we owe to Salomon Bochner.

There is a feature common to both the results which is worth noting. Roughly speaking it may be described as follows: a domain in a complex Lie group which is stable under a real form of the group may be completed by forming certain "averages" in the complex group.

In this note we shall give one reasonably general form of this phenomenon, which more or less includes the case of Reinhardt domains.

We shall follow the usual convention of letting German alphabetic characters denote the Lie algebra or its elements corresponding to the group denoted by the associated ordinary alphabetic character.

Let G be a connected complex Lie group, $\mathfrak{G}$ its Lie algebra. Let $j, j^2 = -$identity, be the endomorphism of $\mathfrak{G}$ giving the complex structure. Let $\mathfrak{H}$ be a real form of $\mathfrak{G}$, so that $\mathfrak{G} = \mathfrak{H} \oplus j\mathfrak{H}$, and H the corresponding analytic subgroup of G.

We want to investigate the envelope of holomorphy of domains D contained in G which are H stable; i.e., $Dh = D$ for all $h \in H$. We make the following assumption governing the situation: *H is a maximal compact subgroup of G.* This assumption has a number of useful consequences, mostly well known, which we now present. For these purposes, let $\hat{G}$ be the universal covering group of G, Γ the fundamental group of G, so that $G \cong \hat{G}/\Gamma$. Let π be the natural map from $\hat{G}$ to G, let $\hat{H}$ be $\pi^{-1}(H)$ and $\hat{H}_0$ be the connected component of the identity in $\hat{H}$.

[1] This research was partially supported by NSF GP 8129.

There are natural homeomorphisms of the identification spaces as indicated:

$$\frac{G}{H} \cong \frac{\hat{G}}{\hat{H}} \cong \frac{\dfrac{\hat{G}}{\hat{H}_0}}{\dfrac{\hat{H}}{\hat{H}_0}}.$$

By a theorem of E. Cartan and others, G/H is Euclidean; on the other hand $\hat{H}/\hat{H}_0$ is discrete. Since Euclidean space is simply connected, we have

LEMMA 1. $\hat{H}_0 = \hat{H}$.

Clearly $\hat{H}$ is then a covering space for H. But $\hat{G}$, viewed as a fibre bundle over $\hat{G}/\hat{H}$, fibre $\hat{H}$, is equivalent to a trivial bundle, since $\hat{G}/\hat{H}$ is Euclidean. $\hat{G}$ being simply connected, it follows that

LEMMA 2. *$\hat{H}$ is the universal covering space of H.*

There is an involutory automorphism σ of $\mathfrak{G}$ given by $\sigma(\mathfrak{h}_1 + j\mathfrak{h}_2) = \mathfrak{h}_1 - j\mathfrak{h}_2$ for $\mathfrak{h}_1, \mathfrak{h}_2 \in \mathfrak{H}$. This automorphism induces an involutory automorphism of $\hat{G}$, whose subgroup of fixed points has connected component of identity equal to $\hat{H}$. Since $\hat{H} \supset \Gamma$, the automorphism, still denoted σ, may be viewed as an automorphism of $\hat{G}/\Gamma \cong G$, fixing $\hat{H}/\Gamma \cong H$. Since H also is compact, we have

LEMMA 3. *G/H is a Riemannian symmetric space* (*see* [1]).

The transvections of G/H are given, of course, by $\exp j\mathfrak{h}$, $\mathfrak{h} \in \mathfrak{H}$, and every element in G may be written uniquely as a transvection times an element of H.

H being compact, it is well known that H decomposes into the direct sum of a semisimple algebra $\mathfrak{S}$, and $\mathfrak{C}$, the center of $\mathfrak{H}$. $\mathfrak{C}$ is the Lie algebra of C, the connected center of H, which must be a torus, and $\mathfrak{S}$ is the Lie algebra of S, a connected semisimple subgroup of H. Every element in H may be written, though not uniquely, in the form sc, $s \in S$, $c \in C$.

For future purposes, we shall suppose that a basis $\mathfrak{c}_1, \mathfrak{c}_2, \ldots, \mathfrak{c}_k$ of $\mathfrak{C}$ has been chosen so that $\exp \sum \lambda_i \mathfrak{c}_i =$ identity implies and is implied by $\lambda_i \in \mathbf{Z}\ \forall i$.

Now let $h \to r(h)$ be a finite d-dimensional representation of H; for our purposes we suppose the representation is already unitary, so that $r(h) \in U(d)$.

There is induced a representation of the algebra $\mathfrak{H}$, which we still denote by $\mathfrak{h} \to r(\mathfrak{h})$.

The representation of $\mathfrak{H}$ can be extended to give a representation R of $\mathfrak{G}$ by setting

$$R(\mathfrak{h}_1 + j\mathfrak{h}_2) = r(\mathfrak{h}_1) + ir(\mathfrak{h}_2)$$

for $\mathfrak{h}_1, \mathfrak{h}_2 \in \mathfrak{H}$. The representation R of $\mathfrak{G}$ now induces a representation $\hat{R}$ of $\hat{G}$. $\hat{R}$, restricted to $\hat{H}$, is simply the lift of r to $\hat{H}$. Since a representation of a connected group is completely determined by the corresponding representation of the Lie algebra, it follows immediately that $\hat{R}$ is trivial on Γ. Hence $\hat{R}$ may be viewed as a representation of $\hat{G}/\Gamma \cong G$, and thus we have:

LEMMA 4. *Every representation of H extends to a representation of G.*

It is worth noting that the entries in the representation $\hat{R}$ are holomorphic functions on G. In fact $\hat{R}$ is easily seen to be the unique holomorphic extension of r. Also note that in the extension, transvections are represented by Hermitian definite matrices.

Now let r be a representation of H in $U(d)$ and R the representation of G in $GL(d, \mathbf{C})$ extending r. $GL(d, \mathbf{C})$ has an analytic structure J arising from the fact that $U(d)$ is a real form. The associated involutory automorphism τ of $GL(d, \mathbf{C})$ is the familiar one mapping a matrix to its conjugate transpose inverse. We have by definition the equations

$$R(j\mathfrak{g}) = JR(\mathfrak{g})$$

and

$$R(\sigma g) = \tau R(g).$$

The first of these equations asserts that R is a holomorphic map of G to $GL(d, \mathbf{C})$. From the second equation we can prove

LEMMA 5. *$R(G)$ is closed in $GL(d, \mathbf{C})$.*

Let $k_v = R(g_v)$ be a sequence of elements in $GL(d, \mathbf{C})$ converging to k in $GL(d, \mathbf{C})$. Write $g_v = p_v h_v$ with $\sigma p_v = p_v^{-1}$, $\sigma h_v = h_v$; i.e., p_v is a transvection and $h_v \in H$. Write $b_v = a_v b_v$ with $\tau a_v = a_v^{-1}$, $\tau b_v = b_v$. Then $R(p_v) = a_v$ and $R(h_v) = b_v$. And it follows that the sequences a_v and b_v converge separately, to a and b, respectively. Since H is compact, $R(H)$ is compact, and it follows that $b \in R(H) \subset R(G)$.

Let $p_v = \exp \mathfrak{p}_v$, so $a_v = \exp R(\mathfrak{p}_v)$. It follows that $R(\mathfrak{p}_v)$ converges to an element in the Lie algebra, say $\mathfrak{u}$. Since the map of Lie algebras is a surjection, there is a $\mathfrak{p} \in j\mathfrak{H}$ such that $R(\mathfrak{p}) = \mathfrak{u}$; then $R(\exp \mathfrak{p}) = \exp \mathfrak{u} = a$, completing the proof.

Now let r be a faithful representation of H in $U(d)$. We claim that

LEMMA 6. *R is a faithful representation of G.*

Suppose $R(g) =$ identity. Write $g = pk$ with $\sigma p = p^{-1}$, $\sigma h = h$. Then each of $R(p)$ and $R(h)$ are the identity. Since $R(h) = r(h)$ we have $h =$ identity. Now let $p = \exp j\mathfrak{h}_1$; $R(p) = \exp ir(\mathfrak{h}_1)$. $ir(\mathfrak{h}_1)$ is Hermitian, and since exponentiation is one-one on Hermitian matrices, $r(\mathfrak{h}_1) = 0$, which implies that $\mathfrak{h}_1 = 0$, implying p is the identity.

By assembling the last few results we can show

THEOREM 1. *G is a Stein manifold.*

Let R be the extension to G of a faithful representation of H. By Lemma 6, the entries of R give sufficient holomorphic functions to separate points.

Let $\mathfrak{h}_1, \mathfrak{h}_2, \ldots, \mathfrak{h}_n$ be a basis of $\mathfrak{H}$ and put $z_v = x_v + iy_v$, $z = (z_1, z_2, \ldots, z_n) \in \mathbf{C}^n$. For z in a sufficiently small neighborhood N of the origin in $\mathbf{C}^n$, the map of N to G given by

$$z \to g = \exp \sum_v (x_v + y_v j)\mathfrak{h}_v$$

gives a complex analytic chart at the identity in G. We have

$$\begin{aligned} R\big(\exp \textstyle\sum (x_v + y_v j)\mathfrak{h}_v\big) &= \exp \textstyle\sum z_v r(\mathfrak{h}_v) \\ &= (f_{\alpha\beta}(z))_{\alpha,\beta}. \end{aligned}$$

$\sum z_v r(\mathfrak{h}_v)$ belongs to $M(d, \mathbf{C})$, the ring of d-dimensional matrices over the complex numbers, which may be identified as before with the Lie algebra of $GL(d, \mathbf{C})$. Since the exponential map of $M(d, \mathbf{C})$ into $GL(d, \mathbf{C})$ has a holomorphic inverse in a neighborhood of the identity, and since the representation r is faithful, it follows readily that for z sufficiently close to the origin, z is a holomorphic function of the entries $f_{\alpha\beta}(z)$. Thus a subset of the $f_{\alpha\beta}(z)$ give holomorphic coordinates near the identity in G. An analogous argument takes care of arbitrary points in G.

Finally, the topology of $GL(d, \mathbf{C})$ may be given as the topology it inherits as a subspace of $M(d, \mathbf{C})$. Since $R(G)$ is closed in $GL(d, \mathbf{C})$, it follows that $R(G)$ topologized as a subspace of $GL(d, \mathbf{C})$ is homeomorphic to G. Let K be a compactum in G. Let α be an upper bound for the moduli of all of the entries of $R(g)$ and $\dfrac{1}{\det R(g)}$ for $g \in K$. Then the set A of $u = (u_{ij}) \in GL(d, \mathbf{C})$ such that $|u_{ij}| \leqq \alpha$ and $\left|\dfrac{1}{\det u}\right| \leqq \alpha$ is compact in $GL(d, \mathbf{C})$. $A \cap R(G)$ is also compact. Hence the holomorphically convex hull of K is compact, proving that G is holomorphically convex, and completing the proof that G is Stein.

REMARK. It may be shown that if G is a closed subgroup of $GL(d, \mathbf{C})$, G stable under τ, then $H = G \cap U(d)$ is maximal compact in G, and H is a real form of G.

Let Δ be the Casimir operator of S, and put

$$\nabla = -\frac{1}{4\pi^2} \sum \mathfrak{c}_i^2.$$

Both Δ and ∇ are in the center of the universal enveloping algebra of H or G. We shall regard Δ and ∇ as differential operators for the manifold H by viewing the elements of the Lie algebra as left-invariant vector fields. Both operators are then formally self-adjoint.

Let $h \to r(h)$ now be an irreducible representation of H, inducing representation $\mathfrak{h} \to r(\mathfrak{h})$ of $\mathfrak{H}$. Restricted to $\mathfrak{S}$, the representation remains irreducible, since the elements of the center $\mathfrak{C}$ of $\mathfrak{H}$ are all represented by scalar matrices. While restricted to $\mathfrak{C}$, the representation is simply the representation of $\mathfrak{C}$ induced by a representation of the torus C.

The representation r of $\mathfrak{H}$ is completely determined by its restriction to $\mathfrak{S}$ and $\mathfrak{C}$, but not every representation of $\mathfrak{H}$ gives rise, of course, to one of H.

A representation r of C is completely determined by a k-tuple $K = (a_1, a_2, \ldots, a_k)$ of integers, namely:

$$r\left(\exp \sum \lambda_v \mathfrak{c}_v\right) = \exp\left(2\pi i \sum_v \lambda_v a_v\right).$$

On the other hand, a representation of $\mathfrak{S}$, if $\mathfrak{S}$ is of rank l, is determined by an l tuple $L = (b_1, b_2, \ldots, b_l)$ of non-negative integers as follows. Every irreducible representation is determined by its highest weight ω, and there exist l fundamental weights $\omega_1, \omega_2, \ldots, \omega_l$ such that every highest weight ω is of form

$$\omega = \sum b_i \omega_i.$$

Thus to every irreducible representation of H we may assign uniquely a pair (L, K) consisting of an l-tuple of non-negative integers, and a k-tuple of integers. We shall parametrize the representations of H by all such pairs, and remind the reader to suppress in ensuing computations the pairs which do not correspond to any representation.

Now let $h \to r(h)$ correspond to pair (L, K). First of all we see that

$$\nabla r(h) = \left(\sum a_v^2\right) r(h) = \chi(K) r(h)$$

and as a straightforward computation shows

$$\Delta r(h) = r(\Delta) r(h),$$

where $r(\Delta)$ is the matrix representing Δ in the extension of the representation to the universal enveloping algebra. $r(\Delta)$ is clearly a scalar matrix, $r(\Delta) = \eta$ identity, where $\eta = \eta(\omega)$ is a constant depending on the highest weight. η has been computed by Freudenthal (see [2]) as follows. Let $\rho = \frac{1}{2}$ sum of the positive roots of H. Then

$$\lambda(\omega) = (\omega, \omega) + 2(\omega, \rho),$$

where the symmetric bilinear form $(\cdot, \cdot)$ is positive-definite. If $\omega = \sum b_i \omega_i$, $L = (b_1, b_2, \ldots, b_l)$, then we denote $\lambda(\omega)$ by $\lambda(L)$ and note that $\lambda(L)$ is an inhomogeneous quadratic form in the b_i whose part of weight 2 is positive-definite. Moreover, it is known that $\lambda(L) \geqq 0$, and $\lambda(L) = 0$ only for $L = 0$.

The degree $d = d_L$ of the representation corresponding to the weight ω has been given by Weyl: $d = \dfrac{\Pi(\alpha, \rho + \omega)}{\Pi(\alpha, \rho)}$, where the products are taken over the positive roots α. d is a polynomial in L.

Let D be an H-stable domain in G. Then $E = \pi(D)$ is a domain in G/H, and $D = \pi^{-1}(E)$. We call E *convex* if it contains the geodesic segment connecting any two of its points. D is called convex if $\pi(D)$ is convex. If E is an arbitrary domain in G/H, the convex hull $\hat{E}$ of E is the intersection of all convex domains containing E. The convex hull of D is defined to be $\pi^{-1}(\widehat{\pi(D)})$.

The convex hull of E may be constructed as follows. Let E_{n+1} be the union of all geodesic segments connecting points of E_n and put $E_0 = E$. Then the E_n are an increasing sequence of open sets whose union is $\hat{E}$.

There is an analogous construction for the hull of an open, H-stable, set in G.

Now let f be a real function defined on G/H. f is called *convex* if for any geodesic segment $x(t)$, $0 \leqq t \leqq 1$, we have

$$f(x) \leqq (1 - t)f(x(0)) + tf(x(1)).$$

If f is defined on G, we say f is convex if $f(gh) = f(g)$ for $h \in H$, and viewed as a function on G/H it is convex as above.

Clearly, if f is a continuous convex function on G, the set $\{g \mid f(g) < 1\}$ is a convex domain.

Our first basic result on convexity is as follows. Let r be a representation of H in $U(d)$ and R its extension to G. Let A be an arbitrary complex matrix of dimension d. By A^* we denote the conjugate transpose of A. Then

LEMMA 7. *$f(g) = \operatorname{tr} AR(g)[AR(g)]^*$ is a convex function on G.*

It is clear that $f(gh) = f(g)$ for $h \in H$.

Now let $\lambda > 0$ and $0 \leqq p \leqq 1$, $q = 1 - p$. By the inequality of the arithmetic and geometric mean

$$\lambda^p \cdot 1^q = \lambda^p \leqq q + p\lambda.$$

If B is a Hermitian definite matrix, then $B = \exp C$, for unique Hermitian C, and we denote by B^t the matrix $\exp tC$, for any real t.

The inequality just given implies that $qI + pB - B^p$ is Hermitian definite. Thus if P is also Hermitian positive-semidefinite, we have

$$\operatorname{tr} P \cdot B^p \leqq q \operatorname{tr} P + p \operatorname{tr} P \cdot B.$$

Let α and β be transvections in G, and put in the last $P = R(\alpha)A^*AR(\alpha)$, $B = R^2(\beta)$. β^p is a well defined transvection for any real p. Thus

$$\operatorname{tr} R(\alpha)A^*AR(\alpha)R^{2p}(\beta) \leqq q \operatorname{tr} R(\alpha)A^*AR(\alpha) + p \operatorname{tr} R(\alpha)A^*AR(\alpha)R^2(\beta);$$

this may be rewritten

$$f((\alpha\beta^p\alpha)^{1/2}) \leqq qf(\alpha) + pf((\alpha\beta^2\alpha)^{1/2}).$$

We have now only to note that the geodesic segment connecting the points αH and $(\alpha\beta^2\alpha)^{1/2}H$ in G/H is $(\alpha\beta^{2p}\alpha)^{1/2}H$, $0 \leqq p \leqq 1$, to see that f is convex as asserted.

If F is a continuous function on H, then F has a Fourier expansion

$$F(h) \approx \sum_{L,K} d_L \operatorname{tr} A_{L,K} r_{L,K}(h),$$

where $r_{L,K}$ is the irreducible representation of H parametrized by pair (L, K), d_L is its degree, and

$$A_{L,K} = \int F(h) r_{L,K}(h^{-1})\, d_\mu,$$

where d_μ is Haar measure on H. If the series is uniformly convergent, then it is convergent to F. We always have Plancherel's theorem,

$$\int |F(h)|^2\, d_\mu = \sum_{L,K} d_L \operatorname{tr} A_{L,K} A^*_{L,K}.$$

If we suppose F is a C^∞ function, then since the operators Δ and ∇ are self-adjoint, it follows that the Fourier coefficients of $\Delta^a \nabla^b F$ are $\lambda^a(L)$ $\chi^b(K)\, A_{L,K}$.

Let D be an H-stable connected domain in G, and let F be a holomorphic function on D. For fixed $g \in D$, we consider $F(gh)$ as a function on H, and take its Fourier series

$$F(gh) \approx \sum d_L \operatorname{tr} [A_{L,K}(g) r_{L,K}(h)].$$

Now $A_{L,K}(g)$ has entries which are readily seen to be holomorphic functions on D, and satisfies in addition $A_{L,K}(gh) = A_{L,K}(g) r_{L,K}(h)$ for $h \in H$. This suggests:

LEMMA 8. *There is a constant matrix $A_{L,K}$ such that $A_{L,K}(g) = A_{L,K} R_{L,K}(g)$.*

A function holomorphic on D is completely determined by its value on the set $g_0 H$ for any $g_0 \in D$. Consider then

$$B(g) = A_{L,K}(g_0) R^{-1}_{L,K}(g_0) R_{L,K}(g).$$

$B(g_0 h) = A_{L,K}(g_0 h)$ for any $h \in H$, so that $B(g) = A_{L,K}(g)$ everywhere.

Let U be an open, relatively compact, H-stable set in D.

LEMMA 9. *The Fourier series for $F(gh)$ is absolutely uniformly convergent for $g \in \hat{U}$, $h \in H$.*

The Fourier coefficients of $\Delta^a \nabla^b F(gh)$ are $\lambda^a(L)\chi^b(K)A_{L,K}R_{L,K}(g)$. Since $\Delta^a \nabla^b F(gh)$ is uniformly bounded for $g \in U$, we have, by Plancherel's theorem, for suitable constant m,

$$d_L \lambda^{2a}(L)\chi^{2b}(K) \operatorname{tr} [A_{L,K}R_{L,K}(g)R^*_{L,K}(g)A^*_{L,K}] \leqq m,$$

true for arbitrary L, K and $g \in U$. From Lemma 7 it follows that the same inequality is true for $g \in \hat{U}$. Then with the aid of the Cauchy–Schwarz inequality we obtain

$$|\operatorname{tr} A_{L,K}R_{L,K}(g)| \leqq \frac{m^{1/2}}{\lambda^a(L)\chi^b(K)}.$$

(To take care of the terms with $\chi(K) = 0$, or $\lambda(L) = 0$, the same treatment but with $b \neq 0$, or $a \neq 0$, is used.)

Now by selecting a and b sufficiently large, it is easy to see that the series

$$\sum_{L,K} d_L |\operatorname{tr} A_{L,K}(g)|$$

is uniformly convergent by Weierstrass's M-test.

From the last result, we can readily derive our principal result:

THEOREM 2. *$F(g)$ extends to be a holomorphic function on $\hat{D}$.*

By the previous result $F(g)$ extends to a holomorphic function on $\hat{U}$.

Let $U_0 \subset U_1 \subset \cdots \subset U_n \subset \cdots$ be an increasing sequence of open, relatively compact, H-stable, open sets converging to D.

For any H-stable open set M we denote by $M^0 = M, M^1, M^2, \ldots$ the increasing sequence of open sets, as described earlier, converging to $\hat{M}$. With this notation, we claim that the increasing sequence of open sets U_n^n converges to $\hat{D}$. What we prove, by induction on v, is that if $g \in D^v$, then $g \in U_n^n$ for some n. This is obvious for $v = 0$. Suppose $g \in D^{v+1}$. Then $\pi(g)$ is on the geodesic segment joining $\pi(p)$ and $\pi(q)$ for some p and $q \in D^v$. By the induction both p and $q \in U_n^n$ for some n. But then $g \in U_n^{n+1} \subset U_{n+1}^{n+1}$, completing the induction.

Now let $g \in \hat{D}$. Then $g \in U_n^n$ for some n, and since U_n^n is open, a neighborhood of g is $\subset U_n^n \subset \hat{U}_n$. Since F extends to be holomorphic on $\hat{U}_n$, it extends to be holomorphic at g, completing the proof of the theorem.

It is tempting to conjecture that if E is convex, then E is holomorphically convex. Since G is a Stein manifold, it would be sufficient to prove that E is pseudoconvex. A possible procedure might be as follows. If R is any holomorphic representation of G in $GL(d, C)$ then $f(g) = \operatorname{tr} AR(g)R^*(g)A^*$ is plurisubharmonic, since it is the sum of squared moduli of holomorphic functions. The set $\{g|f(g) < 1\}$, which we call a half-space, is then holomorphically convex. Hence it would suffice to prove that every convex set is the intersection of the half-spaces containing it.

CORNELL UNIVERSITY

REFERENCES

1. HELGASON, S., *Differential Geometry and Symmetric Spaces.* New York: Academic Press, Inc., 1962.
2. JACOBSON, N., *Lie Algebras.* New York: Interscience Publishers, Inc., 1962.

Automorphisms of Commutative Banach Algebras

STEPHEN SCHEINBERG

This presentation consists of a few observations related to joint work with Herbert Kamowitz. In [2] we showed that if T is an automorphism of a semisimple commutative Banach algebra and if $T^n \neq I$ (all n), then the spectrum of T, $\sigma(T)$, must contain the unit circle. Examples were given to show that this containment can be proper. Section 1 that follows contains more complicated examples of such $\sigma(T)$ plus the theorem that $\sigma(T)$ is necessarily connected. In [1] we studied derivations and automorphisms of a particular radical algebra. Section 2 concerns another class of radical algebras, those of formal power series. These algebras behave like semisimple algebras in that all derivations and automorphisms are necessarily continuous.

1.

Let R be a region which is the interior of its closure and which is bounded away from 0 and ∞. Assume that both $\{|z| < 1\} - \bar{R}$ and $\{|z| > 1\} - \bar{R}$ are semigroups under multiplication. Finally suppose R contains an annulus adjacent to the unit circle.

THEOREM 1. *There is an automorphism T of a semisimple commutative Banach algebra for which $\sigma(T) = \bar{R}$.*

PROOF. It is easy to see that $R = \bigcup R_n$, where each R_n satisfies the requirement just given and, further, ∂R_n is rectifiable. If T_n can be constructed for R_n, then $T = \sum \oplus T_n$ on the direct sum of the algebras will have $\sigma(T) = \bar{R}$. Thus we may assume ∂R is rectifiable. The annulus that R contains separates the plane into two components Ω_0 and Ω_∞, containing 0 and ∞, respectively. Let $\Gamma_0 = \partial R \cap \bar{\Omega}_0$ and $\Gamma_\infty = \partial R \cap \bar{\Omega}_\infty$.

Let A be the family of all bounded analytic functions on R. The multiplication on A will be the Hadamard product; each member of A has a Laurent series on the annulus:

$$\left(\sum a_n z^n\right) * \left(\sum b_n z^n\right) = \sum a_n b_n z^n.$$

To see that $A * A \subseteq A$, write

$$f = f_0 + f_\infty = \sum_0^\infty + \sum_{-\infty}^{-1}$$

for each f in A. Clearly f_0 is analytic on $R_0 = \Omega_0 \cup R$ and $\sup_{R_0} |f_0| \leqq$ const $\sup_R |f|$. The analogous statement applies to f_∞.

Now $f * g = f_0 * g_0 + f_\infty * g_\infty$ and a standard (easy) computation shows that $f_0 * g_0(z) = (2\pi i)^{-1} \int_{\Gamma_\infty} f_0(z/w) g_0(w) w^{-1}\, dw$ and $f_\infty * g_\infty(z) = (2\pi i)^{-1} \int_{\Gamma_0} f_\infty(z/w) g_\infty(w) w^{-1}\, dw$, with suitable orientation on Γ_0, Γ_∞. In these integrals $g_0(w)$ and $g_\infty(w)$ are defined a.e.; the semigroup condition is just what is needed to ensure $z/w \in R_0$ when $z \in R_0$ and $w \in \Gamma_\infty$, and similarly with 0 and ∞ interchanged.

It is now evident that A becomes a Banach algebra when $\|f\|$ is defined to be (a suitably large constant) $\sup_R |f|$. It is also clear that A is semisimple, since the homomorphisms $\phi_k(\sum a_n z^n) = a_k$ separate the members of A [i.e., $\phi_k(f) = (2\pi i)^{-1} \int_{|z|=1} f(z) z^{(-k+1)}\, dz$]. Let T be the shift $f(z) \to zf(z)$. T is obviously an automorphism of A with $\sigma(T) = \bar{R}$.

Examples with disconnected interior may be obtained by a slight modification of this construction. Let G be an open subset of $\{2 < |z| < 3\}$ with rectifiable boundary such that ∂G meets $\{|z| = 2\}$ in a totally disconnected set of positive length. Let A be the family of all bounded analytic functions on $R = \{1 < |z| < 2\} \cup G$ such that the boundary values on the two parts agree a.e. on the common boundary. A is complete with respect to uniform convergence and is an algebra under Hadamard product, since every member of $A * A$ is analytic on $\{1 < |z| < 4\}$. As before, $\|f\| = \text{const } \|f\|_\infty$ makes A into a Banach algebra. The condition on the boundary means that if f vanishes on $\{1 < |z| < 2\}$ then it must vanish on all of R; thus A is semisimple. As before $Tf(z) = zf(z)$ defines an automorphism with spectrum $\bar{R}$.

This procedure can be repeated indefinitely and, together with the previous construction, will produce fairly complicated sets. If $\sigma(T)$ properly contains the unit circle, must it contain an annulus, must it be the closure of its interior, and must it have interior?

THEOREM 2. *If T is an automorphism of a semisimple commutative Banach algebra and $T^n \neq I$ (all n), then the spectrum of T is connected.*

PROOF. Via the Gelfand map we may assume that the algebra A in question is a (not necessarily closed) subalgebra of $C(X)$, that $\|f\| \geqq \|f\|_\infty$ and that T acts as a homeomorphism of $X: Tf(x) = f(Tx)$.

Since $\sigma(T)$ contains the unit circle (see [2]), if it were disconnected it could be separated by a rectifiable path γ lying in the interior or in the exterior of the unit disc. By taking T^{-1} in place of T if necessary, we may assume that γ lies in $\{|z| > 1\}$.

Put $S = (2\pi i)^{-1} \int_\gamma \lambda(\lambda - T)^{-1} \, d\lambda$. Then $\sigma(S) = \sigma(T) \cap$ (inside of γ) and $\sigma(T - S) = \sigma(T) \cap$ (outside of γ). For any $f \in A$, $x \in X$, $Sf(x) = (2\pi i)^{-1} \int_\gamma (1 - \lambda^{-1}T)^{-1} f(x) \, d\lambda = (2\pi i)^{-1} \int_\gamma \sum \lambda^{-n} f(T^n x) \, d\lambda = \sum_0^\infty f(T^n x) \cdot \left[(2\pi i)^{-1} \int_\gamma \lambda^{-n} \, d\lambda\right]$. This is justified since $|\lambda|$ is uniformly >1 for $\lambda \in \gamma$. If γ surrounds 0, this string of equalities gives $Sf(x) = f(Tx)$, implying $S \equiv T$. If γ does not surround 0, we have $Sf(x) = 0$, implying $S \equiv 0$. Either case contradicts the assumption that $\sigma(T)$ meets both the inside and outside of γ.

2.

Given a sequence $\alpha_n > 0$, let

$$A_\alpha = \left\{ x = \sum_1^\infty x_n z^n : \|x\| = \sum |x_n| \alpha_n < \infty \right\}.$$

A_α is a Banach space isomorphic to l^1. It becomes a Banach algebra under formal power-series multiplication iff $\alpha_{n+m} \leqq \alpha_n \alpha_m$. A_α is a radical Banach algebra iff also $\alpha_n^{1/n} \to 0$. Since z generates A_α, many questions are easy to handle.

Recall that a derivation of an algebra is a linear operator D such that $D(xy) = (Dx)y + x(Dy)$. It is easy to see that a continuous (i.e., bounded) derivation D is determined, once Dz is known, by $Dx = (Dz)x'$, where x' is the differentiated formal power series. (It is obvious for polynomials, which are dense.)

THEOREM 3 (Loy [4]). *A derivation D on the Banach algebra A_α is necessarily bounded.*

PROOF (see [4]). By the closed graph theorem, it is enough to show that $(Df)_m$, the mth coefficient of Df, is continuous in f, for each m. By induction, if m is the first integer for which this fails, f_n can be found for which

$\|f_n\| \to 0$ and $(Df_n)_m \to \infty$ very fast. From this one can, therefore, contradict $D(\sum z^n f_n) \in A_\alpha$, because its coefficients will be too large.

THEOREM 4. *These are equivalent for any A_α, where k is any integer $\geqq 1$:*

(i) *There is a derivation D of A_α such that $Dz = c_k z^k + \cdots$ $(c_k \neq 0)$.*
(ii) *There is a derivation D of A_α such that $Dz = z^k$.*
(iii) $\alpha_{n+(k-1)} = 0(\alpha_n/n)$ *as $n \to \infty$.*

PROOF. (ii) $\Rightarrow$ (i) trivially.

(i) $\Rightarrow$ (iii) by simply looking at the norms of $Dz^n = nz^{n-1}(Dz)$, since D is bounded.

(iii) $\Rightarrow$ (ii) by direct examination of the norm.

COROLLARY 1. *Let $\alpha_n = 1/n!$ Then $Dx = z^2x'$ defines a derivation on A_α.*

If we put $\alpha_n = e^{-nc_n}$, then $c_n \to \infty$ is the condition that A_α be radical. The condition $\alpha_{n+m} \leqq \alpha_n\alpha_m$ becomes the "convexity" condition

$$\frac{n}{n+m} c_n + \frac{m}{n+m} c_m \leqq c_{n+m}.$$

Thus $c_1 \leqq c_2 \leqq \cdots$ ensures that A_α is a Banach algebra. Condition (3) will fail for every k if, say, $c_{n+1} - c_n \leqq 1/n$ and $kc_{n+k} \leqq \log n$ for $n \geqq n(k)$. This follows by a straightforward calculation which is omitted. For example, $c_n = \log\log n$ will give the following corollary:

COROLLARY 2. *Let $\alpha_n = (\log n)^{-n}$. If D is a derivation on A_α, then $D \equiv 0$.*

D. J. Newman [3] proved this result for bounded derivations.

$T_\theta x(z) = x(e^{i\theta}z)$ defines an automorphism of A_α. The proof of Theorem 3 works for automorphisms as well and shows that every automorphism of A_α is bounded. It is then clear that any automorphism T is determined by Tz, by the formula $Tx(z) = x(Tz)$, since this holds for polynomials. Let $Tz = az + \cdots$; it is easy to see that $|a| = 1$, since $|a| > 1$ makes T unbounded and $|a| < 1$ makes T^{-1} unbounded. If D is a derivation, then of course e^D is an automorphism and $e^D \neq$ any T_θ, unless $D \equiv 0$. Indeed, $Dz = c_k z^k + \cdots$ implies $e^D z = z + c_k z^k + \cdots$, since $k \geqq 2$ by condition (iii) of Theorem 4.

THEOREM 5. *A_α admits an automorphism other than the T_θ's $\Leftrightarrow$ A_α admits a derivation other than* 0.

PROOF. $\Leftarrow$ is trivial, as noted previously.

$\Rightarrow$ By replacing T by $T_\theta T$ we may assume $Tz = z + c_k z^k + \cdots$, $c_k \neq 0$. Then $Tz^n = z^n + nc_k z^{n+k-1} + \cdots$. Boundedness of T implies boundedness of $n\alpha_{n+k-1}/\alpha_n$, which is exactly condition (iii) of Theorem 4.

In the ring A of all formal power series without constant term, every automorphism is of the form $Tx(z) = x(Tz)$, and $Tz = az + \cdots$ $(a \neq 0)$. When $a = 1$, T is the exponential of a unique derivation D; namely, $D = \log T = \log\{I + (T - I)\} = \sum_1^\infty (-1)^{n-1} \frac{1}{n} (T - I)^n$, the series converging in the strong operator topology (convergence in A is, of course, convergence in each coefficient). In fact, the series applied to any element of A is eventually constant in each coefficient. D is a derivation by the proof of Theorem 6 of [1].

On A_α any $T = e^D$ must send z to $z + \cdots$, since $Dz = c_k z^k + \cdots$ with $k \geqq 2$. The series for $D = \log T$ need not converge in the strong topology; e.g., let D be as in Corollary 1 of Theorem 4. The question of whether every automorphism T of A_α sending z to $z + \cdots$ is the exponential of a derivation is exactly the question of whether the formal $D = \log T$, defined on all formal power series, sends A_α into A_α. This may not always be the case.

Theorem 2 has been obtained independently by R. J. Loy (personal communication).

MASSACHUSETTS INSTITUTE OF TECHNOLOGY

REFERENCES

1. KAMOWITZ, H., and S. SCHEINBERG, "Derivations and automorphisms of $L^1(0, 1)$," *Trans. Amer. Math. Soc.*, *135* (1969), 415–427.
2. ———, "The spectrum of automorphisms of Banach algebras," *J. Funct. Anal.*, *4* (1969), 268–276.
3. NEWMAN, D. J., "A radical algebra without derivations," *Proc. Amer. Math. Soc.*, *10* (1959), 584–586.
4. LOY, R. J., "Continuity of derivations in topological algebras of power series," *Bull. Austral. Math. Soc.*, *1* (1969), 419–424.

Historical Notes on Analyticity as a Concept in Functional Analysis

ANGUS E. TAYLOR

1. Introduction

This is an essay—one of a projected series—on certain aspects of the history of functional analysis. The emphasis of this essay is on the way in which the classical theory of analytic functions of a complex variable was extended and generalized and came to play a significant role in functional analysis. One can perceive two lines of development: (i) the extension of the classical theory to cases in which the function of a complex variable has its values in a function space or in an abstract space, and (ii) the development of a theory of analytic functions from one general space to another (where by a "general space" we mean a function space or an abstract space). The essay also seeks to place the history of these developments in proper perspective in the narrative of the development of functional analysis as a whole.

Before dealing explicitly with the main subject of the essay it is desirable to sketch some relevant details of the history of functional analysis. The explicit emergence of the subject as a distinct and separate branch of mathematics may perhaps be considered as beginning with a series of five notes published by Vito Volterra in 1887 (see Volterra [55], [56], [57], [58] and [59]). The ideas underlying functional analysis were of course much older. At the 1928 International Congress in Bologna Jacques Hadamard mentioned Jean Bernoulli's problem of the curve of quickest descent and the pioneering work of the Bernoullis and Euler in the calculus of variations as the true and definitive foundation of "le calcul fonctionnel" (see Hadamard [24]). Volterra took the significant step of focussing attention on functions for which the independent variable was a function or a curve. This is not to say that Volterra was the first to study operations by which functions are transformed into other functions (or other entities). He was not. The "functional operations" considered by mathematicians prior to the work of Volterra were studied, however, from a somewhat different point of view, more algebraic than analytic. Mathematicians examined the manipulations which could be performed with functional operations, viewed as symbols. (See Pincherle [2] and references cited therein.) In particular, Volterra seemed to be making a new venture in subjecting his "functions of lines" to analysis comparable to that applied in calculus and the classical theory of functions.

In our current terminology, Volterra considered the domain of one of his functions to be a class of functions (in the classical sense) or a class of curves or surfaces. Volterra's functions were numerically valued; thus the ranges were sets of numbers. Volterra drew his examples from boundary-value problems for partial differential equations and from the calculus of variations.

In Volterra's time the name "functional" was still in the future. In the first of his 1887 notes, he referred to "functions which depend on other functions." The title of one of the later notes in this series was "Sopra le funzioni dipendenti da linee." This gave rise to the term "fonctions de lignes" which persisted in its Italian, French, and English forms for a considerable period. It is asserted by Maurice Fréchet and Paul Lévy that the noun "functional" (fonctionelle) originated with Hadamard (see Fréchet [14], p. 2, and Lévy [32], p. 8). Since Hadamard's pioneering work [22] on a general method of representing linear functionals, the term "functional" has rather generally been reserved for a numerically valued function whose argument varies over a class of functions or some abstract set. However, such names as "operazioni funzionali" and "le calcul fonctionnel" were being used before 1900 in connection with operations which map functions into other functions (see Pincherle [41] and [42]). In [41], Pincherle has the following to say with reference to "le calcul fonctionnel": "On reunirait sous ce titre les chapitres de l'analyse ou l'élément variable n'est plus le nombre, mais la fonction considerée en elle-meme." The name "analyse fonctionnelle" was introduced by Lévy, according to Fréchet (see [14], p. 3).

In the early work on functional analysis certain algebraic and analytical operations were available in the nature of the explicit situation, there being no "abstract space" under consideration. It was possible to imitate, to a degree, the formulation of concepts such as continuity of a functional and uniform convergence of a sequence of functionals without a fully explicit treatment of metric and topological notions in the underlying class of functions. It was even possible to calculate such things as "variations" or differentials and functional derivatives without a fully explicit truly general definition. Most commonly these things were done merely by using absolute values and uniformity ideas, assuming the members of the underlying class of functions to be bounded and continuous. It was the work of Fréchet which led to abstraction and the explicit introduction of metrical and topological concepts into the general setting. With the greater generality and abstraction introduced by Fréchet and by E. H. Moore the name "general analysis" gained some currency. Many writers have tried to maintain a restriction on terminology, using "functional" always, as Hadamard had done, for a numerically valued function. In spite of this, "functional analysis" has come to mean not only functional calculus in the sense of Hadamard (l'étude des fonctionnelles), but also general analysis in the sense of Fréchet and Moore—that is, the study of functions (transformations) of a very general character, mapping one set onto another. The sets may be abstract or they may be composed of mathematical objects having a certain amount of structure: continuous functions, linear operators, matrices, measurable sets, and so forth.

These historical notes present some of the results of an attempt to search out the pioneering work on analytic functions in general analysis, that is, of functions from one set to another which are analytic in a suitable sense as an extension of the concept of complex analytic functions of a complex variable.

The essay is divided into four parts, dealing respectively with generalizations of the concept of a polynomial, analytic functions of a complex variable with values in a general space, analytic functions from one general space to another, and the role in spectral theory of abstractly valued analytic functions of a complex variable.

2. Polynomial operations

One line of development of the generalization of the notion of analyticity follows the Weierstrassian point of view, which places the power series at the center of the theory. For this it is necessary to have a functional analysis counterpart of the monomial function defined by the expression $a_n z^n$ (a_n and z complex, n a natural number).

The initial steps in generalizing the concept of a polynomial suitably for functional analysis seem to have been taken by Fréchet. His first application was not to a generalization of analytic functions, but to a generalization of the Weierstrass theorem on approximation of continuous functions by polynomials. In a paper published in 1909 [8] he considers how ordinary real-valued polynomials of one real variable may be characterized as continuous functions such that the application of certain differencing operations to these functions leads to an identically vanishing result. The starting point is the observation that f (real and continuous) is a first degree polynomial in one real variable if

$$\Delta_2 f = f(x + y) - f(x) - f(y) + f(0) \equiv 0.$$

For a polynomial of degree n the corresponding identity is $\Delta_{n+1} f \equiv 0$, where $\Delta_{n+1} f$ is a certain sum involving terms $\pm f(x_{i_1} + \cdots + x_{i_k})$ and $\pm f(0)$, where $(i_1, \ldots, i_k)$ is a combination of integers chosen from the aggregate $(1, 2, \ldots, n + 1)$ and $k = 1, 2, \ldots, n + 1$. For example,

$$\begin{aligned}\Delta_3 f = f(x_1 + x_2 + x_3) - f(x_2 + x_3) - f(x_3 + x_1) - f(x_1 + x_2) \\ + f(x_1) + f(x_2) + f(x_3) - f(0).\end{aligned}$$

In this 1909 paper Fréchet extends the considerations to real functions of several real variables and then to real functions (i.e., functionals) of an infinite sequence of real variables. For this latter case the nature of continuity and of the domain of definition of the function f are not discussed very clearly or generally. Here a continuous functional for which $\Delta_{n+1} f \equiv 0$ for some n is called a "fonctionnelle d'ordre entier n."

In the following year (1910) Fréchet published a paper [9] on continuous (real-valued) functionals defined on the class of real continuous functions which we now denote by $C[a, b]$, where $[a, b]$ is a finite real interval. Here he carries over to this different setting the concept of a "fonctionnelle d'ordre entier n" from the 1909 paper. The definition of such a functional U calls for U to be continuous and for $\Delta_{n+1} U$ to vanish identically. Fréchet generalizes Hadamard's 1903 theorem on the representation of linear functionals by showing that if U is a continuous functional on $C[a, b]$ it can be represented in the form

$$U(f) = \lim_{n \to \infty} [u_n^{(0)} + U_n^{(1)}(f) + \cdots + U_n^{(r_n)}(f)],$$

where

$$U_n^{(r_k)}(f) = \int_a^b \cdots \int_a^b u_n^{(r_k)}(x_1, \ldots, x_{r_k}) f(x_1) \cdots f(x_{r_k})\, dx_1 \cdots dx_{r_k}.$$

Here $u_n^{(0)}$ is a number and $u_n^{(r_k)}(x_1, \ldots, x_{r_k})$ is a continuous function depending only on U and the indices (but not on f); it may be taken to be a polynomial in $x_1, \ldots, x_{r_k}$. The convergence to $U(f)$ of the sum is uniform on sets in $C[a, b]$ which are compact (in the sense of the term "compact" in use at that time as introduced by Fréchet). The functional $U_n^{(r_k)}$ is "entier, d'ordre r_k."

In this paper Fréchet also shows that if U is entire and of order n, then $U(y_1 f_1 + \cdots + y_p f_p)$ is a polynomial of degree at most n in the real variables $y_1, \ldots, y_p$. Here $f_1, \ldots, f_p$ are members of $C[a, b]$. He also observes that U is representable (uniquely) in the form

$$U(f) = U_0 + U_1(f) + \cdots + U_n(f),$$

where U_0 is constant and U_k is entire and of order k as well as homogeneous of degree k.

In the last part of this 1910 paper Fréchet turns to a definition of holomorphism. Suppose $g \in C[a, b]$. Then a functional U is called holomorphic at $f = g$ if there is a representation (necessarily unique)

$$U(f) = U_0 + U_1(f - g) + U_2(f - g) + \cdots + U_n(f - g) + \cdots$$

converging suitably under certain restrictions, where U_n is entire, of order n, and homogeneous of degree n. The representation is to be valid when $\max |f(x) - g(x)| < \epsilon$, for a certain positive ϵ, and the convergence is to be uniform with respect to f when f is confined to a compact subset of the functions which satisfy the inequality. Fréchet observes that Volterra had already considered particular instances of series representations of this type.

Fréchet observes that if U is holomorphic at $f = g$, then $U(g + tf)$ is a holomorphic function of t at $t = 0$. Here he comes close to an alternative approach to the establishment of a theory of analytic functionals. He notes, however, that U can have the property that $U(tf)$ is holomorphic in t at $t = 0$ without U being holomorphic at $f = 0$. The example he gives is this: $U(f) = \max f(x) + \min f(x)$. This functional U is homogeneous of the first degree: $U(tf) \equiv tU(f)$. But U is not entire of order one, and hence is not holomorphic at $f = 0$. Fréchet is here considering real scalars exclusively.

The next stage in the development is apparently due to R. Gateaux, who was clearly strongly influenced by both Fréchet and Hadamard. Gateaux was killed in September 1914, soon after the beginning of World War I, and his manuscript work remained unpublished until 1919. He made decisive contributions to the theory of analytic functions in the framework of functional analysis. At this point I take note only of his work in polynomial operations. This work is presented in two different contexts (see [17] and [18]). In [17], the manuscript of which dates from March 1914, Gateaux considers the space (which he calls E'_ω) of all complex sequences $(x_1, x_2, \ldots)$ with *écart*

$$E(x, x') = \sum_{n=1}^{\infty} \frac{1}{n!} \frac{|x_n - x'_n|}{1 + |x_n - x'_n|}.$$

A "polynomial of degree n" is a functional P which is defined and continuous on a certain set D in E'_ω and such that $P(\lambda z + \mu t)$ is an ordinary polynomial of degree n in λ and μ for each z and t in D. Here Gateaux is using *as a defining property* something which Fréchet had observed as a property possessed by his entire functionals of order n. In another posthumous paper [18] we find Gateaux considering functionals defined on a space of continuous functions. Citing Fréchet, he speaks of a functional as entire of order n if its differences of order $n + 1$ vanish identically while some difference of order n does not so vanish. He then points out that this definition is not satisfactory in the complex case. As an example he cites $U(z) = x$, where $z = x + iy$ (x and y real) for which $\Delta_2 U \equiv 0, \Delta_1 U \not\equiv 0$. Here U is continuous, but it is not suitable to regard it as a "polynomial" in z. In view of this, Gateaux uses the following definition: a continuous functional U is called entire, of order n, if $U(\lambda z + \mu t)$ is a polynomial of degree n in λ and μ whenever z and t are in the given space of continuous functions.

At this point Gateaux introduces the concepts of variations for a functional:

$$\delta U(z, t) = \left[\frac{d}{d\lambda} U(z + \lambda t)\right]_{\lambda=0},$$

$$\delta^2 U(z, t) = \left[\frac{d}{d\lambda} \delta U(z + \lambda t, t)\right]_{\lambda=0}.$$

These are what have subsequently become known as Gateaux differentials. We know that $\delta U(z, t)$ is not necessarily continuous or linear in t, but Gateaux does not discuss these points carefully. Instead, he says with reference to $\delta U(z, t_1)$, "C'est une fonctionnelle de z et de t_1 qu'on suppose habituellement linéaire, en chaque point z, par rapport à t_1." He is discussing the case of real scalars at this point. Again, citing Fréchet, Gateaux obtains the following formulas when U is an entire functional which is homogeneous and of order n:

$$U(z + \lambda t) = U(z) + \cdots + \frac{\lambda^k}{k!} \delta^k U(z, t) + \cdots + \frac{\lambda^n}{n!} \delta^n U(z, t)$$

and

$$U(t) = \frac{1}{n!} \delta^n U(z, t).$$

In a 1915 paper Fréchet [10] points out that the problem of determining the representation of a functional of the second order is reducible to the problem of representing a bilinear functional. This is an early allusion to the relation (which emerges in later studies by various authors) between entire functionals of order n and multilinear functionals.

The next major step, and a definitive one, was taken by Fréchet. In 1925 he published a note [13], later expanded into a paper [15] published in

1929, entitled "Les polynomes abstraits." The paper is a natural and direct outgrowth of his papers of 1909 and 1910, but the setting is more general. Fréchet is now considering functions from one abstract space to another. His spaces are of a general type which he calls "espaces algèbrophiles." These form a class more extensive than the class of normed linear spaces, but less extensive than the class of topological linear spaces as currently defined. The scalars are real, and the characterization of a function from one such space to another as an abstract polynomial is by means of continuity and the identical vanishing of differences, just as in the 1910 paper. The principal result is expressed thus: "Tout polynome abstrait d'ordre entier n est la somme de polynomes abstraits d'ordre $h = 0, 1, \ldots, n$, chaque polynome d'ordre h étant homogène et de degré h, et la décomposition étant unique." By 1929, of course, the theory of abstract linear spaces was much further advanced than in 1910. The difference between the 1910 and the 1929 papers is not in the basic characterization of abstract polynomials, but in the development of concepts and a framework and technique for dealing suitably with linearity and continuity as they enter into this particular problem.

The next six years saw the completion of the theory of abstract polynomials. This completion consisted in tidying up the proofs, putting them into their ultimate general form, and completing the linkage between several different ways of founding the theory. In particular, the difference between the complex and the real case was clarified.

In his 1932 California Institute of Technology doctoral dissertation R. S. Martin [34] used the following definition: A function $f(x)$ from one normed vector space to another is a polynomial of degree n if it is continuous and if $p(x + \lambda y)$ is a polynomial of degree at most n in λ (with vector coefficients) and of degree exactly n for some x, y. This definition applies for the case of either real or complex scalars. Martin was interested in a general theory of analytic functions. He was the student of A. D. Michal. At this time Michal was giving lectures on abstract linear spaces, and he developed an approach to polynomials by using multilinear functionals. These lectures were not published, but references to the work of this period are found in Michal [37].

In 1935 the Polish mathematicians S. Mazur and W. Orlicz published an important paper [36] systematizing the theory quite completely for the case of linear spaces with real scalars. The first of their results was announced to the Polish Mathematical Society in 1933. Their work was independent of that of Martin, Michal, and his group. They separated out the part of the development of the theory which is purely algebraic. In the algebraic portion of the theory they begin with multi-additive operations. If $U^*(x_1, \ldots, x_k)$ is defined for $x_1, \ldots, x_k$ in X, with values in Y, and is additive in each x_i, $U(x) = U^*(x_1, \ldots, x_k)$ is said to be rationally homogeneous of degree k (provided it does not vanish identically). An operation U of mth degree is one which has a representation

$$U(x) = U_0(x) + U_1(x) + \cdots + U_m(x),$$

where $U_m(x) \not\equiv 0$ and each U_k is either identically zero or is rationally homogeneous of degree k. It is then shown how an operation U which is rationally homogeneous of degree k determines and is uniquely determined by a symmetric k-additive operation U^* such that $U^*(x_1, \ldots, x_k)$ becomes $U(x)$ if one puts $x_1 = x_2 = \cdots = x_k$. U^* is obtained from U by differencing, and there is a "multinomial theorem" expressing $U(t_1x_1 + \cdots + t_kx_k)$ as a sum of terms in which the coefficient of each term of the form $t_1^{\nu_1} \cdots t_k^{\nu_k}$ is a multiple of

$$U^*(\underbrace{x_1, \ldots, x_1}_{\nu_1}, \ldots, \underbrace{x_k, \ldots, x_k}_{\nu_k}).$$

Finally, operations of degree m are characterized by both the Fréchet approach and the Gateaux approach. The case of complex scalars is not considered.

Passing beyond the purely algebraic part of the theory, Mazur and Orlicz deal with operations mapping one space of type (F) into another. (Here the nomenclature is that of Banach (see [1], p. 35). It is important for some but not for all of the results that the spaces be complete as metric spaces.) They add the requirement of continuity: a continuous k-additive operation is called k-linear, a continuous operation which is rationally homogeneous of degree k is called a homogeneous polynomial of degree k, and a continuous operation of degree m is called a polynomial of degree m. It is then shown that this concept of a polynomial coincides, for mappings from one (F)-space to another, with the concepts as variously introduced by Fréchet, Gateaux, and Martin.

Finally, for spaces with complex scalars, I. E. Highberg [25], another student of A. D. Michal, showed that for a mapping f from one complex algèbrophile space to another the following two sets of conditions are equivalent:

(i) f is continuous; $f(x + \lambda y)$ is a polynomial of degree at most n in λ for each x, y and of degree exactly n for some x, y.

(ii) f is continuous; f has a Gateaux differential at each point; $\Delta_{n+1}f \equiv 0$ and $\Delta_n f \not\equiv 0$.

The requirement of Gateaux differentiability in the second condition is superfluous in the real case but not in the complex case.

3. Analytic functions of a complex variable

As I pointed out earlier, Fréchet's work of 1910 includes consideration of the notion of a functional U defined on a set in the space $C[a, b]$ and holomorphic at a point g of that space. In particular he noted that, for such a U, $U(g + tf)$ is a holomorphic function of t at $t = 0$. Fréchet was dealing with the case of real scalars, and thus in this situation t and $U(g + tf)$ are real. The essence of the situation is that $U(g + tf)$ is expressible as a power series in t. The concept of an analytic function of a complex variable, with values in a function space, does not appear to come in for consideration at this period either in Fréchet's work or in that of Gateaux. In this essay I cannot deal fully with the question of just how this concept developed. Analytic dependence on a complex parameter appears at many places in the study of differential and integral equations. In Ivar Fredholm's famous 1903 paper [16], for instance, the solution of the equation

$$f(s) - \lambda \int_a^b k(s, t)f(t)\, dt = g(s)$$

is presented in the form

$$f(s) = g(s) + \frac{\lambda}{d(\lambda)} \int_a^b D(s, t; \lambda)g(t)\, dt,$$

provided $d(\lambda) \neq 0$. This is in a context where f and g are members of $C[a, b]$. Here $d(\lambda)$ is an entire analytic function of λ and $D(s, t; \lambda)$ is a continuous function of s, t as well as being an entire analytic function of λ. However, there is no suggestion at this stage of regarding the analytic dependence on λ in terms of the conceptualization of an analytic mapping from the complex plane into a function space.

If we turn to the work of F. Riesz [43] ten years later, we find explicit recognition of what was latent in the work of Fredholm, though not in precisely the same context. Riesz was studying "linear substitutions" in the theory of systems of linear equations in an infinite number of unknowns. He considered, in particular, substitutions of the type called *completely continuous*, acting in the class of infinite sequences which later came to be denoted by l^2 (the classical prototype of a Hilbert space). If A is such a substitution and E is the identity substitution, Riesz's studies led him to the assertion (see [43], p. 106) that the inverse substitution $(E - \lambda A)^{-1}$ is a meromorphic function of λ. In this conclusion we perceive Reisz's conception of $(E - \lambda A)^{-1}$ as a function of the complex variable λ with values in the class of bounded linear substitutions on l^2. However, there is no explicit discussion here of exactly what it means in general for such a function to be analytic at a particular point or to have a pole at a particular point. There is only a discussion of the particular function $(E - \lambda A)^{-1}$. This is couched in terms of what occurs when one looks at certain related linear substitutions (and their inverses) involving only a finite number of unknowns. These systems are dealt with by using determinants of finite order. A bit further on (see [43], pp. 114–119) Riesz examines $(E - \lambda A)^{-1}$ for the case of an arbitrary bounded linear substitution A (i.e., not merely the completely continuous case). Riesz shows that the class of regular points λ (those for which $E - \lambda A$ is appropriately invertible) form an open set in the complex plane and that $(E - \lambda A)^{-1}$ is analytic at each regular point μ in the following sense: $(E - \lambda A)^{-1}$ can be represented as a power series in $\lambda - \mu$ with certain coefficients which are bounded linear substitutions. The series converges in a well defined sense when λ is sufficiently close to μ. The mode of convergence is that of what is sometimes called the uniform topology of the bounded linear substitutions.

Further references to the work of Riesz will be made later when we come to the consideration of the role of analyticity as a concept in spectral theory. However, it should be mentioned at this point that Riesz observed (see [43], pp. 117–119) that it is possible to integrate $(E - \lambda A)^{-1}$ and other such functions along contours in the complex plane and to make effective use of the calculus of residues in the study of linear substitutions.

In a paper published in 1923 Norbert Wiener [60] pointed out that Cauchy's integral theorem and much of the classical theory of analytic functions of a complex variable remain valid for functions from the complex plane to a complex Banach space (which did not then regularly carry that name). Wiener made a few observations about applications. In particular, he pointed out that some of the work of Maxime Bôcher [2], on complex-valued functions $f(x, z)$ of a real variable x and a complex variable z, can be regarded as a study of an analytic function of z with values in the function space $C[a, b]$. (See Taylor [51], p. 655, for a slight refinement of this.) In another paper Taylor [50] extended Wiener's work to obtain some rather surprising things in connection with an analytic function of z with values in $L^p(a, b)$.

In the 1930's A. D. Michal and his students began to use Wiener's observations in the study of analytic mappings from one complex Banach space to another. In 1935 A. E. Taylor discovered that the following definition leads to a satisfactory general theory of such mappings: a function f with values in a complex Banach space Y is called analytic in a neighborhood N of a point x_0 in the complex Banach space X if f is continuous in N and if, for each x in N and each u in X, $f(x + \lambda u)$ is analytic (i.e., differentiable) as a function of the complex variable λ in some neighborhood of $\lambda = 0$. This definition and the theory flowing from it are natural generalizations of the work of Gateaux [18]. The details of Taylor's work are given in [45] and [47]. That Gateaux's work could be generalized in this manner was also observed by L. M. Graves (see [20], pp. 651–653). Taylor also studied some of the divergences from the classical theory which appear in the case of vector-valued analytic functions of a complex variable.

During the 1930's the research of American mathematicians was increasingly directed to the study of Banach spaces. The theory of analytic functions began to play a systematic role in spectral-theoretic studies of linear operators. This will be dealt with separately in a later part of this essay. Another interesting and important development occurred as a result of studying the concept of analyticity in the context of various alternative modes of convergence. In 1937 Nelson Dunford and Taylor independently discovered very closely related results, both of which depend essentially on what has come to be known as the principle of uniform boundedness. Dunford's result, published in 1938 (see [3], p. 354) is that, if f is a function from an open set D in the complex plane to a complex Banach space X, then f is analytic on D, in the sense of being differentiable at each point of D with respect to the strong topology of X, provided that $x^*(f(z))$ is analytic on D for each continuous linear functional x^* defined on X. This result, which has enormous usefulness, was initially very surprising because this is a case in which the weak convergence of difference quotients implies their strong convergence. The result of Taylor relates to the dependence of a bounded linear operator on a complex parameter. If A_λ is such an operator (from one complex Banach space X to another such space Y) for each complex λ in the open set D, Taylor showed (see [50], p. 576) that A_λ is differentiable on D as an operator-valued function provided that $A_\lambda x$ is differentiable on D as a vector-valued function for each vector x in X. That is, convergence of the difference quotients in the strong topology of Y implies convergence in the uniform topology of the space of operators. This result was also unexpected and surprising. The result of Dunford can be deduced from that of Taylor, and vice versa.

The use in functional analysis of analytic functions from the complex plane to a function space or an abstract space has developed and ramified enormously since the 1930's. Since this essay treats only the early stages of the subject I forego further details and examples. For a report on the subject as of 1943 see Taylor [51]. For a striking application to obtain a famous classical theorem by recognizing it as a functional analysis instance of another classical theorem see [52] (this is also dealt with in [54], pp. 211–212).

4. Analytic functions of a vector variable

In a 1909 paper David Hilbert sketched the start of a theory of numerical analytic functions of a countable infinity of complex variables (see [26], pp. 67–74). He used the Weierstrassian approach; that is, he dealt with functions represented by a "power series" the successive terms of which are: a constant, a linear form, a quadratic form, a ternary form, and so on. A definition of analytic functions based on this notion of a power series appears in the work of Helge von Koch [30] in 1899. This work deals with infinite systems of differential equations; von Koch imitates the classical pattern of argument used to obtain analytic solutions of differential equations. Hilbert considers analytic continuation and the composition of analytic functions. He gives an example to show that analytic continuation can give rise to uncountably many branches of a function.

As was pointed out earlier in this essay, Fréchet in 1910 introduced a definition of what it means for a functional to be holomorphic at a point in the space $C[a, b]$. The basic idea is that of an expansion in a series of homogeneous polynomials of ascending degree. Fréchet's work is much clearer than that of Hilbert, perhaps mainly because the questions surrounding proper definitions of convergence and of domains of definition of forms and power series are obscure when one deals with a countable infinity of complex variables without an adequate consideration of the structure of the space composed of the sequences $(x_1, x_2, \ldots)$. Fréchet's 1910 paper marked out quite clearly the lines along which the theory was to develop, and the work of Gateaux exploited Fréchet's beginning in brilliant fashion. Doubtless there were other forerunners of the definitive work of these pioneers. For example, Volterra had considered infinite series in which integrals of the form

$$\int_a^b \cdots \int_a^b K(x_1, \ldots, x_n) f(x_1) \cdots f(x_n)\, dx_1 \cdots dx_n$$

play the role of the homogeneous polynomial of degree n on $C[a, b]$.

At this stage there was lacking to Fréchet and Gateaux something fully comparable to the Cauchy point of view, according to which a function is characterized as analytic on a suitable set if it is differentiable on each point of the set. The early workers in functional analysis borrowed the concept of a "variation" from the calculus of variations. Gateaux used variations in his theory of analytic functionals (see [18]) to characterize a functional U as being holomorphic on a certain domain of definition if it is continuous and admits a first variation $\delta U(z, t)$ at each point z of the domain. (The definition of this variation was given earlier in this essay, in the discussion of polynomials.)

The variation $\delta U(z, t)$ came to be called a Gateaux differential. However, it lacked certain properties which are desirable in a differential. Fréchet made a definitive contribution with his 1925 paper [12] in which he shows how to define the concept of a differential of a function which maps one abstract space (of suitable type) on another. Fréchet's differential has strong properties; it has become the standard instrument of differential calculus in modern Banach space theory. In his paper Fréchet ascribes to Hadamard a basic principle: the fundamental fact about the differential of a function in calculus is that it depends linearly on the differentials of the independent variables. He refers to Hadamard's book [22] on the calculus of variations, where Hadamard asserts that the methods of differential calculus can be extended to functionals $U(y)$ for which the first variation is a linear functional of the variation of y. Actually, Fréchet's concept of the differential is a direct extension to the abstract situation of the well known definition introduced into calculus by O. Stolz. The line of Fréchet's thinking as early as 1912 with respect to the Stolz definition is clearly shown in [24] and in other papers mentioned therein. However, the tools were not yet available to formulate the definition adequately in the abstract space context which Fréchet came to in 1925.

The Fréchet concept of the differential is stronger than the Gateaux concept. More precisely, if f is a function from a normed linear space X to a normed linear space Y, if f is defined in a neighborhood of the point x_0 and has a Fréchet differential at x_0, then f has a Gateaux differential at x_0 and it coincides with the Fréchet differential. This proposition becomes false, however, if we interchange the names Fréchet and Gateaux in it. Some of the disparity between the two concepts is revealed by the observation that f may have a Gateaux differential and yet be discontinuous, while f must be continuous at a point if it has a Fréchet differential at that point. Because of this great difference between the two differentials it is remarkable that the following theorem is true. (It was discovered independently by L. M. Graves and Angus E. Taylor, but Graves has priority in the time of discovery. (See [20], as well as [45] and [46]. Both Graves and Taylor knew the work of Gateaux.) Suppose that X and Y are complex Banach spaces and that f is a function with values in Y which is defined on an open set D in X. Then f has a Fréchet differential at each point of D if and only if it is continuous and has a Gateaux differential at each point of D. Moreover, under these conditions f has Fréchet differentials of all orders at each point of D, f is holomorphic at each point of D, and the power series expansion of f in the neighborhood of a point x_0 in D is convergent within the largest spherical neighborhood centered at x_0 which lies wholly in D. The homogeneous polynomials in $x - x_0$ which form the terms of the power series are expressible as multiples of the Fréchet differentials of

f at x_0 according to the appropriate generalization of the Taylor series. The proof of this theorem utilizes and exposes the core of the theory of analytic mappings from one complex Banach space to another. Gateaux's work, building on the original power series conception of Fréchet, set the whole development in motion, but Gateaux's theory of analytic functionals made no reference to or use of the Fréchet concept of a differential.

One consequence of all this is that there is a simple definition of analyticity completely analogous to the classical definition from the Cauchy standpoint: f is analytic on an open set if it has a Fréchet differential at each point of the set (both domain and range in complex Banach spaces).

A few years prior to the discoveries by Graves and Taylor the power series approach to abstract analytic functions was under intensive study by Michal and his students, especially Clifford and Martin. They did not use the methods based on complex variables and Cauchy's formulas, as Gateaux had done, but relied entirely on the power series approach. (See [34], [38], and [39], especially p. 71 in [39].)

In a posthumously published monograph by Michal are given some historical notes about his involvement with the theory of polynomials and abstract analytic functions (see [37], pp. 35–39). Taylor was Michal's student. His work, originally motivated by an interest in a long paper of Luigi Fantappiè [7] on a rather different theory of analytic functionals, led into the consideration of $f(x + \lambda y)$ as a function of the complex variable λ, and from this, using Cauchy's integral formulas, Taylor discovered the linkage between the Gateaux and Fréchet differentials. It was apparent that Fantappiè's work did not lend itself to treatment in the framework of Banach spaces. It has subsequently become clear that a more general theory of topological linear spaces is necessary, and there have been extensive developments based on the foundations laid by Fantappiè. (See, for instance, [31], pp. 375–381, and references therein to Grothendieck, Sebastiao e Silva, and others.)

The state of subsequent development of the theory of analytic functions of a vector variable is partially indicated, with many references, in the massive book of Hille and Phillips ([28], Chapter 3, especially pp. 109–116). See also Dunford and Schwartz ([6], pp. 520–526) for an important application. Significant contributions were made in three papers by Max Zorn [61], [62], [63]. There are great untouched areas in the theory of analytic functions of a vector variable, especially in the study of such things as manifolds of singularity, convergence sets of power series, and envelopes of holomorphy. This is not surprising, since these subjects are so deep and complicated in the case of functions of a finite number of complex variables.

5. The role of analytic functions in spectral theory

I made reference earlier in this essay to the work of Riesz in showing that $(E - \lambda A)^{-1}$ depends analytically (and in some cases meromorphically) on λ. This work of Riesz foreshadows a wealth of important developments, but the full scope of what was latent in Riesz' book did not become evident for many years. Taylor, in a paper [49] published in 1938, showed that if T is a closed linear operator in a complex Banach space, the inverse $(\lambda I - T)^{-1}$, if it exists in a suitable sense for at least one λ, is an operator-valued analytic function on the (necessarily open) set of λ's for which it is defined. He also showed that if T is bounded and everywhere defined there must be some value of λ for which $(\lambda I - T)^{-1}$ does not exist; that is, the spectrum of T is not empty. This result hinges on the use of Liouville's theorem for vector-valued analytic functions. The special case of this theorem on the spectrum, for operators in a Hilbert space, had been known earlier. Marshall Stone [44] obtained the result without using vector-valued analytic functions. He did this by applying linear functionals to $(\lambda I - T)^{-1}x$.

At about this same time and a few years later work was going on in the study of complete normed rings (later to be known as Banach algebras) with the use of analytic functions as a tool. Ring elements $\lambda e - a$ and their inverses were investigated by Gelfand [19], Lorch [33], Mazur [35], and Nagumo [40]. It is interesting to note that, although Gelfand defined the concept of an analytic function of the complex variable λ with values in a complete normed ring, he developed the theory of such functions by reducing the arguments to the numerical case through the use of linear functionals. All of this work in normed rings has many parallels with the more general spectral theory of linear operators.

Spectral theory of linear operators is the functional analysis counterpart of the theory of eigenvalues of linear transformations in spaces of finite dimension. The spectrum of a linear operator T acting in a Banach space corresponds to the set of eigenvalues of the linear transformation in the finite-dimensional case. The spectrum of T consists of all values of λ for which $\lambda I - T$ fails to have an inverse in a suitable sense. When the Banach space is finite-dimensional, the spectrum is a finite set of eigenvalues and $(\lambda I - T)^{-1}$ is a rational function of λ. Much of spectral theory, for a particular operator acting in a Banach space, may be viewed as the study of the operator-valued function $(\lambda I - T)^{-1}$, called *the resolvent* of T, and of the behavior of this function of λ both globally and locally. A great deal of information about T itself can be extracted from this study.

The calculus of residues and Cauchy's integral formula play important roles in spectral theory. Integrals of the form

$$\frac{1}{2\pi i}\int_C f(\lambda)(\lambda I - T)^{-1}\, d\lambda$$

over suitable closed contours C in the plane, where the complex-valued function f is analytic on a neighborhood of the spectrum of T, are used to define operators which can be used in significant ways. The heuristic basis for the definition of an operator denoted by $f(T)$ is the symbolic "Cauchy's formula"

$$f(T) = \frac{1}{2\pi i}\int_C \frac{f(\lambda)}{\lambda - T}\, d\lambda.$$

For the finite-dimensional case (i.e., for the study of finite square matrices) there are nineteenth-century anticipations of this sort of thing in the work of Frobenius and others, most explicitly in an 1899 paper by Poincaré. For references to this early work see Dunford and Schwartz ([6], pp. 606–607) and Taylor ([51], p. 662, [53], p. 190). Systematic exploitation of the calculus of residues and of the symbolic operational calculus based on the Cauchy integral formula, as applied to general spectral theory, starts with Riesz and picks up again in the years around 1940. The work of investigators of normed rings, referred to earlier, is part of this general development. Dunford [4], [5] and Taylor [51], [53], [54] dealt explicitly with operator theory. (See also Einar Hille [27]. For further references see Hille and Phillips [28], pp. 164–183.)

The subsequent development and application in functional analysis of the ideas and methods of classical analytic function-theory have been varied and rich. I conclude this essay by mentioning just one example of such development. Analytic function-theory methods have been used to deal with perturbations of operators and their spectra. Perturbation theory goes back a long way in mathematics, of course. The pioneering work in studying perturbations of operators in Hilbert space seems to be that of F. Rellich, dating from 1937. Numerous subsequent investigators have made use of the symbolic operational methods of Dunford and Taylor. For references to the work of Rellich, B. v. Sz. Nagy, F. Wolf, and others see Kato [29].

UNIVERSITY OF CALIFORNIA
BERKELEY

REFERENCES

1. BANACH, STEFAN, *Théorie des operations linéaires*, Warsaw, 1932.
2. BÔCHER, MAXIME, "On semianalytic functions of two variables," *Annals of Mathematics*, (2), *12* (1910), 18–26.

3. DUNFORD, NELSON, "Uniformity in linear spaces," *Transactions of the American Mathematical Society, 44* (1938), 305–356.
4. ———, "Spectral theory," *Bulletin of the American Mathematical Society, 49* (1943), 637–651.
5. ———, "Spectral theory, I: Convergence to projections," *Transactions of the American Mathematical Society, 54* (1943), 185–217.
6. DUNFORD, N., and J. T. SCHWARTZ, *Linear Operators, Part I: General Theory.* New York: Interscience Publishers, Inc., 1958.
7. FANTAPPIÈ, LUIGI, "I funzionali analitici," *Memorie della R. Accademia Nazionale dei Lincei,* (6) 3 fasc., *11* (1930), 453–683.
8. FRÉCHET, MAURICE, "Une définition fonctionnelle des polynomes," *Nouvelles Annales de Mathématiques,* (4) *9* (1909), 145–162.
9. ———, "Sur les fonctionnelles continues," *Annales Scientifiques de l'École Normale Supérieure,* (3) *27* (1910), 193–216.
10. ———, "Sur les fonctionelles bilinéaires," *Transactions of the American Mathematical Society, 16* (1915), 215–234.
11. ———, "Sur la notion de différentielle totale," *Comptes Rendus du Congrès des Sociétés Savantes en 1914,* Sciences, 5–8, Paris, 1915.
12. ———, "La notion de différentielle dans l'analyse générale," *Annales Scientifiques de l'École Normale Supérieure,* (3) *42* (1925), 293–323.
13. ———, "Les transformations ponctuelles abstraites," *Comptes Rendus Acad. Sci. Paris, 180* (1925), 1816.
14. ———, *Les Espaces Abstraits et leur théorie considérée comme introduction à l'analyse générale.* Paris: Gauthier-Villars, 1928.
15. ———, "Les polynomes abstraits," *Journal de Mathématiques Pures et Appliquées,* (9) *8* (1929), 71–92.
16. FREDHOLM, IVAR, "Sur une classe d'équations fonctionnelles," *Acta Mathematica, 27* (1903), 365–390.
17. GATEAUX, R., "Fonctions d'une infinité des variables indépendantes," *Bullétin de la Societé Mathématique de France, 47* (1919), 70–96.
18. ———, "Sur diverses questions de calcul fonctionnel," *Bullétin de la Societé Mathématique de France, 50* (1922), 1–37.
19. GELFAND, I. M., "Normierte Ringe," *Matem. Sbornik, 9* (51) (1941), 3–24.
20. GRAVES, L. M., "Topics in the functional calculus," *Bulletin of the American Mathematical Society, 41* (1935), 641–662.
21. HADAMARD, JACQUES, "Sur les opérations fonctionnelles," *Comptes Rendus Acad. Sci. Paris, 136* (1903), 351.
22. ———, *Leçons sur le calcul des variations.* Paris: Hermann, 1910.
23. ———, "Le calcul fonctionnel," *l'Enseignement Mathématique, 14* (1912), 1–18.
24. ———, "Le développement et le rôle scientifique du calcul fonctionnel," in *Atti del Congresso Internazionale dei Matematici,* Bologna, 3–10 Settembre 1928 (VI) Tomo I, pp. 143–161.
25. HIGHBERG, IVAR, *Polynomials in Abstract Spaces,* unpublished doctoral dissertation, California Institute of Technology, 1936.
26. HILBERT, DAVID, "Wesen und Ziele einer Analysis der Unendlichvielen unabhängigen Variabeln," *Rendiconti del Circolo Matematico di Palermo, 27* (1909, 59–74.
27. HILLE, EINAR, "Notes on linear transformations, II: Analyticity of semigroups," *Annals of Mathematics,* (2) *40* (1939), 1–47.
28. HILLE, EINAR, and R. S. PHILLIPS, *Functional Analysis and Semi-Groups,* rev. ed. Providence, R.I.: American Mathematical Society, 1957.
29. KATO, TOSIO, *Perturbation Theory for Linear Operators.* Berlin: Springer-Verlag, 1966.
30. VON KOCH, HELGE, "Sur les systèmes d'ordre infini d'équations differentielles," *Öfversigt af Kongl. Svenska Vetenskaps—Akademiens Forhandlingar, 61* (1899), 395–411.

31. Köthe, Gottfried, *Topologische Lineare Räume*, I. Berlin: Springer-Verlag, 1960.
32. Lévy, Paul, "Jacques Hadamard, sa vie et son oeuvre—Calcul fonctionnel et questions diverses," in Monographie N° 16 de *l'Enseignement Mathématique—La Vie et l'oeuvre de Jacques Hadamard*. Genève, 1967, pp. 1–24.
33. Lorch, E. R., "The theory of analytic functions in normed Abelian vector rings," *Transactions of the American Mathematical Society*, *54* (1943), 414–425.
34. Martin, R. S., *Contributions to the Theory of Functionals*, unpublished doctoral dissertation, California Institute of Technology, 1932.
35. Mazur, S., "Sur les anneaux linéaires," *Comptes Rendus Acad. Sci. Paris*, *207* (1938), 1025–1027.
36. Mazur, S., and W. Orlicz, "Grundlegende Eigenschaften der polynomischen Operationen" (erste Mitteilung), *Studia Mathematica*, *5* (1935), 50–68.
37. Michal, Aristotle D., *Le Calcul Différentiel dans les espaces de Banach*. Paris: Gauthier-Villars, 1958.
38. Michal, A. D., and A. H. Clifford, "Fonctions analytiques implicites dans des espaces vectoriels abstraits," *Comptes Rendus Acad. Sci. Paris*, *197* (1933), 735–737.
39. Michal, A. D., and R. S. Martin, "Some expansions in vector space," *Journal de Mathématiques Pures et Appliquées*, *13* (1934), 69–91.
40. Nagumo, M., "Einige analytische Untersuchungen in linearen metrischen Ringen," *Japanese Journal of Mathematics*, *13* (1936), 61–80.
41. Pincherle, Salvatore, "Mémoire sur le calcul fonctionnel distributif," *Mathematische Annalen*, *49* (1897), 325–382.
42. ———, "Funktional-Gleichungen und Operationen," in *Encyklopädie der Mathematischen Wissenschaften mit Einschluss ihrer Anwendungen*, II_1, Heft 6, Leipzig, 1906, pp. 761–817.
43. Riesz, F., *Les Systèmes d'équations linéaires à une infinité d'inconnues*. Paris: Gauthier-Villars, 1913.
44. Stone, M. H., *Linear Transformations in Hilbert Space*. New York: American Mathematical Society, 1932.
45. Taylor, Angus E., *Analytic Functions in General Analysis*, unpublished doctoral dissertation, California Institute of Technology, 1936.
46. ———, "Sur la théorie des fonctions analytiques dans les espaces abstraits," *Comptes Rendus Acad. Sci. Paris*, *203* (1936), 1228–1230.
47. ———, "Analytic functions in general analysis," *Annali della R. Scuola Normale Superiore di Pisa*, *6* (1937), 277–292.
48. ———, "On the properties of analytic functions in abstract spaces," *Mathematische Annalen*, *115* (1938), 466–484.
49. ———, "The resolvent of a closed transformation," *Bulletin of the American Mathematical Society*, *44* (1938), 70–74.
50. ———, "Linear operations which depend analytically on a parameter," *Annals of Mathematics*, *39* (1938), 574–593.
51. ———, "Analysis in complex Banach spaces," *Bulletin of the American Mathematical Society*, *49* (1943), 652–669.
52. ———, "New proofs of some theorems of Hardy by Banach space methods," *Mathematics Magazine*, *23* (1950), 115–124.
53. ———, "Spectral theory of closed distributive operators," *Acta Mathematica*, *84* (1951), 189–224.
54. ———, *Introduction to Functional Analysis*. New York: John Wiley & Sons, Inc., 1958.
55. Volterra, Vito, "Sopra le funzioni che dipendono da altre funzioni," Nota I. *Rendiconti della R. Accademia dei Lincei*, Series IV, vol. III (1887), pp. 97–105.
56. ———, Nota II, *ibid.*, pp. 141–146.
57. ———, Nota III, *ibid.*, pp. 153–158.
58. ———, "Sopra le funzioni dipendenti da linee," Nota I, *ibid.*, pp. 225–230.
59. ———, Nota II, *ibid.*, pp. 274–281.

60. WIENER, NORBERT, "Note on a paper of M. Banach," *Fundamenta Mathematica, 4* (1923), 136–143.
61. ZORN, MAX, "Characterization of analytic functions in Banach spaces," *Annals of Mathematics*, (2) *46* (1945), 585–593.
62. ———, "Gateaux differentiability and essential boundedness," *Duke Mathematical Journal, 12* (1945), 579–583.
63. ———, "Derivatives and Fréchet differentials," *Bulletin of the American Mathematical Society, 52* (1946), 133–137.

$\mathscr{A}$-Almost Automorphic Functions

WILLIAM A. VEECH[1]

T will denote a locally compact, σ-compact, Abelian group, and $\mathscr{C} = \mathscr{C}(T)$ will be the algebra of bounded, complex-valued, uniformly continuous functions on T. Bochner defines $f \in \mathscr{C}$ to be *almost automorphic* if from every sequence $\{\beta_m\} = \beta \subseteq T$ may be extracted a subsequence $\{\alpha_n\} = \alpha$ $(\alpha_n = \beta_{m_n})$ such that (a) the limit $T_\alpha f(t) = \lim_n f(t + \alpha_n)$ exists for each t, and (b) the equation

$$(1)\qquad \begin{aligned} T_{-\alpha}T_\alpha f(t) &= \lim_n T_\alpha f(t - \alpha_n) \\ &= f(t) \end{aligned}$$

holds identically.

The limits in (a) and (b) are not required to be uniform (but they are automatically locally uniform), and, in fact, because T is σ-compact, (a) places no restriction at all on f, so long as $f \in \mathscr{C}$. With condition (b) it is a different matter altogether, and the functions which satisfy it are precisely the functions $f \in \mathscr{C}$ which are continuous in the "Bohr topology" on T, the topology on T induced by the algebra $\mathscr{A}$ of continuous Bochner–von Neumann almost periodic functions (see [6]).

Here we shall consider a condition which is weaker than (1). We say $f \in \mathscr{C}$ is *$\mathscr{A}$-almost automorphic* if the totality $S = \{T_{-\alpha}T_\alpha f\}$ is norm precompact (for $\|g\|_\infty = \sup_t |g(t)|$) and if a second more technical condition is satisfied. Theorem 2 which follows gives at least the coarse structure of the class (actually algebra) of $\mathscr{A}$-almost automorphic functions.

Proposition 1 which follows enables us to reformulate both Bochner–von Neumann almost periodicity and $\mathscr{A}$-almost automorphy in such a way that there is no explicit requirement of compactness, and this may be of independent interest.

We begin with some notation. A set $B \subseteq \mathscr{C}$ is a *T-algebra* if it is a uniformly closed, self-adjoint, translation-invariant algebra containing the

[1] Research supported by NSF grants GP-5585 and GP-7952X.

constants. Each $f \in \mathscr{C}$ is contained in a smallest T-algebra which we denote by B_f. B_f is generated by polynomials in the translates of f and its conjugate.

Given $f \in \mathscr{C}$, f_t is to be the tth translate of f:$f_t(s) = f(s + t)$. Let $X_f^0 = \{f_t\}_{t \in T}$, and note that X_f^0 is precompact precisely when $f \in \mathscr{A}$. X_f^0 is *always* precompact in the topology of local uniform convergence, and we let $X_f \subseteq \mathscr{C}$ be its closure in this weaker topology. Because T is σ-compact X_f is metrizable. There is an isometric isomorphism between B_f and $C(X_f)$ given by $h \leftrightarrow H$, where $h(t) = H(f_t)$, $t \in T$.

X_f is a translation invariant subset of $\mathscr{C}$, and therefore there is a natural action of T on X_f yielding a *flow* (T, X_f). We say f is *minimal* if the associated flow is minimal (every point has a dense orbit) or, equivalently, if for every $T_\alpha f$ there is a sequence β such that $T_\beta T_\alpha f = f$.

If $T_\alpha f$ exists, it is easy to see that $T_\alpha h$ exists for each $h \in B_f$, and, moreover, if $T_\beta T_\alpha f = f$, then $T_\beta T_\alpha h = h$, $h \in B_f$. We conclude therefore that each $h \in B_f$ is minimal if f is minimal. In general we say that a T-algebra is minimal if each of its elements is minimal. We remark also that this property of $h \in B_f$ characterizes B_f. If h is such that $T_\alpha h$ exists whenever $T_\alpha f$ exists, then $h \in B_f$. [For then $H(f_t) = h(t)$ extends to be continuous on X_f.]

LEMMA 1. *Suppose $f \in \mathscr{C}$ is such that whenever $T_\alpha f$ exists there is a sequence β such that $f \in B_{T_\beta T_\alpha f}$. Then f is minimal.*

PROOF. By a well known Zorn's lemma argument there exists a minimal set $M \subseteq X_f$ for (T, X_f). If $m \in M$, say $m = T_\alpha f$, then $(T, X_{T_\alpha f})$ is simply (T, M). Thus $T_\alpha f$ is minimal, as is $T_\beta T_\alpha f$ whenever the limit exists. Thus, by our assumption on f, $f \in B_g$ for some minimal function g, and f is minimal.

If α is a sequence such that $T_\alpha f$ exists for some f, then T_α may be regarded as a linear operator on B_f. Evidently the range of this operator is contained in $B_{T_\alpha f}$ and contains a dense subset of the latter. Generally it is true that $\|T_\alpha h\|_\infty \leqq \|h\|_\infty$, but if f is minimal, equality holds. (For then there exists β with $T_\beta T_\alpha h = h$, $h \in B_f$, and the reverse inequality follows from this.) It follows in particular that T_α is onto when f is minimal.

LEMMA 2. *Let $f \in \mathscr{C}$ be minimal. If $f \in B_{T_\alpha f}$ for some sequence α, there exists a sequence β such that $T_\alpha T_\beta f = f$.*

PROOF. By the remark preceding the lemma, T_α is onto, and therefore $f = T_\alpha g$ for some $g \in B_f$. Since g is then minimal, we can find β with $T_\beta f = T_\beta T_\alpha g = g$, and *a fortiori* $T_\alpha T_\beta f = f$. The lemma is proved.

We now fix $f \in \mathscr{C}$ and consider the significance of the existence in X_f of a set S with the following properties:

(a) S is closed.
(b) $f \in B_{T_\alpha f}$ if $T_\alpha f \in S$.
(c) For each $T_\alpha f \in X_f$ there exists a sequence β such that $T_\beta T_\alpha f \in S$.
(d) If $T_\alpha f \in S$, and if β is a sequence such that $T_\alpha T_\beta f = f$, then $T_\beta f \in S$.
(e) If $T_\alpha f$, $T_\beta f \in S$, and if $T_\alpha T_\beta f$ exists, then $T_\alpha T_\beta f \in S$.

LEMMA 3. *Suppose f and $S \subseteq X_f$ enjoy properties* (b) *to* (d) *just given. Then f is minimal, and $T_\alpha f \in B_f$ for all $T_\alpha f \in S$.*

PROOF. Using assumptions (b) and (c), minimality is a consequence of Lemma 1. If $T_\alpha f \in S$, then because f is minimal we can apply Lemma 2 to find a sequence β such that $T_\alpha T_\beta f = f$. By (d), $T_\beta f \in S$, and in particular $f \in B_{T_\beta f}$. It follows then that $T_\alpha f \in B_{T_\alpha T_\beta f} = B_f$, and the lemma obtains.

Continuing to suppose f and S satisfy (b) to (d), we apply Lemma 3 as follows. Suppose $x \in X_f$ and $y \in S$ are given, say $x = T_\alpha f$ and $y = T_\beta f$. By Lemma 3, $T_\alpha T_\beta f$ exists, and one sees readily that it depends only upon x and y. It therefore makes sense to define a binary operation $x \circ y$, $x \in X_f$, $y \in S$, by

$$x \circ y = T_\alpha T_\beta f. \tag{2}$$

In this way we obtain a map from $X_f \times S$ to X_f.

LEMMA 4. *Let $f \in \mathscr{C}$ and $S \subseteq X_f$ enjoy properties* (b) *to* (e) *described earlier. Then*

(α) *S inherits the structure of a group when* (2) *is restricted to $S \times S$.*

(β) *For each $y \in S$, the map $x \to x \circ y$ defines a homeomorphism of X_f which is without fixed points unless $y = f$.*

(γ) *The actions of T and S commute with one another; i.e., $t(x \circ y) = (tx) \circ y$, $t \in T$, $x \in X_f$, $y \in S$.*

PROOF. We first prove (α). If $x, y \in S$, then by (e) $x \circ y \in S$. If $x = T_\alpha f$, $y = T_\beta f$, $z = T_\gamma f \in S$, we may choose δ so that $T_\delta f = T_\alpha T_\beta f$. Then $T_\delta h = T_\alpha T_\beta h$, $h \in B_f$, and in particular $T_\delta T_\gamma f = T_\alpha T_\beta T_\gamma f$. The left side represents $(x \circ y) \circ z$, while the right represents $x \circ (y \circ z)$. Thus the operation is associative. (We have not used here that $x \in S$.) Using minimality, (d), (e), and Lemma 2, we obtain that $f \in S$, and this is an identity for S. A final application of minimality, (d), and Lemma 2 yield the existence of right inverses, and therefore S is a group. (α) is proved.

For (β) we fix $y = T_\beta f \in S$, and note that because $T_\beta f \in B_f$, there is a function $G_\beta \in C(X_f)$ such that $G_\beta(f_t) = T_\beta f(t)$. If $T_{\alpha_n} f$, $n = 1, 2, \ldots$, is a sequence of elements of X_f which is convergent to some $x = T_\alpha f \in X_f$, then for each $t \in T$ we have

$$T_\alpha T_\beta f(t) = G_\beta(tx) = \lim_n G_\beta(tT_{\alpha_n} f)$$
$$= \lim_n T_{\alpha_n} T_\beta f(t).$$

This implies continuity. Because $(x \circ y) \circ y^{-1} = x \circ (y \circ y^{-1}) = x = x \circ (y^{-1} \circ y) = (x \circ y^{-1}) \circ y$, it is true that $x \to x \circ y$ is a homeomorphism. If $x \circ y = x$, then $T_\alpha T_\beta f = T_\alpha f$, where $T_\alpha f = x$, $T_\beta f = y$. Now since $T_\beta f \in B_f$, the function $T_\beta f - f$ is minimal, and therefore $T_\alpha(T_\beta f - f) \equiv 0$ implies $T_\beta f = f$. (β) is proved. (γ) is trivial, and the lemma is proved.

PROPOSITION 1. *Suppose $f \in \mathscr{C}$ and $S \subseteq X_f$ enjoy the properties* (a) *to* (e). *Then*

(i) *The operation* (2) *and the topology of X_f give S the structure of a compact topological group.*

(ii) (2) *is continuous from $X_f \times S$ to X_f, and (X_f, S) is a transformation group.*

(iii) (X_f, S) *is equicontinuous, strongly effective ($x \circ y = x$ for some x implies $y = f$), and the actions of S and T commute.*

(iv) *S is norm compact, as is $Sh = \{T_\alpha h | T_\alpha f \in S\}$ for any $h \in B_f$.*

PROOF. S is closed and hence compact in X_f, and therefore (i) will follow from (ii) and Lemma 4. If $y \in S$ is fixed, we know from Lemma 4 that (2) is continuous in x. Now we fix x and prove continuity in y. It is here that metrizability plays an essential role.

Suppose $x \in X_f$ and $x = T_\alpha f$. Then by definition $x = \lim_n f_{\alpha_n} = \lim_n x_n$, and $x \circ y = \lim_n x_n \circ y$, $y \in S$. Now $y \to x_n \circ y$ is continuous on S for each n, and therefore $y \to x \circ y$ is of the first Baire class on S. It follows from Baire's theorem that for some y_0 in S (2) is continuous at y_0. Now suppose $y \in S$ is arbitrary, and let $y = \lim_n y_n$, $y_n \in S$. Certainly $y_n \circ y^{-1} \to f$, and so from the continuity at y_0, $\lim_n x \circ y_n \circ y^{-1} \circ y_0 = x \circ y_0$. Then by left continuity we obtain

$$\lim_n x \circ y_n = \lim_n x \circ y_n \circ y^{-1} \circ y_0 \circ y_0^{-1} \circ y$$
$$= x \circ y_0 \circ y_0^{-1} \circ y = x \circ y$$

and therefore (2) is continuous at y. We have now that (2) is continuous in each of the variables separately, and therefore by the Ellis separate continuity theorem (or in this case a fairly easy category argument) (2) is continuous on $X_f \times S$. As stated previously, (i) and (ii) are established.

Statement (iii) is obvious and (iv) is true because the induced action of S on B_f has norm compact orbits. Each orbit is Sh for some $h \in B_f$.

Our reformulation of Bochner–von Neumann almost periodicity is

THEOREM 1. *A necessary and sufficient condition for $f \in \mathscr{C}$ to belong to $\mathscr{A}$ is that $f \in B_{T_\alpha f}$ whenever $T_\alpha f$ exists.*

PROOF. If $f \in \mathscr{A}$, and if $T_\alpha f$ exists, then it exists uniformly. Moreover, as is noted in (1) $\lim_n \|T_\alpha f(t - \alpha_n) - f(t)\|_\infty = 0$. That is, $f \in B_{T_\alpha f}$. For the converse we set $S = X_f$ in Proposition 1 and note that properties (a) to (e) are trivial.

We will introduce a slightly more general terminology. Let $F(t) = [f_1(t), f_2(t), \ldots]$ be a vector-valued function on T whose coordinate functions belong to $\mathscr{C}$. There may be a finite or an infinite number of coordinates. B_F will be the T-algebra generated by the coordinates and symbols like F_t, $T_\alpha F$, and X_F have obvious definitions. If F and G are two such functions, $F \in B_G$ will mean $B_F \subseteq B_G$. X_F is metrizable, and B_F is isometrically isomorphic with $C(X_F)$ by the usual correspondence.

DEFINITION 1. A function $f \in \mathscr{C}$ is *$\mathscr{A}$-almost automorphic* if the totality $H_f = \{T_{-\alpha}T_\alpha f | T_\alpha f$ and $T_{-\alpha}T_\alpha f$ exist$\}$ is norm precompact, and if H_f generates a minimal T-algebra.

REMARK. If f is almost automorphic, then f is obviously $\mathscr{A}$-almost automorphic.

Let (T, X) be a minimal flow where X is compact metrizable and T, as usual, is Abelian. For each $x \in X$ define $E(x)$ to be the set of $y \in X$ for which there exists a sequence $\alpha = \{\alpha_n\}$ such that $\lim_n \alpha_n x = y'$ exists and $\lim_n - \alpha_n y' = y$. We recall the following

THEOREM A. (See [7], Theorems 1.1–1.2.) *With notations as given previously the relation $y \sim x$ if $y \in E(x)$ is a closed, invariant equivalence relation. If $Y = X/\sim$, the natural flow (T, Y) is minimal and equicontinuous. Moreover, $\sim$ is contained in any relation with the latter property.*

Now let f be $\mathscr{A}$-almost automorphic, and let B be the T-algebra generated by H_f. Since H_f is norm precompact, it has a dense subset $f_1, f_2, \ldots$, and we set $F = (f_1, f_2, \ldots)$. Clearly, $B = B_F$. By assumption B is minimal, and therefore (T, X_F) is a minimal flow. In what follows a *bitransformation group* shall be a triple (T, X, S) which satisfies properties (i) to (iii) of Proposition 1 [with (T, X) minimal]. This terminology is due to Ellis.

THEOREM 2. *Given $f \in \mathscr{C}$ the following statements are equivalent:*

(i) *f is $\mathscr{A}$-almost automorphic.*

(ii) *There exists* $F = (f_1, f_2, \ldots)$ *such that* $f \in B_F$ *and* $F \in B_{T_{-\alpha}T_\alpha F}$ *whenever the limits exist.*

(iii) *There exists a bitransformation group* (T, X, S), *a point* $x_0 \in X$ *such that* $Sx_0 = E(x_0)$, *and* $\varphi \in C(X)$ *such that* $f(t) = \varphi(tx_0)$.

PROOF. (i) $\rightarrow$ (ii). We associate F with the $\mathscr{A}$-almost automorphic function f just given. To see that $F \in B_{T_{-\alpha}T_\alpha F}$ when the limits exist we shall apply Theorem A to (T, X_f) which, since f is minimal, satisfies the hypotheses of that theorem. Since the coordinates of F are dense in H_f $(= E(f))$, the existence of $T_{-\alpha}T_\alpha F$ implies the existence of $T_{-\alpha}T_\alpha h$, $h \in H_f$. Secondly, because $\sim$ is transitive, we have $T_{-\alpha}T_\alpha h \in H_f$, $h \in H_f$. B_F is a minimal T-algebra by assumption, and therefore $T_{-\alpha}T_\alpha\colon H_f \rightarrow H_f$ is a norm isometry. Now H_f is closed in X_f because $\sim$ is closed, and being norm precompact, it follows that H_f is norm closed (and so compact). An isometry of a compact metric space into itself must be onto, and therefore $T_{-\alpha}T_\alpha$ is onto. If β is any sequence such that $T_{-\beta}T_\beta F$ exists, then clearly $T_{-\beta}T_\beta F \in B_F$, and *a fortiori* $T_{-\alpha}T_\alpha T_{-\beta}T_\beta F \in B_{T_{-\alpha}T_\alpha F}$. If f_n is the nth coordinate function, we may choose β so that $T_{-\alpha}T_\alpha T_{-\beta}T_\beta f_n = f_n$ [since $E(f_n) = E(f)$]. Therefore $f_n \in B_{T_{-\alpha}T_\alpha F}$, and since n is arbitrary, $F \in B_{T_{-\alpha}T_\alpha F}$.

(ii) $\rightarrow$ (iii). We take $X = X_F$, $S = E(F)$, $x_0 = F$, and $\varphi(tF) = f(t)$. The latter function extends to be continuous on X because $f \in B_F$. We verify properties (a) to (e) for F and S. (b) and (c) are trivial, and therefore, by Lemma 1 and Theorem A, (a) is true. (e) is a consequence of the transitivity of $\sim$ and (d) is a consequence of symmetry. Now (iii) is a consequence of Proposition 1. (The change from f to F is purely formal.)

(iii) $\rightarrow$ (i). Because $Sx_0 = E(x_0)$, we have $Sf = H_f$. Moreover, if $B = \{g \in \mathscr{C} \mid g(t) = G(tx_0), G \in C(X)\}$, then B is a minimal T-algebra, and $SB \subseteq B$. Thus, in particular $H_f \subseteq B$, and H_f generates a minimal T-algebra. The theorem is proved.

We conclude with some further remarks.

Let f be $\mathscr{A}$-almost automorphic, and suppose $E(f) = X_f$. Then $f \in \mathscr{A}$, and f is almost automorphic. That is, $E(f) = \{f\}$, which, coupled with the preceding, implies f is constant. It follows that if f is a nonconstant $\mathscr{A}$-almost automorphic function, $X_f/\sim$ is not trivial, and B_f contains nonconstant almost periodic functions.

If f is $\mathscr{A}$-almost automorphic, and if (T, X, S) is as in Theorem 2(iii), there is a natural quotient flow $(T, X/S)$. If $p\colon X \rightarrow X/S$ is canonical, and if $y_0 = px_0$, then $E(x_0) = Sx_0$ implies $E(y_0) = \{y_0\}$. That is, y_0 is an "almost automorphic" point, and (T, X) is a group extension (Furstenberg's terminology) of an almost automorphic flow. Conversely, if (T, X) is a group extension of an almost automorphic flow (T, Y), and if $y_0 \in Y$ is almost automorphic, then $E(x_0) = Sx_0$ if $px_0 = y_0$. The $\mathscr{A}$-almost automorphic functions are precisely the orbit functions, $f(t) = \varphi(tx_0)$, coming from group extensions of almost automorphic flows, subject to the restriction that px_0 be almost automorphic.

If f is $\mathscr{A}$-almost automorphic, and if $T_\alpha f$ is $\mathscr{A}$-almost automorphic whenever the limit exists, it is possible to choose (T, X, S) so that $(T, X/S) = (T, X/\sim)$. Then, in the terminology of [5], f is almost periodic with respect to the almost periodic functions. This result is similar in spirit to the statement that f is almost periodic if and only if $T_\alpha f$ is almost automorphic whenever the limit exists (see [6], Theorem 3.1).

Using Theorem 2 it is easy to see that $\mathscr{A}$-almost automorphic functions on T form a T-algebra. For example, if f and g are $\mathscr{A}$-almost automorphic, and if F and G are associated as in Theorem 2(ii), set $H = (f_1, g_1, f_2, g_2, \ldots)$ (with an obvious convention if one or both of the sets is finite). Then $f + g$, $f \cdot g \in B_H$, and $H \in B_{T_{-\alpha}T_\alpha H}$ whenever the limits exist. A similar argument shows the class to be closed under uniform limits. Self-adjointness and translation invariance are trivial.

When T is the group of real numbers in its usual topology, the Bebutov–Kakutani theorem (see [4]) implies that a separable minimal T-algebra has the form B_f for some $f \in \mathscr{C}$. Thus, if g is $\mathscr{A}$-almost automorphic, there exists $f \in \mathscr{C}$ with $g \in B_f$ and $f \in B_{T_{-\alpha}T_\alpha f}$ whenever the limits exist. We do not know if this is true in general.

The functions $f \in \mathscr{C}$ for which $f \in B_{T_{-\alpha}T_\alpha f}$ whenever the limits exist are precisely those functions for which (T, X_f) is a (minimal) group extension of an almost automorphic flow.

One can carry out a "Fourier analysis" similar in spirit to that of [5] for the $\mathscr{A}$-almost automorphic functions.

University of California
Berkeley
Institute for Advanced Study

REFERENCES

1. Bochner, S., "A new approach to almost periodicity," *Proceedings of the National Academy of Sciences of the U.S.A., 48* (1962), 2039–2043.
2. ———, "Beiträge zur Theorie der fastperiodischen Funktionen, I," *Math. Ann., 96* (1926), 119–147.
3. Ellis, R., "Locally compact transformation groups," *Duke Math. J., 24* (1957), 119–126.
4. Kakutani, S., "A proof of Bebutov's theorem," *Journal of Differential Equations, 4* (1968), 194–201.
5. Knapp, A., "Distal functions on groups," *Trans. Am. Math. Soc., 128* (1967), 1–40.
6. Veech, W., "Almost automorphic functions on groups," *Am. J. Math., 87* (1965), 719–751.
7. ———, "The equicontinuous structure relation for minimal Abelian transformation groups," *Am. J. Math., 90* (1968), 723–732.